Laboratory Manual for Biotechnology and Laboratory Science

This manual explores the basic laboratory skills and concepts essential for a career in biotechnology and other laboratory sciences. Written by four biotechnology instructors, all with over 25 years of teaching experience, it incorporates instruction, exercises, and laboratory activities that the authors have been using and perfecting for years. These exercises and activities help students understand the fundamentals of working in a biotechnology laboratory. As students build skills through an organized and systematic presentation of materials, procedures, and tasks, they also explore overarching themes that relate to all biotechnology workplaces.

Features

- Provides clear instructions and step-by-step exercises to make learning the material easier for students.
- Emphasizes fundamental laboratory skills that prepare students for the industry.
- Builds students' skills through an organized and systematic presentation of materials, procedures, and tasks.
- Updates reflect recent innovations and regulatory requirements to ensure students stay up to date.
- Introduces skills important for careers in forensic, clinical, quality control, environmental, and other testing laboratories.

T0262830

Laboratory Manual for Biotechnology and Laboratory Science: The Basics, Revised Edition

Authored By

Lisa A. Seidman
Faculty Emeritus, Madison Area Technical College

Mary Ellen Kraus
Faculty, Madison Area Technical College

Diana Lietzke Brandner
Former Laboratory Coordinator, Madison Area Technical College

Jeanette Mowery
Faculty Emeritus, Madison Area Technical College

CRC Press
Taylor & Francis Group
Boca Raton London New York

CRC Press is an imprint of the
Taylor & Francis Group, an **informa** business

Revised edition published 2023
by CRC Press
6000 Broken Sound Parkway NW, Suite 300, Boca Raton, FL 33487-2742

and by CRC Press
4 Park Square, Milton Park, Abingdon, Oxon, OX14 4RN

© 2023 Taylor & Francis Group, LLC

First edition published by Pearson Education, Inc. publishing as Benjamin Cummings 2011

CRC Press is an imprint of Taylor & Francis Group, LLC

ISBN: 978-1-032-41993-0 (hbk)
ISBN: 978-1-032-41991-6 (pbk)
ISBN: 978-1-003-36074-2 (ebk)

DOI: 10.1201/9781003360742

Typeset in Warnock Pro
by Deanta Global Publishing Services, Chennai, India

Contents

Preface

The authors of this manual are faculty in the Biotechnology Laboratory Technician Program at Madison Area Technical College which prepares students to become laboratory professionals. Since the inception of this program, we have been challenged to define the nature of the work for which we are preparing students, and the knowledge, skills, and attitudes that our students need to achieve to be successful. This manual is the distillation of our conversations, interactions with talented colleagues, and 25-plus years of rewarding experiences teaching students.

When we first developed our program, we began, as do many undergraduate laboratory teachers, with an experiment or two to introduce students to the scientific method. Undergraduate experimentation teaches students the important lesson that science is not a dreary collection of facts but is an engaging process and a logical way of inquiring about nature. There is, however, a problem inherent in much undergraduate research: experimentation that is performed badly is unlikely to answer any scientific questions at all. In an attempt to mitigate this problem, colleges sometimes hire professional staff who maintain and calibrate equipment, prepare materials and solutions, and perform other essential tasks behind the scenes in rooms that students never enter. The result is that beginning students do not learn the basic skills that make laboratory investigation meaningful and they do not appreciate the challenges of making things "work" in the laboratory. We found that this customary approach is not the most effective in preparing professional biotechnologists. Instead, we now begin with the concept that producing quality laboratory work requires understanding fundamental principles and mastering fundamental techniques. We developed this laboratory manual to be an early stepping stone for students on the path to becoming professionals.

This manual was originally developed for associate degree college students. We knew that these individuals would be called on to perform basic laboratory tasks. Over time, however, we have instructed teachers, laboratory professionals, and individuals with bachelor of science degrees trying to enter the job market. Many of these individuals never had the opportunity to learn and practice laboratory fundamentals, despite having academic degrees and experience. We have realized that the same fundamental skills, knowledge, and attitudes are essential for any laboratory professional. These fundamentals are critical to the success of cutting-edge research scientists who probe the inner workings of nature. They are essential for scientists who develop ideas into practical products. The same fundamentals apply to laboratory analysts who analyze samples in forensic, clinical, quality control, and other testing laboratories. The way in which the basics "play out" may vary in different workplaces, yet the fundamentals remain, well … fundamental. Thus, this manual aims to systematically build students' basic skills, introduce the fundamental principles underlying these basic tasks, and explore overarching themes that relate to all laboratory workplaces.

The goal of this manual is to help students establish a coherent, integrated understanding of laboratory work. It is therefore organized in a particular way. At the first level of organization are the basic tasks or skills that students explore, such as working safely and maintaining a laboratory notebook. These tasks are the subject of individual exercises. At the next level, the exercises are organized into units, such as safety and documentation, which have unifying fundamental principles. At the highest level of organization, certain themes of quality work, such as the importance of reducing variability, are integrated into every activity and discussion.

Perhaps the methods covered in this manual (such as how to prepare a solution) seem less glamorous than such tasks as editing DNA. However, we have learned from our students that developing a deep understanding of the fundamentals of laboratory work is not only essential, it is also rewarding.

Moving forward the frontiers of science requires a solid foundation

Companion Textbooks

This laboratory manual includes introductory information to introduce each unit. However, more in-depth discussions can be found in the textbook *Basic Laboratory Methods for Biotechnology: Textbook and Laboratory Reference*, by Lisa A. Seidman, Cynthia J. Moore, and Jeanette Mowery (CRC Press 2022, ISBN 9780367244880).

Students who would benefit from support in performing math calculations may like the textbook *Basic Laboratory Calculations for Biotechnology*, by Lisa A. Seidman (CRC Press 2022, ISBN 9780367244804). The calculations text is a friendly guide to the most common math procedures in biotechnology settings.

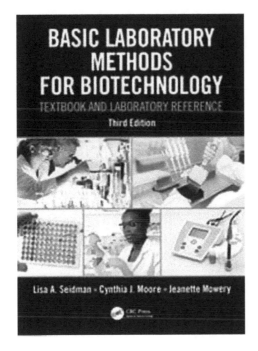

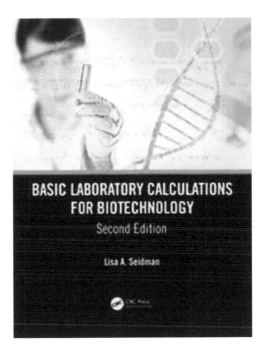

Acknowledgments

As is always the case when writing a book, many people contributed to it and we are grateful for their input. We thank our many colleagues who encouraged us to prepare this manual. Thank you to Elaine Johnson and Linnea Fletcher for their friendship and continuous support. We thank our excellent reviewers for their ideas and expertise: Kristine M. Snow, Fox Valley Technical College; Josephine Pino, Portland Community College–Rock Creek Campus; David M. Brooks, East Central College; Jonathan Morris, Manchester Community College; Jean L. Schoeni, TRAC Microbiology; and Thomas C. Tubon Jr., Madison Area Technical College. We thank Tracy M. Theobald, our student intern extraordinaire, who meticulously tested the exercises and provided invaluable feedback. We thank Alex V. Sheyn, a student in art and graphic design, for his creativity and talent in imagining and providing special art work. Sandra Bayna, a creative biotechnology student, also provided art work. We acknowledge the many contributions of our students who suffered through earlier, less than perfect drafts and whose successes always inspire us. We thank our deans for their continued support: Joy McMillan, David Shonkwiler, and John Stransky. We also thank the staff at Benjamin Cummings for their skills and contributions, including Gary Carlson, Lindsay White, and Camille Herrera. The current revision was supported by our excellent editors at CRC Press, Barbara Knott and Hilary Lafoe, and by the production staff at Deanta Global Publishing Services.

This manual is based in part on work funded by the National Science Foundation Advanced Technology Education Initiative, under grant numbers 9752027 and 0101093. Any opinions, findings, conclusions, or recommendations expressed in this material are those of the authors and do not necessarily reflect the views of the National Science Foundation.

Authors

Lisa A. Seidman earned her PhD from the University of Wisconsin and has taught for more than 30 years in the Biotechnology Laboratory Technician Program at Madison Area Technical College. She is presently serving as emeritus faculty at the college.

Mary Ellen Kraus has been a faculty member in the Biotechnology Laboratory Technician Program at Madison Area Technical College for more than 25 years. She earned her BS in Biochemistry from the Pennsylvania State University and her PhD in biochemistry from Cornell University.

Diana Lietzke Brandner earned her MS in Biotechnology Education from the University of Wisconsin–Madison. She was a lead laboratory coordinator in the Biotechnology Laboratory Technician Program at Madison Area Technical College for more than 30 years.

Jeanette Mowery earned her PhD in Biomedical Science from the University of Texas Health Science Center–Houston. She has taught for more than 25 years in the Biotechnology Laboratory Technician Program at Madison Area Technical College and is currently serving as emeritus faculty at the college.

Introduction

A. WHAT IS THIS MANUAL ABOUT?

The purpose of this laboratory manual is to help you systematically develop your skills and your understanding of the basic laboratory methods used by biotechnologists and other laboratory professionals. Let's then consider three introductory questions: What is biotechnology? What is a "laboratory"? What do we mean by "basic" methods?

Biotechnology is sometimes defined as any application that uses organisms, cells, or materials derived from organisms and cells to make a product of value to humans. Biotechnology encompasses ancient arts such as bread making, beer brewing, and animal breeding. When people now speak of biotechnology, however, they usually are not referring to these long-existing applications, but rather to sophisticated methods of manipulating genetic information and cells, methods that blossomed in the later part of the 20th century. "Modern" biotechnology includes genetic manipulation of cells (Figure I.1), regenerative medicine (e.g., the use of stem cells to treat disease), advanced forms of diagnostics, bioinformatics, aspects of nanotechnology, and an ever-expanding array of sophisticated applications. To encompass all these applications and techniques, from the ancient to the emerging, we broadly define **biotechnology** to be *the transformation of knowledge acquired by basic biological research into goods and products of value to people.*

The term "laboratory" perhaps brings to mind a room with counters festooned with beakers, flasks, and esoteric apparatus. In such a room, people conduct experiments to explore the intricacies of nature. But must a laboratory be a room? If ecologists manipulate environmental variables in a pond, is the pond a laboratory? What about psychologists studying human interactions in a social setting? Can a party be a "laboratory"? Let's approach the question from another perspective: What is the purpose of a laboratory? Perhaps a laboratory is not so much a special place, but rather a setting with a particular purpose. We could explore this question in more detail, but, for now, we will define a **laboratory** as *a setting whose purpose is to allow investigators to produce knowledge, information, or data.* The product of a laboratory is not tangible, like a drug, widget, or gizmo, but is rather a product of the mind. People in laboratories may indeed produce tangible items, such as photographs or instrument recordings, but these tangible items are produced in order to obtain information and knowledge.

Biologically oriented laboratories may be classified into types:

- **Basic biology research laboratories.** *The product of these laboratories is knowledge about how living systems work.*
- **Research and development laboratories.** *The product of these laboratories is transformation of knowledge from basic research into a practical product.*
- **Testing laboratories.** *The product of these laboratories is data or information about samples that can be used to make decisions.*

What is **basic**? The website Thesaurus.com includes the following synonyms for "basic": *central, chief, indispensable, key, main, necessary, principal, underlying, and vital.* Clearly the basics are important! The basics are those things that all laboratory professionals should know and practice, regardless of where they work. Basic laboratory methods may also be

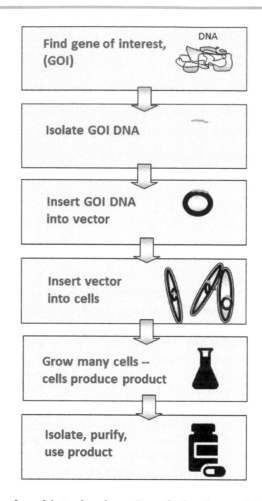

FIGURE I.1   **The modern biotechnology "revolution" was driven by discoveries of methods that allow scientists to move genetic information, DNA, from one organism into another.** Biotechnologists can now use cells as "factories" to produce protein products of tremendous value, such as insulin, anticancer therapeutics, cleaning products, and enzymes for food production.

called "good" laboratory practices (lowercase, glp). Good laboratory practices are the prerequisite for "doing good science." Good laboratory practices ensure that laboratory professionals produce a **good** product— knowledge, information, or data that is *trustworthy*, *reproducible*, and *usable*. We have now used the term "good" multiple times in three sentences; this is a word we will come back to over and over again as we explore what it means in the context of laboratory work.

B. ORGANIZATION OF THIS LABORATORY MANUAL

This laboratory manual is about basic skills, knowledge, and attitudes that you will need to become a professional in the challenging and exciting biotechnology laboratory workplace. This manual is organized as follows:

■ **Laboratory Exercises and Classroom Activities** provide you with experiences and challenges relating to specific skills and techniques. Laboratory Exercises contain procedures that you perform in the laboratory, while the Classroom Activities do not require specialized laboratory facilities.
■ The Laboratory Exercises and Classroom Activities are organized into **units**, which *share fundamental principles*. These principles are introduced at the beginning of each unit.

- Certain **overarching themes of good laboratory practice (glp)** emerge throughout all the units as basic skills are introduced. These overarching themes include the following:
  - Documentation provides the framework for glp.
  - Laboratory professionals plan their work.
  - Good laboratory practices lead to consistency and a reduction in the variability of results.
  - Problems and mistakes must be anticipated and prevented whenever possible and corrected when they occur.
  - Good laboratory practices include the validation and verification of processes, methods, and results.
  - Results must be properly analyzed and interpreted.
  - Laboratory professionals work in a safe way to protect themselves, their colleagues, their equipment, and their environment.

Each laboratory exercise is followed by a laboratory meeting in which the class will explore their results together. There are questions provided to direct this classroom discussion. Collaboration and replication of results by other people are critical pieces of the scientific process, so this discussion step is important.

C. USING THIS LABORATORY MANUAL

Science is facts; just as houses are made of stone, so is science made of facts; but a pile of stones is not a house, and a collection of facts is not necessarily science.

—Jules Henri Poincaré (1854–1912), French mathematician

Just as science is more than a pile of facts, working in the laboratory is more than learning to perform a series of steps. Working in a laboratory is certainly something you do with your hands—and is rewarding in that way. However, to build something other than a pile of useless data, laboratory work must also be something you do with your mind. Thinking as a professional is a recurrent theme in this manual, reiterated in the planning you will perform, the discussions you will have with your colleagues, and the analyses that you will perform on your data.

Laboratory professionals prepare for their work. Most of the laboratory exercises in this manual, therefore, begin with questions to guide your preparation before coming to class. The background information that precedes each laboratory exercise is also intended to help you prepare. By understanding and preparing for laboratory work, you avoid the trap of working in a "cookbook" way, that is, following directions without really understanding why you are doing each step. Some students prepare for lab by putting tables and preliminary information into their laboratory notebooks. This may be essential if your time in the laboratory is limited. If you prepare by writing in your laboratory notebook before class, always be sure to distinguish between what you plan to do and what you *actually* do—do not be surprised if they are not the same! This issue is discussed in more detail in Unit II, Documentation in the Laboratory.

Class discussion is an important tool for developing the thought processes of a laboratory professional. Scientists discuss their work with colleagues; in this process their understanding of their work deepens and evolves. Scientists also synthesize their results and thoughts into reports and other communications. Unless your instructor tells you otherwise, when you have completed each laboratory exercise, you should hand in copies of your laboratory notebook pages and answers to the Lab Meeting/Discussion questions, typed on a word processor as part of this synthesizing process.

D. A NOTE ABOUT REGULATED WORKPLACES

Many biotechnologists work in facilities that are involved in the production of pharmaceuticals, medical products, or foods. All these products are **regulated by the government**, meaning that, *by law, these products must be designed, manufactured, labeled, and* *distributed according to certain requirements to ensure their quality.* It is beyond the scope of this manual to explore these regulations. There are, however, many sources of information about regulatory affairs online, in books, and in journals. For more information, consult the resources in the bibliography at the end of the manual.

Although regulatory affairs are not the primary focus of this laboratory manual, they may affect the everyday work of a laboratory biotechnologist and certainly cannot be ignored. The requirements of working in a regulated environment are, therefore, mentioned where relevant throughout this manual. Some terms you should know are:

- **Current Good Manufacturing Practices (CGMP):** *The regulations that affect the manufacture, labeling, and distribution of pharmaceuticals; enforced by the United States Food and Drug Administration (FDA).*
- **Good Laboratory Practices (GLP):** *The regulations that affect laboratory studies relating to pharmaceutical products before the products are tested in humans; enforced by the FDA.*
- **Good Clinical Practices (GCP):** *The regulations that affect studies of medical products that involve human subjects; enforced by the FDA.*
- **Good laboratory practices (glp):** *A term sometimes used to refer to any good work practices in a laboratory, whether or not that laboratory is regulated by the government.*
- **GXP:** *Stands for GMP, or GLP, or GCP.*
- **ISO 9000:** *A family of quality standards administered by the International Organization for Standardization to help ensure that an organization's products are of good quality.* Organizations comply with ISO 9000 voluntarily.

Safety in the Laboratory

DOI: 10.1201/9781003360742-1

UNIT INTRODUCTION

We can all agree that we should work safely in the laboratory, but doing so requires knowledge, planning, and commitment. Working safely is an active process of anticipating threats to our health and the integrity of our equipment, reagents, and laboratory space, and taking action to minimize those threats. The consequences of failing to work safely in the laboratory can be severe—there have been incidents that have resulted in the loss of life and destruction of workspaces. The purpose of this safety unit is to help you begin to acquire the skills and knowledge that will allow you to avoid harm by working safely—but not fearfully.

Safety is a vast topic, and this unit cannot be comprehensive, nor does it take the place of workplace training programs. Working through this unit will, however, introduce you to a safety mindset and will provide you with a framework of knowledge as you take responsibility for your own safety throughout your career.

This unit contains a series of classroom activities and laboratory exercises. Because this unit should be understood prior to working in the laboratory, the first activities are "dry" in that they do not require laboratory equipment.

This unit is separated into three parts, each with its own introduction to fundamental principles:

1. **Creating a Safe Workplace** addresses broad issues and the resources, practices, and governmental agencies that support safety in the workplace.
2. **Working Safely with Chemicals** introduces specific practices that should be followed whenever working with chemicals in the laboratory.
3. **Working Safely with Biological Hazards** introduces special practices that relate to the use of biohazardous agents in the laboratory.

The ideas, activities, and laboratory exercises that make up this unit are summarized in Figure 1.1.

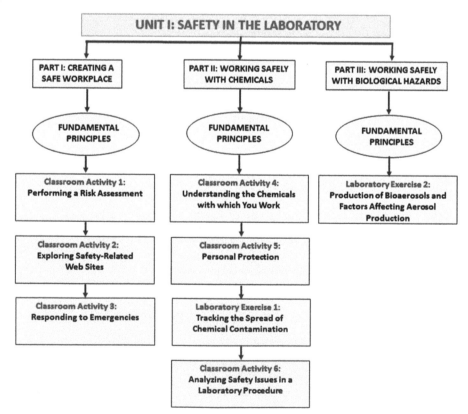

FIGURE 1.1   Unit overview. This unit will provide you with information, classroom activities, and laboratory exercises to guide you towards a safety mindset.

SAFETY PART 1: CREATING A SAFE WORKPLACE

FUNDAMENTAL PRINCIPLES

Working safely may get old, but so do those who practice it.

—Author unknown, http://www.quotegarden.com/safety.html

Imagine a group of hikers, all of whom want to reach the summit of a mountain and return safely but who disagree on the route. One hiker argues that the shortest route up the mountain, which requires clambering over rocks and scaling cliff faces, is reasonably safe, and is best because it is fastest. Another hiker, however, thinks it is safer for the group to stay on well-maintained hiking trails. The group must decide what hazards are likely to be encountered on each route, how much risk they are willing to accept, and what path to choose. Suppose that the group decides to use the established hiking trails, reasoning that this will improve their chances of reaching the summit and returning safely. These hikers could just start to hike and hope that, because they stay on the trail, they will be safe. A better approach, however, would be to consider and plan for things that could still go wrong. They might consider the possibility of sunstroke, dehydration, falls, insect bites, and so on. The hikers could then manage these dangers by bringing enough water, wearing proper clothing, carrying a first-aid kit, and taking similar measures. With planning and preparation, the hikers can maximize their chances of reaching and returning from the summit safely—but they still acknowledge that there are inherent hazards on the journey.

Like the hikers, laboratory professionals recognize that there are inherently **hazardous** situations and materials in a laboratory that *pose potential threats to their well-being or the safety of others.* Each of these laboratory hazards has an associated **risk**, *which is the likelihood that the hazard will result in harm.* Biotechnology laboratories contain numerous sorts of hazards, including the following:

- **Chemicals** that are flammable, explosive, or toxic
- **Biological agents** that can cause disease, injury, or death
- **Electrical/electronic equipment**
- **Extremes of heat and cold**
- **Pressurized gas cylinders**
- **Special equipment** (e.g., centrifuges and autoclaves) that can be dangerous
- **Sharps** (e.g., broken glass, razor blades, scalpel blades)
- **Ultraviolet light sources** that can cause serious burns and damage to exposed skin or eyes

Laboratory professionals anticipate these hazards and develop safety programs that greatly reduce risk, with the understanding that no workplace or activity is risk free.

Classroom Activity 1: Performing A Risk Assessment

OVERVIEW

One of the key steps in developing an effective laboratory safety program is to perform a series of risk assessments. A **risk assessment** *is an analysis of a procedure or situation in which the hazards and their associated risks are identified.* Personnel then develop plans to manage and minimize the risks. Risk assessments are not unique to laboratories; the process is applicable to numerous situations and settings. In this activity you will perform the following tasks:

- Practice identifying, assessing, and addressing hazards in familiar situations.
- Learn the difference between a *hazard* and a *risk*.
- Learn to anticipate consequences of unsafe conditions.
- Develop strategies to address unsafe conditions.

CLASSROOM ACTIVITIES

You should be able to complete this activity in approximately 45 minutes, with additional time allotted for class discussion.

Part A: Individual Risk Assessment of a Familiar Place

A.1. Identify a place that you know well. Each person should list and describe his or her place on the worksheet at the end of this activity.

    A.1.1. Envision this place and identify three hazards present. List them on the worksheet provided. Describe the risk(s) associated with each hazard. An example is shown on the worksheet.

    A.1.2. For each hazard, discuss a remedy to reduce the associated risk and indicate the level of risk reduction accomplished by this remedy, as demonstrated in the example.

A.2. Discuss your individual answers with two or three classmates.

Part B: Analysis of an Accident

B.1. Recall an accident that you have experienced that was caused, at least in part, by unsafe conditions or practices. Describe this accident on the worksheet.

B.2. Discuss the accidents you recalled with three or four other students. Discuss the safety issues that played a role in each incident and whether any remedies were employed to prevent the accident from reoccurring.

B.3. Each group should share one or two stories with the entire class.

RISK ASSESSMENT WORKSHEET

Part A: Individual Risk Assessment of Familiar Places

<u>Example</u>	<u>Your Situation 1:</u>
Situation: *Home*	**Situation:**
Hazard: *Frayed cord in living room*	**Hazard:**
Risk: *May cause fire or shock*	**Risk:**
Remedy: *Replace cord*	**Remedy:**
Level of risk reduction: *High*	**Level of risk reduction:**
(High reduction is almost complete elimination of risk.)	

Your Situation 2:

Situation:

Hazard:

Risk:

Remedy:

Level of risk reduction:

Your Situation 3:

Situation:

Hazard:

Risk:

Remedy:

Level of risk reduction:

Part B: Analysis of an Accident

Explain the accident/situation:

What inadequate safety measures played a role?

What remedies were employed afterwards (if any) to prevent such an incident from happening in the future?

DISCUSSION QUESTIONS

1. How does the risk assessment process contribute to workplace safety?
2. What are the **root causes** of the accidents you discussed? For instance, suppose an individual was injured when he tripped over a broom that a coworker left on the floor. The broom is the immediate cause of the accident, but something else might be the *root* cause. Perhaps the facility is too crowded and there was no place to put away the broom. Perhaps everyone is so overworked that they have no time to put anything away in its proper storage location. Perhaps the individual who left the broom on the floor was poorly trained and did not know where to put it. You can see that the broom itself is not the root cause of the problem. If the root cause of the accident was poor supervision, overcrowding, or overwork, then simply putting the broom in the closet will not solve the safety problem in the long term.
3. At the conclusion of the class discussion, write a brief paragraph summarizing any insights that you gained regarding workplace safety by completing this activity.

Classroom Activity 2: Exploring Safety-Related Government Websites

OVERVIEW

The purpose of this classroom activity is to explore how government shares the responsibility with employers and employees to create a safe workplace. You will visit the websites of two government agencies involved in safety: the Centers for Disease Control (CDC) and the Occupational and Safety Health Administration (OSHA). In this activity, you will perform the following tasks:

- Examine the content of each website.
- Evaluate the types of content available.
- Compare the missions of these two organizations and the materials and resources that they distribute.
- Begin to study safety practices that should be used when working with hazardous chemicals.

Note that the topics highlighted in this activity are a starting point for exploring these websites. You are encouraged to delve into additional topics other than those indicated in this activity.

BACKGROUND

A. Shared Responsibility for Workplace Safety

Now that you have completed a sample risk assessment in Classroom Activity 1, you probably have a better appreciation of the time, effort, and thought required to analyze just a single procedure or situation. If the staff of every laboratory had to perform a new risk assessment for every protocol, procedure, or situation that they might encounter, it is hard to imagine that anyone would ever be ready to begin laboratory work. Fortunately, the responsibility for safety in the workplace is shared:

- **Federal governmental agencies** create **regulations** *that must be followed, by law,* for managing hazards in the laboratory.
- **Employers** are required to comply with governmental regulations to create and maintain a safe work environment. Employers must provide training, create plans to respond to possible emergencies, designate safety officers, maintain libraries with hazardous chemical information, provide safety equipment, and so on.
- **Employees** are required to work in a way that is consistent with their institution's safety program to maintain a safe work environment.

B. Governmental Agencies and Safety

Five federal government agencies have responsibilities relevant to safety in the biotechnology workplace:

- **OSHA** *is the major agency dealing with worker (including laboratory personnel) safety.*
- **The CDC** *researches and deals with biohazardous agents.* Although we primarily think of the CDC's role in combating disease, the CDC also provides information about many aspects of workplace safety.
- **The Environmental Protection Agency (EPA)** *regulates disposal of waste. It is also concerned with release of genetically modified organisms into the environment.*
    - **The U.S. Department of Transportation (DOT)** *deals with transportation of hazardous materials.*
    - **The Nuclear Regulatory Commission (NRC)** *controls radioactive materials.*

C. The "Right-to-Know" Law—You Have a Right *and* an Obligation to Know

Let us return once more to our hikers. We determined that a safe hike to and from the summit required planning. We identified hazards that the hikers could expect to encounter, for example, extremes of temperature. The risk associated with this hazard could be reduced by packing the appropriate clothing. Of course, the "appropriate" clothing depends upon the temperatures that the hikers encounter. Will it be hot, cold, or, perhaps, both? These hikers need information to make a wise decision about packing, for instance, a reliable weather forecast. Creating and maintaining a safe work environment similarly requires access to information.

In 1983, recognizing that personnel need information to work safely, OSHA announced the **Federal Hazard Communication Standard (HCS)**, more commonly known as the **"right-to-know" law**. *This law regulates the use of hazardous materials in the workplace and emphasizes the responsibility of employers to provide safety information to their employees.* The right-to-know law requires employers to clearly identify hazards in the workplace, label chemicals, provide information on safe handling and storage of chemicals, and provide worker training programs. The right-to-know law did not originally target laboratory settings, but its scope was extended in 1990 to cover laboratory workplaces. One of the major requirements of this extension was that all institutions that oversee laboratories file a **Chemical Hygiene Plan (CHP)**. The **CHP** *is an extensive written manual that contains procedures and policies regarding safety issues including waste disposal, laboratory monitoring, staff safety training, record keeping, laboratory protective clothing, and accident and emergency response.* In laboratories that have more than three workers, there must be a **laboratory safety officer**, *whose role is to oversee safety-related issues in the laboratory and ensure that the CHP is followed.*

In 2012, the HCS was aligned with the Globally Harmonized System of Classification and Labelling of Chemicals (GHS), providing a standardized approach to hazard identification, chemical labeling, and safety data sheets (formerly known as material safety data sheets, MSDS).

OSHA's right-to-know law is based on the premise that knowledgeable employees provide the foundation for workplace safety. As an employee, you have both the right and the obligation to be informed, and you will have many sources of information available. Catalogs, reference manuals, and websites help laboratory personnel make informed safety plans. Institutions generally require new employees to complete safety training where hazards specific to that laboratory site are explained. The CHP and the safety officer are sources of information specific to a particular facility. The websites of several government agencies are important resources for safety information. Two of these websites, those of OSHA and CDC, are explored in the following Classroom Activities.

CLASSROOM ACTIVITIES

Part A: Exploration of the CDC Website

A.1.   Enter the CDC website by either typing http://www.cdc.gov into your browser or by searching for "Centers for Disease Control and Prevention" using an internet search engine.

   A.1.1.   Explore some of the content of the CDC website, taking notes on the following:

   A.1.1.1.   What is the mission of the CDC? (Hint: Enter "CDC Mission" into the search box for the CDC website or type http://www.cdc.gov/about/organization/mission.htm into your internet browser.)

   A.1.1.2.   How is this agency organized? (Hint: Check the "leadership" menu.)

   A.1.1.3.   What resources does the CDC provide related to laboratory safety? (Hint: Type http://www.cdc.gov/Workplace/ into your internet browser.) List at least five different topics covered by the CDC that you think relate to laboratory safety. (You will find many resources relating to COVID-19, but there is also information about other hazards on the website.)

A.2. Explore the following specific topics by entering these search terms into the CDC website search box. Examine the information that appears and, for each term, list two links that you find to be useful:
    A.2.1. Chemical safety
    A.2.2. Eye safety
    A.2.3. Protective clothing

A.3. Write a brief review of the CDC website directed toward a fellow student beginning his or her first laboratory job working with hazardous substances. Discuss the following topics:

    A.3.1. What is the CDC?
    A.3.2. How is the CDC organized?
    A.3.3. What does the CDC do?
    A.3.4. What is the CDC's website address?
    A.3.5. What information does the website provide concerning chemical safety?
    A.3.6. What three areas do you consider a "must-read" for the new laboratory worker?

Part B: Exploration of the OSHA Website

B.1. Enter the OSHA website by either typing http://www.osha.gov into your browser or by searching for "OSHA" using an internet search engine.
    B.1.1. Explore some of the content of the OSHA website, taking notes on the following:
        B.1.1.1. What is the mission of OSHA? (Hint: Find the "About OSHA" section.)
        B.1.1.2. Find the OSHA publication called "Laboratory Safety Guidance." Based on the table of contents, what hazards does OSHA recognize as being significant in laboratories?

B.2. Explore the following specific topics by entering the following search terms into the OSHA website search box. Examine the information that appears and list two links for each term that you find to be useful:
    B.2.1. Chemical safety
    B.2.2. Laboratory safety
    B.2.3. Eye safety
    B.2.4. Protective clothing

B.3. Write a brief review of the OSHA website directed towards a fellow student beginning his or her first laboratory job working with hazardous chemicals. Discuss the following topics:
    B.3.1. What is OSHA?
    B.3.2. What is OSHA's website address?
    B.3.3. What information does the website provide?

B.4. How do the resources relating to chemical and laboratory safety that are available at the OSHA website compare to the information at the CDC website?

Classroom Activity 3: Responding to Emergencies

OVERVIEW

The purpose of this activity is to consider possible laboratory emergencies and how to prepare for them. There is no way to completely eliminate risk in a laboratory environment, so it is important to plan for emergencies. In this activity you will perform the following tasks:

- Create a map identifying locations of emergency equipment.
- Identify evacuation routes.

BACKGROUND

Situations in which laboratory safety has been compromised can be classified as either accidents or emergencies. Accidents, which are unexpected occurrences that result in harm to the person or the workplace, are far more common than emergencies. Commonplace examples include collisions between coworkers, cuts or burns, and tripping over laboratory stools. Accidents are more likely to happen when workers have not adequately prepared for their work, are tired, or are not concentrating on the job at hand. Emergencies differ from accidents in that an immediate response is required to prevent further harm to people or the workplace.

One of the most important ways to plan for emergencies is to know what resources are available (see Figure 1.2). Standard emergency equipment in laboratories includes

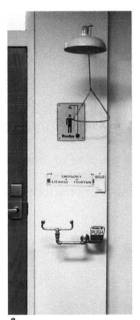
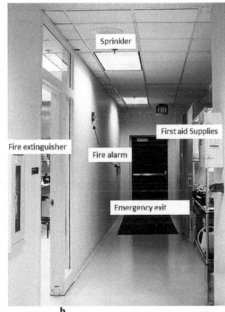

a   b   c

FIGURE 1.2   Looking around the laboratory at emergency equipment. (a) This picture contains two devices for emergencies: a safety shower and eyewash station. (b) The laboratory door is on the left of this hallway. Fire extinguishers, such as the one near the door, are located in hallways outside the laboratory so people must exit the laboratory before beginning to fight the fire. A first-aid kit is mounted on the wall across from the fire extinguisher. What other emergency features do you see in this photo? (c) This wall-mounted fire blanket smothers fires by eliminating oxygen. To use: put on fire resistant gloves, if available. Remove the blanket and unfold it. Hold the blanket with your hands wrapped in its top edge to protect them. Cover the flames with the blanket or throw it over the fire. You can wrap a person in a fire blanket if his or her clothing is on fire. This particular fire blanket is coiled in the cabinet in such a way that a person whose clothing is on fire can hook their arm through the strap that extends from the cabinet and pull. The blanket will uncoil and the person can wrap his or her body into it.

fire extinguishers, eyewash stations, fire blankets, emergency showers, and first-aid kits. Discussion of the location and use of emergency equipment should occur before performing work in the laboratory.

A laboratory fire is a type of emergency that unfortunately sometimes occurs—even in a teaching laboratory. Laboratory fires are especially dangerous because some chemicals commonly used in laboratories are extremely flammable or even explosive. Also, the release of chemicals into the environment, as may happen in a fire, may injure anyone in the vicinity and poses a special danger to firefighters. Training in how and when to use fire extinguishers is often available through institutional safety departments, and it is strongly recommended that everyone take advantage of this training.

Some emergencies require evacuation of the laboratory. Sometimes the nature of the emergency may make the nearest exit inaccessible. It is therefore important to be aware of all possible exits from the laboratory space and to plan alternate evacuation routes. It is also important to plan for everyone to meet at a prearranged location in the event of evacuation so that you know everyone has safely left the laboratory.

CLASSROOM ACTIVITIES

Part A: Mapping Emergency Laboratory Equipment

A.1.   Obtain a sheet of graph paper.

A.2.   Sketch the floor plan of your laboratory in the context of the building in which it is located. Be sure to indicate hallways and exit doors to the outside of the building.

A.3.   Create a map of your laboratory space, including lab benches, doors, windows, sinks, large pieces of equipment, and similar items.

A.4.   Locate the following emergency equipment and clearly label each item on your map:

    A.4.1.   Fire extinguisher

    A.4.2.   Eyewash station

    A.4.3.   Emergency shower

    A.4.4.   Fire blanket

    A.4.5.   First-aid kit

A.5.   Take a tour of your laboratory with your instructor or lab director who will instruct you in how to use the safety equipment.

A.6.   Use YouTube (or a similar website) to find videos on how to use the following safety equipment:

    A.6.1.   Fire extinguisher

    A.6.2.   Eyewash station

    A.6.3.   Emergency shower

    A.6.4.   Fire blanket

    A.6.5.   First-aid kit

Part B: Determining Evacuation Routes

B.1.   Using colored ink, clearly label the various evacuation routes from the laboratory. Be sure that your route extends all the way to the outside of the building.

B.2.   Identify an outside meeting site for gathering in the event of an evacuation.

B.3.   Your instructor may stage a mock emergency and evacuation of the laboratory. Different individuals may need to use different evacuation routes for the same emergency depending upon their location in the laboratory relative to the emergency site. Meet at the identified gathering site.

SAFETY PART 2: WORKING SAFELY WITH CHEMICALS

What thing is not a poison? All things are poison and nothing is without poison. It is the dose only that makes a thing not a poison.

—Paracelsus, 16th century physician, alchemist, and scientist

A. TYPES OF HAZARDOUS CHEMICALS

i. **What Are Hazardous Chemicals?**

Hazardous chemicals *are those that have the potential to cause injury or harm to people, the workplace, or the environment.* Many chemicals are used in laboratories. While you cannot be expected to remember every chemical and its associated hazards, it is possible to be familiar with the general categories of hazardous chemicals. The safety practices that are used with one chemical in a category are generally applicable to other chemicals in the same category.

ii. **Flammable Chemicals**

Flammables *are chemicals that are easily ignited or burned.* (Note that "flammable" and "inflammable" are synonymous terms.) Alcohols and organic solvents are examples of flammable chemicals routinely used in biology laboratories.

Flammable chemicals are generally **volatile**, *meaning they evaporate quickly at room temperature.* This is a concern because vapors can invisibly spread across a laboratory, suddenly igniting if they come in contact with a flame and spreading fire rapidly across a large space.

Some flammable chemicals are more easily ignited than others. The relative flammability of a chemical is expressed by its **flash point**, which is *the temperature at which enough vapor is emitted that the chemical will burn in the presence of an ignition source.* Chemicals with lower flash points are more flammable than those with higher flash points. For example, the flash point of 95% ethanol is 13°C (55°F), acetic acid is 40°C, and that of glycerol is 160°C. This means that ethanol will burn at room temperature (or even cooler) if there is an open flame, but acetic acid and glycerol will not.

iii. **Corrosive Chemicals**

Corrosive chemicals *are capable of destroying human tissue (and some types of equipment) on contact.* Biotechnologists frequently use corrosive chemicals, the most common of which are acids and bases. While people generally are aware of the dangers of spilling acids and bases on their skin or splashing them in their eyes, they may not realize that corrosive chemicals are also harmful by inhalation.

iv. **Reactive Chemicals**

Most chemicals can undergo chemical reactions to some extent, but those that are hazardous in the laboratory have the potential to participate in violent chemical reactions. These violent reactions may result in the generation of excessive heat, hazardous chemical products, and in extreme cases, explosions.

B. TOXIC CHEMICAL HAZARDS

i. **Definitions**

Toxic chemicals *are biological poisons.* Toxics are a category of hazardous chemical, just as corrosives and flammables are categories.

Toxic chemicals may affect any system of the body, such as the nervous system, kidneys, and liver. Many common laboratory chemicals are toxic at some level of exposure. A skull-and-crossbones emblem on the label denotes a chemical that is very toxic.

Acute toxic agents *are those that can be damaging in a short period of time, even with a single exposure.* **Chronic toxic agents** *are those that are able to accumulate in the body over multiple exposures.* Chronic agents can be dangerous over time, even if a person uses only small quantities of the substance at any given time.

ii. **Mutagens, Carcinogens, and Teratogens**
  ■ A **carcinogen** *is a type of toxic chemical that is known to cause cancer in animals and is assumed to also cause cancer in humans.*
  ■ A **mutagen** *is a type of toxic chemical that alters the genetic makeup of a cell.* The effects of mutagenic chemicals are believed to be cumulative and occur despite natural cellular DNA repair processes. All mutagens are also suspected to be carcinogens. Certain chemicals commonly used in biotechnology laboratories are suspected mutagens and/or carcinogens; ethidium bromide is an example.
  ■ A **teratogen** *is a type of toxic chemical that is known to cause defects in a developing fetus.* It is best for women who are pregnant or planning to become pregnant to completely avoid working with teratogenic chemicals. If this is not possible, such women should take special care when planning their work.

iii. **Routes of Entry for Toxic Chemicals**
    There are four routes by which chemicals may enter the body:

• **Ingestion**, *through the mouth*	• **Inhalation**, *through the lungs*
• **Injection**, *directly through the bloodstream*	• **Skin contact**, *by absorption through the skin or eye contact*

Most exposures in the laboratory occur through either skin contact or inhalation because it is usually easy to avoid ingesting or injecting chemicals.

iv. **Symptoms of Exposure**
    Although you will obviously avoid exposing yourself or others to toxic substances, it is still imperative for you to be aware of the physical signs indicating accidental exposure. The symptoms of toxicity vary depending upon the exposure (acute or chronic), the dose, and the type of agent. Common symptoms of toxic chemical exposure are shown in Table 1.1.

Most of the symptoms of acute exposure are like those you would experience when coming down with the flu. However, if you move to fresh air for 15 minutes and these symptoms subside, then the symptoms return when you go back to the laboratory, or if multiple laboratory workers suffer the same symptoms, you should consider the possibility of chemical exposure.

TABLE 1.1   Symptoms of Exposure to Toxic Chemicals

Symptoms of Acute Exposure	Symptoms of Chronic Exposure
• Headache	• Rashes or other dermatitis
• Dizziness	• Unusual body odor
• Nausea	• Unusual taste in the mouth or breath odor
• Coughing	
• Difficulty breathing	• Unusual urine or skin color
• Irritation of the eyes, nose, or throat	• Numbness or tremors
• Difficulty speaking	
• Uncoordinated movement	

v. **Assessment of Toxicity**

Toxicity is relative; everything, even water, can be toxic. Many life-saving medications are toxic if an overdose is taken. Most toxic chemicals are those that require a relatively small level of exposure (or smaller dose) to cause harm.

The relative toxicity of a chemical is commonly reported as its **lethal dose-50% (LD_{50})**, which is *the amount of chemical that has been shown to cause death in 50% of laboratory animals tested*. Chemicals that are designated as "highly toxic" have an LD_{50} of less than 500 mg/kg when administered by ingestion or injection, and less than 1 g/kg when administered by absorption through the skin.

The toxicity of a volatile or gaseous chemical is reported by its **lethal concentration-50% (LC_{50})**, which is *the concentration of the chemical in the air that will kill 50% of laboratory animals tested*. This concentration is expressed in parts per million (ppm). A "highly toxic" gas will have a LC_{50} of less than 2000 ppm.

C. LABELS AND SAFETY DATA SHEETS*

As a new employee in a laboratory work situation, you will probably not be familiar with all the chemicals in use. Fortunately, the right-to-know law requires that you have access to safety information.

Reading the label of a chemical stock bottle is your first step toward working safely with it. Even if you have used a chemical before (perhaps especially if you have done so), you should read the label before opening the bottle. People tend to get comfortable when performing familiar tasks, and reading a hazardous chemical label is a quick reminder of necessary precautions.

Manufacturers sometimes label stock bottles using the National Fire Protection Association (NFPA) **Hazard Diamond System** *to provide a quick reference as to a chemical's flammability, health hazard, reactivity, and other special concerns* (see Figure 1.3). The diamond consists of four quadrants: The top quadrant contains *flammability information*. The left quadrant contains information about *health hazards associated with the chemical*. The right quadrant includes information about *reactivity and explosion hazard*. The bottom quadrant is sometimes used to indicate *special safety concerns* relating to a particular chemical. The bottom quadrant may also be used instead to provide information relating to *personal protective equipment*.

Sometimes you will not have access to the stock bottle, but rather you will be working with a reagent made and labeled by a coworker. Everyone in a laboratory needs to be careful to properly label reagents in order to minimize risk to others. Labels should *always* include any relevant safety information. For example, a beaker containing a strongly acidic or basic substance should always be labeled as such. Moreover, such a beaker should always be securely covered and positioned where it cannot be accidentally spilled.

The label on a chemical's bottle provides a quick overview of the hazards associated with a particular chemical. More *comprehensive information* can be obtained by consulting the **Safety Data Sheet (SDS)** for that chemical. The chemical's manufacturer will supply the SDS to the laboratory at the time of purchase. SDSs are discussed in more detail in Classroom Activity 4.

---

* The GHS has been evolving in recent years and SDSs have become easier to interpret with each revision. Recent SDSs use an internationally consistent set of Hazard Statements and Precautionary Statements. They also use specific pictograms to represent common hazards, such as flammability or toxicity. At the time of writing, the most recent version of the GHS can be found at: https://unece.org/transport/documents/2021/09/standards/ghs-rev9.

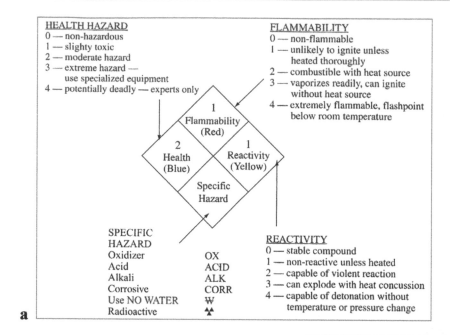

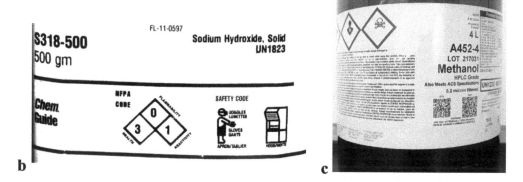

FIGURE 1.3   Reading labels. (a) The Hazard Diamond System provides rapidly interpretable information about a chemical's hazards. On the example shown in this diamond, the chemical is moderately toxic, only slightly flammable, and nonreactive unless heated. (b) You may also see alternative forms of the hazard diamond triangle in which personal protective equipment is identified. Sodium hydroxide is a base that can cause serious injury. Therefore it is rated a 3 for health hazard. Goggles, a lab apron (biologists would probably use a lab coat), gloves, and a fume hood are recommended to avoid injury. (c) A label that conforms to recent GHS requirements. The label has internationally accepted pictograms, seen on the top left, that correspond to specific hazards. The skull and crossbones indicates acute toxicity. The pictogram to the left indicates that there is a specific toxicity hazard. Underneath the pictograms there is detailed text that describes hazards and precautions to use when handling this chemical

Classroom Activity 4: Understanding the Chemicals with Which You Work

OVERVIEW

In this classroom activity, you will practice interpreting SDSs. To do so, you will compare the SDSs of two chemicals. One of these chemicals, sodium azide, is toxic at low exposure levels. The other chemical, sodium chloride (table salt), is relatively nontoxic. Sodium chloride (NaCl) should be familiar to you, but you may never have heard of sodium azide (NaN_3), which is commonly used as a preservative in hospitals and laboratories to prevent bacterial growth.

BACKGROUND

It is a regulatory requirement that the SDSs for all chemicals used in a workplace must be located in a location where employees may reference them. If, for some reason, an SDS is missing from your laboratory, SDSs can readily be found online by using an internet search engine and typing in the chemical's name.

Prior to 2012, there was no standard format for an SDS and different manufacturers varied somewhat in their format. Now the format for an SDS is specified as shown in Table 1.2.

It requires some experience to properly interpret SDSs. This is because they often use the same language to describe very different chemicals. The section on "First-Aid Measures" for sodium azide and sodium chloride are almost identical, for example, even though sodium azide is highly toxic while sodium chloride is ordinary table salt. With more experience, you will realize that sodium chloride is toxic only at high doses that are rarely encountered. Note also that the requirements for SDSs have evolved and newer SDs are easier to interpret than they were in the past.

CLASSROOM ACTIVITIES

1. Read the following SDSs for sodium chloride and sodium azide from Real Fine Chemicals, a fictitious company. Although the company is fictitious, the chemicals and the information about them on the SDSs are real.

2. Compare the NFPA ratings for health, flammability, and reactivity for the two chemicals. Based on these ratings alone, which chemical is most hazardous? Why?

TABLE 1.2   Information Contained in an SDS

Section 1: **Identification** including product identifier, manufacturer, name address, phone number, emergency phone number, recommended use, restrictions on use.	Section 7: **Handling and storage**, incompatibilities.
Section 2: **Hazard(s)** identification, required label elements.	Section 8: **Exposure controls/ personal protection.**
Section 3: **Composition, information on ingredients.**	Section 9: **Physical and chemical properties.**
Section 4: **First-aid measures.**	Section 10: **Stability and reactivity.**
Section 5: **Firefighting measures.**	Section 11: **Toxicological information.**
Section 6: **Accidental release measures**, emergency procedures, containment, cleanup.	Section 12: **Ecological information.**
	Section 13: **Disposal considerations.**
	Section 14: **Transport information.**
	Section 15: **Regulatory information.**
	Section 16: **Other information.**

3.   Look for the words "Hazard Identification" in each SDS.
    3.1.   Compare the two chemicals with respect to emergency measures described.
    3.2.   Describe those emergency measures.
4.   Find the section of the two SDSs that describes toxicity and compare the toxicity of the two chemicals.
    4.1.   What is the LD_{50} for sodium chloride in rats?
    4.2.   What is the LD_{50} for sodium azide in rats?
    4.3.   If an experimenter tested 300 rats and gave them 27 mg/kg of sodium azide, how many rats would be expected to die in the experiment?
    4.4.   What are the words used in the "Hazard Identification" section to describe the toxicity of sodium azide in the United States?
    4.5.   Is sodium azide toxic by inhalation?
        4.5.1.   At what level is inhalation toxicity reported?
        4.5.2.   How could you avoid exposure by inhalation?
    4.6.   What are the routes of entry for sodium azide into the body?
        4.6.1.   Which route(s) do you think is the biggest danger to you as a laboratory technician?
        4.6.2.   How could you protect yourself from this hazard?
5.   Find the section of the SDS that deals with disposal. Compare the language used for NaN_3 and NaCl.
    5.1.   What do you think could happen if sodium azide solution was poured down the drain of the laboratory sink?
    5.2.   How does the SDS direct you to dispose of sodium azide?
6.   Your supervisor asks you to make a 10% stock solution of sodium azide in water. When you get the stock bottle, you notice the skull-and-crossbones on the label.
    6.1.   What concerns do you have about making this solution?
    6.2.   After reading the SDS for sodium azide, what could you do to protect yourself?
    6.3.   What should you do to protect others in the laboratory from exposure to sodium azide?
    6.4.   What should you do to protect the environment from sodium azide?
    6.5.   How would you label the solution that you make?
7.   You are performing an experiment of a certain type for the first time. The procedure you are following calls for incubating your sample with Blocking Buffer. You find the buffer in the refrigerator, and there is a label on the buffer bottle:

    Blocking Buffer
    PBS with 3% BSA and 0.02% azide

Based on the presence of the sodium azide in the solution, what precautions do you think you should take when performing this experiment?
8.   In some cases, you may need to ask your instructor or supervisor where you can find more information about a particular chemical. Other times, you can find out about a particular chemical from online sources. The CDC website has information about sodium azide. Go to the CDC website, and look up sodium azide. Compare the information on the CDC's website to that in the SDS.

REAL FINE CHEMICALS

SAFETY DATA SHEET

SECTION 1: IDENTIFICATION, PRODUCT AND COMPANY INFORMATION

Product Name: Sodium Chloride Realpure grade Product Number: 123456
Company: Real Fine Chemicals, 110 Pharmaceutical Way, Madison, WI 53704
Technical Phone: 111-555-3333   Emergency Phone: 111-333-4444

SECTION 2: HAZARDS IDENTIFICATION

GHS label elements, including precautionary statements:
Not a hazardous substance.

NFPA Rating:
Health: 0   Flammability: 0   Reactivity: 0

Effects of Overexposure: Irritation to skin, eyes, and mucous membranes.

SECTION 3: COMPOSITION/INFORMATION ON INGREDIENTS

Name: Sodium Chloride, meets Realpure testing spec.
CAS: 8647-14-5   Formula: NaCl
Synonyms: Common salt, table salt, etc.   RTECS Number: VZ4725000

SECTION 4: FIRST-AID MEASURES

Eyes: Flush with water 15 min. Ensure adequate flushing by separating eyelids with fingers.
Skin: Flush with clean water.
Inhalation: If inhaled, remove to fresh air.
Oral Exposure: If swallowed, wash out mouth with water, provided person is conscious. Call a physician.

SECTION 5: FIREFIGHTING MEASURES

Flash point: None
Firefighting:

  Protective equipment: Wear self-contained breathing apparatus and protectiveclothing to prevent contact with skin and eyes.
  Specific Hazard: Emits toxic fumes under fire conditions.

SECTION 6: ACCIDENTAL RELEASE MEASURES

Spill Release Procedure: Sweep up, place in bag, and hold for waste disposal. Avoid raising dust.
Ventilate area and wash spill site after material pickup is complete.

SECTION 7: HANDLING AND STORAGE

Avoid inhalation. Avoid contact with eyes, skin, and clothing. Avoid prolonged or repeated exposure.

SECTION 8: EXPOSURE CONTROLS/PERSONAL PROTECTION

Safety shower and eye bath. Mechanical exhaust required.

Personal Protective Equipment:
Respiratory: None required. NIOSH/MSHA-approved respiration device in accordance with exposure of concern.

SECTION 9: PHYSICAL/CHEMICAL PROPERTIES

Appearance and Odor: Solid, white crystalline, no odor
Vapor Pres: 1   SG/density: 2.163 g/cm^3 Molecular Weight: 58.44 amu

SECTION 10: STABILITY AND REACTIVITY

Stability: Stable
Materials to avoid: Strong oxidizing agents
Hazardous Decomposition Products: Sodium/sodium oxides, hydrogen chloride gas
Hazardous Polymerization: Will not occur

SECTION 11: TOXICOLOGICAL INFORMATION

Route of Exposure:

Skin Contact: May cause skin irritation
Skin Absorption: May be harmful if absorbed through the skin,
Eye Contact: Can cause irritation or redness due to abrasion
Inhalation: May cause irritation to mucous membranes
Ingestion: May be harmful if swallowed

Signs and Symptoms of exposure:

Ingestion of large amounts causes vomiting and diarrhea
Dehydration and congestion may occur in internal organs
Chemical, physical, and toxicological properties have not been thoroughly investigated

Toxicity Data:

Oral: Rat, 3000 mg/kg LD$_{50}$
Irritation Data: Skin, Rabbit: 50 mg over 24 hours, mild irritation
Eyes: 100 mg, 24 hours, moderate irritation effect

SECTION 12: ECOLOGICAL INFORMATION

No data available

SECTION 13: DISPOSAL CONSIDERATIONS

By methods consistent with local, state, and federal regulations

…rest of SDS is omitted to save space

REAL FINE CHEMICALS
Safety Data Sheet

SECTION 1: PRODUCT AND COMPANY INFORMATION

Product Name: Sodium Azide, Realpure grade   Product Number 191113
Company   Real Fine Chemicals, 110 Pharmaceutical Way, Madison, WI 53704
Technical Phone: 111-555-3333   Emergency Phone: 111-333-4444

SECTION 2: HAZARDS IDENTIFICATION

Emergency Overview: Highly Toxic (USA), Very Toxic (European Union)
Heating may cause explosion. Very toxic by inhalation, skin contact, and ingestion. Contact with acids liberates very toxic gas. Very toxic to aquatic organisms; may cause long-term adverse effects in the aquatic environment. Readily absorbed through skin. May react with plumbing to form highly explosive metal azides.

Target organs: Nerves, heart

NFPA Rating:
Health: 4   Flammability: 0   Reactivity: 2

SECTION 3: COMPOSITION/INFORMATION ON INGREDIENTS

Name: Sodium Azide, meets Realpure testing spec.
CAS: 26628-22-8   Formula: NaN_3
Synonyms: Azide, Sodium   RTECS Number: VY8050000

SECTION 4: FIRST-AID MEASURES

Oral Exposure: If swallowed, wash out mouth with water provided person is conscious. Call a physician immediately.
Eyes: Flush with water 15 min. Ensure adequate flushing by separating eyelids.
Skin: Remove contaminated clothing/shoes. Wash with soap, mild detergent, and water.
Inhalation: If inhaled, remove to fresh air.

SECTION 5: FIREFIGHTING MEASURES

Explosion Hazards:

  Container explosion may occur under fire conditions. Azide reacts with many heavy metals to form explosive compounds. Move containers from fire area if possible.

Flash point: 300°C

Extinguishing Media:
  Suitable: Dry chemical powder Unsuitable: Water

Firefighting:

  Protective equipment: Wear self-contained breathing apparatus and protective clothing to prevent contact with skin and eyes.
Specific Hazard: Emits toxic fumes under fire conditions

SECTION 6: ACCIDENTAL RELEASE MEASURES

Spill Release Procedure: Don't touch spilled material. Stop leak if you can do so without risk. Use water spray to reduce vapors. For small spills, take up with absorbent material.

SECTION 7: HANDLING AND STORAGE

Handling: Do not breathe dust. Do not get in eyes or on skin or clothing. Avoid prolonged or repeated exposure.
Storage: Keep tightly closed. Store in a cool, dry place.
Incompatible Materials: Azide reacts with many heavy metals to form explosives

SECTION 8: EXPOSURE CONTROLS/PERSONAL PROTECTION

Safety shower and eye bath. Mechanical exhaust required
Personal Protective Equipment:
High Levels: Use supplied air respirator, helmet and hood

SECTION 9: PHYSICAL/CHEMICAL PROPERTIES

Appearance and Odor: Solid, no odor   Vapor Pres: 0.01 hPa
SG/density: 1.85 g/cm^3 Molecular Weight: 65.01 amu
Solubility: In water, complete, colorless 1 M in H_2O

SECTION 10: STABILITY AND REACTIVITY

Conditions of Instability: Heat sensitive.
Conditions to Avoid: Do not grind or subject to frictional heat.
Keep from contact with oxidizing materials. Fire or excessive heat may cause explosive decomposition.
Materials to Avoid: Halogenated solvents. Avoid contact with metals. Avoid contact with acid, acid chlorides.

SECTION 11: TOXICOLOGICAL INFORMATION

Route of Exposure:

  Skin Contact: May cause skin irritation
  Skin Absorption: May be fatal if absorbed through the skin
  Eye Contact: Can cause irritation
  Inhalation: May be fatal if inhaled. Material may be irritating to mucous membranes.
  Ingestion: May be fatal if swallowed

Signs and Symptoms of Exposure:

Exposure may cause nausea, vomiting, and headache. Profound hypotensive effect in animals.
Dehydration and congestion may occur in internal organs.
Chemical, physical and toxicological properties have not been thoroughly investigated
Chronic Exposure
Carcinogen

Toxicity Data:

Oral: Rat, 27 mg/kg LD_{50}
Inhalation: Mouse, 32.4 mg/m^3 LC_{50}

SECTION 12: ECOLOGICAL INFORMATION

Acute Ecotoxicity Tests EC_{50} *Daphnia pulex*   Time: 48 h   Value: 4.2 mg/L

SECTION 13: DISPOSAL CONSIDERATIONS

Contact a licensed professional waste disposal service. Observe local, state, and federal environmental regulations.

…rest of SDS is omitted to save space

Classroom Activity 5: Personal Protection

OVERVIEW

Previous activities in this section focused on understanding the hazards associated with chemicals. We will now start to talk about strategies and rules to minimize the risk of working with hazardous substances, particularly chemicals. In this activity you will consider the rationale behind the most basic rules for the following:

- Personal laboratory hygiene.
- Proper laboratory attire.
- Common types of personal protective equipment.

BACKGROUND

A. Personal Hygiene

Box 1.1 summarizes basic laboratory safety rules that are often referred to as rules of "personal hygiene." While the term might seem odd in this context, following these rules will protect you from the most foolish accidents, such as ingesting a hazardous chemical. Personal hygiene rules include such things as not eating, drinking, or smoking in the laboratory (indeed, as far as safety is concerned, you should not smoke anywhere).

B. Laboratory Attire
  i.   **Clothing**

Dress properly to protect yourself from hazards. Have as little exposed skin as possible, while keeping clothing easy to remove if necessary. Here are some considerations:
- **Shorts, skirts, and dresses** are not ideal laboratory attire because the legs are exposed.
- **Footwear** is an important defense against chemicals that spill or items that are dropped. Shoes that are appropriate for the laboratory have a closed toe; sandals are not acceptable. Rubber-soled shoes are best because they will provide nonskid contact with the floor.
- **Loose clothing and hair** should be avoided because they might become caught on equipment, be exposed to chemicals, or encounter flames.
- **Jewelry** provides a way for chemicals to be trapped in close contact with the skin, so it is common practice to remove watches and rings prior to laboratory work.
  ii. **Lab Coats**

Personal protective equipment (PPE) *is specialized clothing or equipment worn to protect oneself.* A laboratory coat is a type of PPE that is your first line of defense against chemicals that you are using and against chemicals that your colleagues may be using. A laboratory coat can be easily removed in the event of a chemical spill. In addition to protecting skin and clothing from chemical exposure, the laboratory coat—which should never be worn outside the laboratory—reduces the chance that chemical or biological agents will be tracked outside.

Laboratory coats should be completely buttoned and should fit properly. They should not be too loose or have sleeves that are too long because the sleeves and the areas around the waist pockets can get caught on equipment, lab drawers, and the like. Do not roll up the sleeves because the pocket formed by the rolled-up sleeves can catch chemicals or broken glass. Most workplaces provide laundry facilities so that lab coats do not need to be laundered at home. Students, unfortunately, seldom have access to such facilities and must wash theirs at home.

Laboratory coats do not protect against every chemical hazard: they can catch fire. they are not impermeable to corrosives; and they do not protect against ingestion or inhalation of hazardous substances. Nonetheless, they provide valuable protection in the laboratory.

Box 1.1   Personal Hygiene Rules

- While in the laboratory, never eat, drink, smoke, chew gum, apply cosmetics, or insert contact lenses into your eyes.
- Never store your food or drinks in a laboratory refrigerator or use a laboratory microwave to heat food.
- Do not use ice from a laboratory ice machine in your drinks.
- Do not eat or drink from laboratory glassware.
- Wash your hands frequently and always right before leaving the lab.

C. Eye Protection

Eye protection is arguably the single most important type of PPE in the laboratory. Peruse advertisements in scientific publications and you may see people working in laboratories without safety glasses, perhaps to look more attractive for the photo. You, however, should be smarter and not let vanity override your good sense.

Safety glasses, safety goggles, and face shields are three forms of eye protection. All three can be purchased to fit over normal glasses. Each type of eye protection has particular uses:

- **Safety glasses with side shields** are appropriate for most biological laboratory situations. Approved safety glasses will provide protection from chemicals and from minor impact, such as from broken glass. Biotechnologists should look for safety glasses that protect the eyes from ultraviolet (UV) light because UV light is commonly used in biotechnology applications.
- **Safety goggles**, unlike safety glasses, have a seal around the eyes to provide extra protection against large splashes. It is common practice for biologists to use safety glasses for routine work and safety goggles for situations where more protection is needed.
- **Face shields** cover the entire face and neck (see Figure 1.4). They should be worn over safety glasses or goggles. Face shields provide additional protection from impact and when working in situations where the face and neck are at risk as well as the eyes. Biotechnologists who work with unshielded UV light sources should always wear face shields to completely cover their face and neck and should also make sure their wrists and hands are completely covered.

D. Gloves

i.   **Purpose of Gloves**

Gloves, like lab coats and safety glasses, are another essential type of personal protection (see Box 1.2). Gloves can be used to protect your hands against a number of hazards including the following:
- **Extreme heat and extreme cold**, which are generally managed by thick, fabric gloves that resemble oven mitts
- **Sharp objects**, which may be handled using thick rubber gloves, although these gloves will not completely protect against punctures
- **Biological agents**
- **Hazardous chemicals**

Thin, disposable nitrile, vinyl, and latex gloves provide adequate protection against many nonpathogenic or low-risk biological agents and against many of the most common chemicals used in biotechnology laboratories. These thin gloves provide a barrier to protect the skin while still allowing good dexterity.

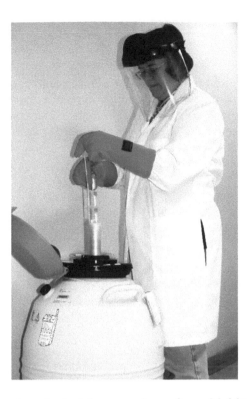

FIGURE 1.4   A laboratory technician wearing a face shield to remove stem cells from storage in liquid nitrogen. Liquid nitrogen is cold enough to change the physical properties of many materials, and it will cause the equivalent of third-degree burns if it contacts the skin.

They are, however, unsuitable for use with higher-risk biological agents and provide no protection against a number of particularly toxic chemicals. Note also that many people are allergic to latex or develop an allergy with frequent exposure. The trend is, therefore, away from using latex gloves.

ii. **Using Gloves Properly**

Gloves used incorrectly can be worse than useless. Why? First, remember that when using gloves to protect yourself, the glove is there to protect your skin from a hazardous substance. As you work with the hazardous substance, you may contaminate the outer surface of the glove. If you then touch any other object—such as a pen, lab coat pocket, doorknob, telephone, keyboard, or pipette—that object will likely become contaminated. You, your colleagues, and other people entering the laboratory are unlikely to know that items in the laboratory are contaminated and may touch them with bare hands. Moreover, contamination that is carelessly spread about the laboratory on gloves is likely to adversely affect later experiments and assays. Therefore gloves must be routinely changed frequently to avoid inadvertently contaminating other objects. Knowing when to take off and change gloves is just as important as knowing when to put them on.

A second issue with gloves is that some hazardous chemicals will readily penetrate gloves made of certain materials. It is thus possible to assume that a glove is providing protection when, in fact, it is not. When in doubt, consult the glove manufacturer's technical information.

A third issue with gloves is that inexpensive ones for routine work occasionally have small holes or imperfections such that your hands may be unknowingly exposed to the agent with which you are working. Where you need more assurance that your hands are completely protected, consider using two pairs of gloves at a time or purchasing more expensive surgical gloves to use underneath less-expensive gloves.

Box 1.2 summarizes issues relating to proper glove use.

Box 1.2   Proper Glove Use

- Change gloves *every time* they might have been contaminated!
- Change gloves frequently even if contamination is not suspected.
- Do not touch doorknobs, light switches, or similar surfaces while wearing gloves.
- Make sure the gloves you are using are appropriate for the hazard(s).
- Check gloves carefully for holes or tears.
- Keep cuts on your skin bandaged.
- Keep nails short.
- Make sure gloves cover wrists.
- Wash hands after glove use and anytime contamination of your skin is suspected.
- Remove gloves properly, as shown in Figure 1.5.

CLASSROOM ACTIVITIES

1. As an individual, provide a rationale for the following safety rules; that is, explain what might happen if the rule is not followed.
    - 1.1.   Do not wear necklaces or neckties in lab.
    - 1.2.   Do not wear loose clothing in lab.
    - 1.3.   Avoid wearing jewelry in lab.
    - 1.4.   Do not launder your lab coat at home (once you are no longer a student).
    - 1.5.   Do not wear open-toed shoes in lab.
    - 1.6.   Keep fingernails short.
    - 1.7.   Do not leave Bunsen burners unattended.
    - 1.8.   Use water baths, not Bunsen burners, when organic solvents are in use.
    - 1.9.   Wear face shields when using UV light.
    - 1.10.   Always dispose of sharps items, like broken glass, in the appropriate container.
2. Determine what type(s) of gloves are available in your laboratory and of what material they are made. Consult the manufacturer's information regarding these gloves. Will these gloves protect you from all chemicals? What are the limitations of these gloves?

3. Compare answers with your classmates.

4. Practice removing gloves as shown in Figure 1.5.

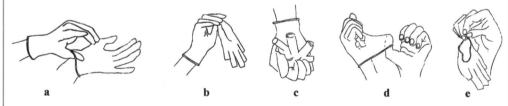

a   b   c   d   e

FIGURE 1.5   Removing gloves without contaminating your skin. (a) Hook a finger on the cuff of one glove, being careful not to contact the skin of your wrist. (b) Pull the glove from the hand inside out. (c) Roll up the removed glove in the palm of the remaining gloved hand. (d and e) Pull off the glove inside out; discard properly. (Artist: Dana Benedicktus)

Laboratory Exercise 1: Tracking the Spread of Chemical Contamination

OVERVIEW

In this laboratory exercise, you will simulate working with a hazardous compound. You will use relatively harmless fluorescent compounds instead of hazardous chemicals so that you can practice safely. Fluorescent compounds glow when exposed to UV light. You will, therefore, be able to find places where you have accidentally dispersed these compounds during your work. Surface contamination will not be so easily detected when working with most hazardous materials; therefore, it is critical that you learn to minimize the chance that contamination will occur.

In the laboratory, you will first prepare solutions of various fluorescent chemicals. You will then dilute these solutions, spot them onto filter papers, and allow the papers to dry. You will observe your filter papers with a handheld UV light to estimate the limit of detection for each fluorescent chemical. Finally, you will use the UV light to discover inadvertent contamination by your "hazardous chemicals."

As you complete this procedure, you will perform the following tasks:

- Observe the ease with which a hazardous chemical may contaminate a workplace.
- Consider the consequences of careless laboratory practice.
- Learn to minimize the likelihood of chemical contamination in the laboratory.
- Practice using PPE.
- Practice using SDSs.
- Work with dilutions.
- Explore the concept of the "limit of detection."

Planning Your Work: Tracking the Spread of Chemical Contamination

1. What is the purpose of diluting the fluorescent compounds?

2. There are some calculations in this exercise. Review these calculations, possibly with the assistance of your instructor and/or other students:

   a. In steps 6–15 of the following Laboratory Procedure, you will weigh out 0.5 g of each fluorescent compound. In step 17, you will dissolve the compounds in 50 mL of water. As a result, you will have

$$\frac{0.5\,\text{g compound}}{50\,\text{mL water}}$$

   Divide both the numerator and denominator of this ratio by 50:

$$\frac{0.1\,\text{g compound}}{1\,\text{mL water}}$$

   The concentration of each compound will thus be 0.01 g/mL. This is the concentration of the **original stock solutions**.

   b. In step 18, you will combine 10 mL of the original stock solution of each fluorescent compound with 10 mL of water. The original stock solution will be at a concentration of 0.01 g/mL. When you remove 10 mL of this stock, you remove $\left(\dfrac{0.01\,\text{g compound}}{1\,\text{mL}}\right)(10\,\text{mL}) = \textbf{0.1 g compound}$.

   You will combine the 10 mL of original stock solution with 10 mL of water for a total volume of 20 mL. The 0.1 g of compound is now in 20 mL, so the concentration will be

$$\frac{0.1\,\text{g compound}}{20\,\text{mL water}}$$

   Dividing the numerator and denominator by 20 gives a concentration of

$$\frac{0.005\,\text{g compound}}{1\,\text{mL water}}$$

   This will be the concentration of each fluorescent compound in the **first dilution**.

   c. In step 19, use the same logic to confirm that the concentration of compound in the second dilution will be 0.0005 g/mL.

   d. In step 20, confirm that the concentration of compound in the third dilution will be 0.00005 g/mL.

   e. In step 21, confirm that the concentration of compound in the fourth dilution will be 0.000005 g/mL.

3. What data will you collect in this exercise? How will you analyze your data?

Safety Briefing

- Fluorescein is hazardous by inhalation, and therefore, it is most hazardous when it is in solid form. Once it is diluted in liquid form, it is less hazardous. Weigh the solid fluorescein in the fume hood to minimize the risk of inhalation.

- Do not look directly into the UV lamp's light! UV light can damage the retina of your eye. Wear UV-safe glasses when using UV lights.

LABORATORY PROCEDURE

1.   Examine stock bottle labels and SDSs for all chemicals prior to use.
2.   Cover work area with bench paper, white if possible.
3.   Put on PPE: lab coat and safety glasses.
4.   Label four 200 mL beakers for your stock solutions: one for fluorescein, one for rhodamine, one for DayGlo, and one for Tide.
     4.1.   Include the chemical name, date, concentration, and your initials.
5.   Put on gloves.
6.   In a fume hood, weigh out 0.5 g of fluorescein (F). Place it in its labeled beaker. Alternatively, your instructor might have already weighed out this chemical for you.
7.   Remove your gloves and save them to check later for contamination.
8.   Put on new gloves.
9.   Weigh out 0.5 g of rhodamine (R). Place it in its labeled beaker.
10. Remove your gloves and save them to check later for contamination.
11. Put on new gloves.
12. Weigh out 0.5 g of DayGlo (D). Place it in its labeled beaker.
13. Remove your gloves and save them to check later for contamination.
14. Put on new gloves.
15. Weigh out 0.5 g of Tide laundry detergent (T). Place it in its labeled beaker.
16. Remove your gloves and save them to check later for contamination.

Note: At this point, your instructor will tell you whether or not to wear gloves.

- If you do wear gloves, be sure to change them every time they may have been contaminated.
- If you do not wear gloves for every step, wash your hands every time they may have been exposed to the chemicals.

17. Prepare four stock solutions.
     17.1.   Add 50 mL of purified water to each beaker.
     17.2.   Stir until each compound is dissolved. Your instructor will provide directions for stirring.
     17.3.   You have now prepared a 0.01 g/mL solution of each of the four chemicals. These *original solutions* are called your **stock solutions.**
18. Prepare the "first dilution" of each "hazardous" chemical stock solution (from step 17.3):
     18.1.   Label four large test tubes (that can hold at least 30 mL), or 50 mL beakers, or reagent bottles as follows:
              • Dilution I-F (where "F" is fluorescein)   • Dilution I-R (where "R" is rhodamine)
              • Dilution I-D (where "D" is DayGlo)   • Dilution I-T (where "T" is Tide)
     18.2.   Place 10 mL of the appropriate stock solution in each labeled vessel.
     18.3.   Add 10 mL of purified water to each tube or beaker and mix thoroughly.
     18.4.   Label each of these four **dilutions 0.005 g/mL**.

19.   Prepare the "second dilution" of each "hazardous" chemical from step 18.4.
    19.1.   Label four 100 or 200 mL beakers or reagent bottles as follows:
        • Dilution II-F • Dilution II-R   • Dilution II-D   • Dilution II-T
    19.2.   Place 10 mL of the appropriate first dilution from step 18 in each labeled beaker or reagent bottle.
    19.3.   Add 90 mL of purified water to each beaker and mix thoroughly.
    19.4.   Label each of these four dilutions 0.0005 g/mL.

20.   Prepare the "third dilution" of each "hazardous" chemical from step 19 as follows:
    20.1.   Label four 100 or 200 mL beakers or reagent bottles as follows:
        • Dilution III-F • Dilution III-R   • Dilution III-D   • Dilution III-T
    20.2.   Place 10 mL of the appropriate second dilution from step 19 in each labeled beaker or reagent bottle.
    20.3.   Add 90 mL of purified water to each beaker and mix thoroughly.
    20.4.   Label each of these four dilutions 0.00005 g/mL.

21.   Prepare the "fourth dilution" of each "hazardous" chemical from step 20.
    21.1.   Label four 100 or 200 mL beakers or reagent bottles as follows:
        • Dilution IV-F • Dilution n IV-R • Dilution IV-D   • Dilution IV-T
    21.2.   Place 10 mL of the appropriate third dilution from step 20 in each labeled beaker or reagent bottle.
    21.3.   Add 90 mL of purified water to each beaker and mix thoroughly.
    21.4.   Label each of these four dilutions 0.000005 g/mL.

22.   This table summarizes these dilutions:

Solution	Dilution of Original Stock Solution	Concentration of Solution (g/mL)
Stock	NA	0.01
Dilution I	1/2	0.005
Dilution II	1/20	0.0005
Dilution III	1/200	0.00005
Dilution IV	1/2000	0.000005

23.   Prepare filter papers.
    23.1.   Label four pieces of filter paper (each about 4" × 4"), one for each compound, using a pencil with your name, date, the exercise title, and the compound's name.
    23.2.   With a pencil, write a list of the dilutions (I to IV) on each filter paper.
    23.3.   Draw a light pencil line 3 cm long next to the name of each solution on each filter paper.
    Example of how one filter paper (for fluorescein) is labeled:
    Dilution I-F   _____
    Dilution II-F _____
    Dilution III-F _____
    Dilution IV-F _____
    23.4.   Place the four labeled filter papers onto lab bench paper.
24.   Apply drops to filter paper as follows:
    24.1.   For fluorescein, with a Pasteur pipette, slowly apply five drops of Dilution I in a row on the line labeled "Dilution I-F."
    24.2.   Repeat for fluorescein for Dilutions II, III, and IV.
    24.3.   Repeat steps 24.1 and 24.2 for the other three compounds.
25.   Allow the filter papers to dry.

26.   Observe the filter papers and record observations on the following worksheet.
    26.1.   Observe the filters with the normal room lights.
    26.2.   Observe the filter papers using a UV lamp. This is best done with the room lights off.
    26.3.   Describe on the following worksheet what you see when observing your filter paper and how it relates to the dilutions you prepared. What is the lowest concentration that you can see with the room lights? What is the lowest concentration you can see with the fluorescent light; therefore, what is the limit of detection?

27.   Use the UV light to check for contamination around your station. Your check should include at least the locations described in the following worksheet.

28.   Clean your work space.
    28.1.   Wipe up all spills and contamination with ethanol and paper towels.
    28.2.   Flush liquid waste down the drain with plenty of water.
    28.3.   Recheck the bench top and other areas for trace fluorescent materials.
    28.4.   Clean the balances and put away the materials.

Tracking the Spread of Contamination Worksheet				
Dilutions/Observations of Locations	Fluorescein	Rhodamine	DayGlo	Tide
Dilution I				
Dilution II				
Dilution III				
Dilution IV				
Lab bench				
Balance				
Pipettes and pipette aids; pipette bulbs				
Wastebasket				
Lab coat				
Gloves				
Floor				
Pen				
Lab notebook				
Fume hood				
Other				

LAB MEETING/DISCUSSION QUESTIONS

1.   What precautions do the SDSs recommend for the materials used in this lab?
2.   What areas exhibited contamination? Do you think that you could minimize the likelihood of having these areas of contamination in the future? What precautions would you need to take?
3.   How would your lab colleagues be affected if the materials used in this lab were highly toxic, pathogenic, or radioactive?
4.   When working with hazardous chemicals in the context of a procedure that you have not previously performed, it can be helpful to perform a "dry run" of the process. During a dry run, the procedure is carried out using a nonhazardous chemical in place of the hazardous one. A dry run allows the analyst to identify particularly dangerous parts of the procedure and to redesign the procedure to minimize risk. This exercise was essentially a dry run. Imagine that you must make Dilutions I–IV with chemicals that are hazardous by inhalation and ingestion and that are not fluorescent or visible. What changes to the procedure, as written, would you make on the basis of your dry-run results?

Classroom Activity 6: Analyzing Safety Issues in a Laboratory Procedure

OVERVIEW

In this activity, you will analyze safety issues involved in a laboratory procedure and develop strategies that could be used to protect people and laboratory equipment while performing this procedure. You will need to use the information provided in this unit, and you will need to consult outside resources, including SDSs and websites. You will not actually perform this procedure in the laboratory; rather, you will focus on the planning process.

As you work on this activity you will perform the following tasks:

- Use SDSs to obtain information.
- Choose appropriate PPE.
- Choose equipment that is compatible with the chemicals involved.
- Determine how to properly dispose of hazardous waste.

BACKGROUND

A. Chemical Compatibility

We have discussed methods to protect yourself and other people when working with hazardous chemicals. Remember, however, that hazardous chemicals may also harm equipment and degrade materials such as test tubes and gloves. It is, therefore, important to choose equipment that is chemically compatible with the chemicals that you are using.

Many organic and corrosive materials that are commonly used in biotechnology laboratories will degrade certain types of plastics. This means that you must make the right choice when selecting plastic tubes and containers for centrifugation, experimental use, and storage. Chemical compatibility charts are available in which you can locate the chemicals that you are using and determine which types of plastics are resistant to those chemicals. You can often find chemical compatibility charts in the catalogs of manufacturers who make plastic items for the laboratory.

The issue of chemical compatibility illustrates a major point regarding laboratory safety: preliminary research and planning will help eliminate hazards, delays, and loss of valuable samples. If you plan your work the day you intend to do it and then realize that your centrifuge tubes will not remain intact during your procedure, then you must postpone your work until the new centrifuge tubes arrive. In a still-worse scenario, you place a valuable and/or hazardous substance in an incompatible container, which therefore disintegrates, resulting in the loss of your material and possible contamination of the laboratory.

B. Chemical Fume Hoods

A **chemical fume hood** *is a specially ventilated, enclosed chemical and fire-resistant work area that provides user access from one side* (see Figure 1.6). A chemical fume hood is one of the most effective tools to reduce personal exposure to airborne chemicals.

In a chemical fume hood, air is drawn from the room, past the work area, and outside the building through a vent. The opening to the work space is covered by a transparent **sash**, *which is a window constructed of impact-resistant materials that can be raised and lowered by the user.* When using a hood, your hands pass under the sash and you look through it at your work so that your eyes and head are protected by the sash.

Analysts generally use fume hoods whenever working with chemicals that are volatile, have an unpleasant smell, or are toxic at low exposure levels. Fume hoods are only effective if they are used correctly; see Box 1.3 for basic rules regarding the use of chemical fume hoods.

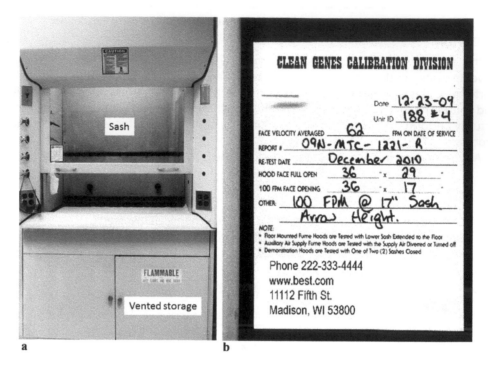

FIGURE 1.6   A Chemical fume hood. (a) The hood is sitting on top of a storage cabinet that is vented through the hood and is therefore appropriate for storing volatile chemicals. A tissue in the hood is blowing, indicating that the hood is turned on. (b) A tag (visible on the bottom left on the sash window) documents that the hood performance has been routinely tested.

Box 1.3   Proper Use of a Chemical Fume Hood

- Never peer under the sash to better see what you are working on—this totally eliminates any protection the hood might provide to you.
- Be sure that fume hoods are checked annually for proper function (see Figure 1.6b).
- Be sure that the hood is working before using it.
- Minimize motions into and out of the hood by loading all required chemicals and equipment prior to beginning work.
- Check that the chemicals in use are compatible with each other and approved for the hood.
- Keep larger equipment (e.g., centrifuges) at least 2 inches off the hood surface to avoid impeding airflow.
- Raise and lower the sash slowly to avoid large changes in the air volume passing through the hood face.
- Keep hand movements in the hood slow.
- Wear face and eye protection and keep your face in front of the hood sash for added protection.
- Keep toxic chemicals at least 5 inches behind the hood sash for maximum hood effectiveness.
- Do not use infectious agents in a fume hood; biological safety cabinets must be used for infectious agents. (Biological safety cabinets are discussed later in this unit.)

C. Chemical Storage and Chemical Disposal

Knowing how to safely store chemicals and determining how to dispose of chemical waste are parts of laboratory work that are frequently overlooked. These are also complex topics because there are many chemicals that have different requirements for safe storage and disposal. These topics are, therefore, outside the scope of this manual. As a student, you will not have to make decisions about storage or disposal, but you should be careful to comply with the instructions from your instructors and laboratory managers. Never pour any chemicals down the sink without consulting your instructor.

SDSs are good sources of information regarding proper storage and disposal of chemicals. The *Flinn Chemical and Biological Catalog Reference Manual*, which is referenced in the bibliography, contains clearly written information regarding chemical storage and disposal. Most institutions have their own processes to ensure that chemicals are stored and disposed of properly.

Box 1.4 summarizes a number of safety considerations relating to hazardous chemicals.

Box 1.4   Working Safely with Chemicals in the Laboratory

General Chemical Safety Rules

- Be knowledgeable about the risks associated with the chemicals you use.
- Wear appropriate personal protective equipment: laboratory coat, safety glasses, and the correct gloves.
- Label clearly all solutions, vessels, and equipment. Identify the name of the substance(s) contained and any associated hazards. Remember to also include your name or initials and the date of preparation on every label.
- Use the minimum amount of chemicals and other hazardous substances required.
- Transport chemicals cautiously. Chemical stock bottles should be transported in trays large enough to contain the contents of the bottle should it break. If glass bottles are not contained within a tray, they should be transported with one hand underneath while wearing the appropriate chemical-resistant gloves. The seam joining the bottom of the glass bottle to the sides is a weak point and, occasionally, the bottom of a glass bottle will fall off of it.
- Close the storage bottle's lid whenever you are not actively using a chemical. Do not place the lid on the bottle without tightening it; a coworker may try to pick up the bottle by the lid and cause a spill.
- Be sure that glass vessels are contained in spill trays if they must be heated because thermal shock can cause glass vessels to break.
- Ensure that chemical spill kits are accessible in laboratories where hazardous chemicals are used.
- Dispose of all chemicals in an appropriate fashion.

Working with Flammable Chemicals

- Keep flammable chemicals away from sources of heat and flame.
- Keep all containers of flammable chemicals tightly closed when not in use and store away from incompatible or reactive materials. Large quantities of flammable chemicals must be stored in an explosion-proof cabinet.
- Work in a well-ventilated area.
- Avoid static electricity.
- If a spill occurs, thoroughly clean your skin first. Then proceed with cleanup of your work space.
- Know how to extinguish a fire should one occur. Use only the appropriate fire extinguisher for the chemicals in use.

Working with Corrosive Chemicals

Corrosive

- Work in a fume hood rated for the corrosive chemical being used.
- Wear the appropriate PPE, including gloves appropriate for the chemical, lab coats, goggles, and a rubber apron if necessary.
- Always add acid to water when diluting or mixing, never water to acid. This will minimize the likelihood of splashing or spilling concentrated acid.
- Any reactions between an acid and base should be performed slowly to control the amount of heat and gas generated.
- Store acids and bases separate from one another in cabinets approved for these chemicals. The bottles should be in trays or on shelves with a lip so that any spillage will be contained.
- Do not use hydrofluoric acid (HF), which is an extremely hazardous chemical. Only persons trained to work with HF should ever work with this chemical.

Working with Reactive Chemicals

Explosive

- *Never* combine chemicals for disposal without following an established set of instructions or researching the reactive properties of the chemicals involved!
- Because of the potentially drastic pressure changes that can occur when working with reactive chemicals, avoid closed glass containers.
- Label all disposal containers so that everyone knows what they contain.
- Never mix unknown chemicals together, especially in closed waste containers.
- Store only compatible chemicals in the same area.

Working with Toxic Chemicals

Poison

- Be aware of the hazards of any toxic substance used.
- Do not rely on odor as a guide to the level of exposure received. Some chemicals are strong smelling at low concentrations that are not harmful. Conversely, some chemicals do not produce an odor even at toxic levels.
- Plan your laboratory work so that you minimize the amount of toxic agent used, as well as the time you are in contact with the toxic substance.
- Monitor yourself for signs of sensitization or toxicity.

Classroom Activities

Here's what I'd like you to do today

Directions from the Laboratory Supervisor

Please analyze this procedure to identify any hazardous materials and to determine what safety precautions are necessary. Do not actually perform the procedure.

PHENOL-CHLOROFORM EXTRACTION OF DNA

1. Add 5 mL of phenol [Sigma P1037] to 5 mL of an aqueous DNA solution. Mix thoroughly.
2. Spin for 1 min. in a clinical (tabletop) centrifuge at setting 4.
3. Remove the aqueous phase to a new tube.
4. Add 5 mL of a 1:1 mixture of phenol:chloroform [Sigma BioUltra 25666] to the aqueous phase. Mix thoroughly.
5. Spin for 1 min. in a clinical (tabletop) centrifuge at setting 4.
6. Remove the aqueous phase to a new tube.
7. To the aqueous phase, add 5 mL of chloroform. Mix thoroughly.
8. Spin for 1 min. in a clinical (tabletop) centrifuge at setting 4.
9. Remove the aqueous phase to a new tube.
10. To the phenol-extracted DNA solution, add 1/10 volume 3M sodium acetate [Sigma S7899]. Add 2 volumes (10 mL) of ethyl alcohol [Sigma 459836].
11. Spin tubes in a high-speed centrifuge for 10 min. at 10,000 X g.
12. Remove aqueous supernatant and store DNA pellet frozen.

Analysis

1. Identification of hazardous chemicals
    1.1. List the chemicals in this procedure.
    1.2. Consult the SDS for each chemical to determine special precautions to be taken when handling this material. You can find the SDS for each chemical online at the Sigma-Aldrich website.
2. Selection of protective equipment and attire
    2.1. At what steps should a fume hood be used?
    2.2. What type of protective wear should be worn?
    2.3. Use a web browser to research the appropriate type of gloves to use. Assume, for example, that Microflex disposable latex and disposable nitrile gloves are already in the lab. Will these work? If not, what gloves would be better?
3. Choose equipment
    3.1. List the equipment required for this procedure.
    3.2. Choose tubes that are compatible with the chemicals in use.
        • Assume that the centrifuge tubes in the lab are made from polycarbonate or polypropylene.
        • Assume that the centrifuge tubes in the lab are from Nalgene.

    Will these tubes be acceptable? Find out by using a web browser to find a chemical compatibility chart for plastics.
4. Research hazardous waste disposal methods
    4.1. Consider how waste should be handled during the experiment.
    4.2. Research methods to dispose of hazardous waste following the procedure, using the *Flinn Chemical and Biological Catalog Reference Manual*.

Your instructor might collect a written report containing information from the sections covered in this activity. Alternatively, you may discuss this analysis as a class.

SAFETY PART 3: WORKING SAFELY WITH BIOLOGICAL HAZARDS

FUNDAMENTAL PRINCIPLES

> LAIs [Laboratory Acquired Infections] are ... of public health concern as an infected worker may present a risk of transmission to his colleagues, relatives, family members or other citizens.
>
> **—Belgian Biosafety Server website**

A. GENERAL CONSIDERATIONS

Biohazards *are biological agents that have the potential to cause harm.* It is not unusual for biotechnologists to work with biological agents, such as bacteria, fungi, and viruses, which may cause laboratory-acquired infections (LAIs). Studies have documented thousands of cases of LAIs resulting in hundreds of deaths. You may be reassured to know, however, that as a student, you are unlikely to work with biological agents that are particularly hazardous.

Not all biohazardous agents are equally dangerous. One way to describe the risk of a particular biohazardous agent is to describe its pathogenicity. A **pathogen** is *an agent that is both infectious* (able to infect a host) *and can cause disease in the host.* **Pathogenicity** *is the ability of a pathogen to cause harmful effects in the host relative to other pathogens.*

The CDC classifies biological agents according to their risk. There are four classes, or **biosafety levels (BSLs)**: BSL-1, BSL-2, BSL-3, and BSL-4. More dangerous organisms are assigned a higher number, which triggers more stringent safety rules for their handling. Less dangerous organisms require less-stringent practices and less specialized equipment.

BSL-1 agents *are nonpathogenic and well-characterized.* Examples of BSL-1 agents include some laboratory strains of the common gut bacterium, *E. coli*; most yeasts; and most plants. BSL-1 agents are often used in teaching laboratories. BSL-1 agents will not cause disease in healthy people and, therefore, require no special equipment or precautions other than Standard Microbiological Practices, as shown in Box 1.5. **Standard Microbiological Practices** *are the most basic rules for handling biohazards.* It is essential to follow these when working with any biological agents in order to reduce the risk of exposure for you, your colleagues, and those outside the laboratory. Following these rules will also help protect the biological material with which you are working from outside contaminants. In teaching laboratories, you will work with agents that pose minimal risk of causing illness, but you must still follow Standard Microbiological Practices.

Box 1.5   Standard Microbiological Practices

- Only trained individuals may enter the laboratory.
- Lab coats and safety glasses must be worn at all times to reduce the risk of exposure by skin or eye contact.
- Wash hands after working with organisms and before leaving the lab. This will reduce the spread of the microbes within the laboratory, decrease the risk that an individual will become infected, and decrease the risk that a worker will accidentally carry a biological agent out of the laboratory.
- Never eat, drink, or smoke in the laboratory. Any of these practices greatly increases the chance of being exposed by ingestion.
- Avoid hand-to-mouth or hand-to-eye contact. Avoid nervous habits such as touching the face, playing with hair, or chewing on pencils or nails.
- Never "mouth pipette" (use your mouth to suck liquids into pipettes).
- Minimize aerosol production. (Aerosols are discussed in Laboratory Exercise 2.)
- Work with biohazardous materials on a clean, hard benchtop and always keep disinfectant handy in the event of a spill.
- Decontaminate the work space after any spill and at the end of each work session.
- Dispose of biohazardous waste properly.
- Use practices to reduce the risk from sharp objects.

B. EXPOSURE TO BIOHAZARDOUS AGENTS

As with toxic chemicals, there are four routes by which an agent may infect a person: inhalation, skin or eye contact, injection, and ingestion. Research indicates that inhalation of airborne agents is the most common route of laboratory infection. Inhalation exposure is usually related to the generation of bioaerosols. **Aerosols** *are very small particles or droplets suspended in the air.* **Bioaerosols** *are small liquid droplets containing a biological agent that are capable of remaining suspended in the air for long periods of time.* Bioaerosols are invisible so people are not aware of them. They are small enough to be easily inhaled. Over time, these droplets settle onto surfaces in the laboratory, at which point they can be picked up on hands or exposed skin and enter the body through skin contact or by ingestion.

Bioaerosols are easily generated when liquids containing the biohazard (e.g., a bacterial culture broth) are mixed, pipetted, spilled, or otherwise agitated. Even opening a vial containing the liquid can release bioaerosols. Because it is so easy to generate bioaerosols, it is important to use methods that reduce or eliminate their production and to use containment (discussed later) whenever bioaerosol generation cannot be avoided. It is also essential to properly disinfect surfaces and to properly dispose of biohazardous materials. Laboratory Exercise 2 explores aerosol production using fluorescent compounds.

C. CONTAINMENT

i. **What Is Containment?**

The goal of **containment** *is to protect the worker, the workplace, and the environment from the dangerous effects of a biological agent by physically containing the hazard.* Containment is achieved by the following:

- **Equipment** (e.g., biological safety cabinets and special tightly sealed centrifugation containers)
- **PPE** (e.g., self-contained breathing apparatus)
- **Practices** (e.g., avoiding the production of bioaerosols)
- **Laboratory features** (e.g., special air-handling systems)

ii. **Biological Safety Cabinets**

Biological safety cabinets (BSCs) *superficially look similar to chemical fume hoods, but biological safety cabinets protect people and the outside environment from bioaerosols; chemical fume hoods do not.* BSCs contain special filters that remove bioaerosols from the air and efficiently remove nearly all particles, microbes, and most viruses.

BSCs come in three classes: Class I, Class II, and Class III. You are likely to use a Class II biological safety cabinet, at school or in the workplace, when handling cultured cells (see Figure 1.7). **Class II BSCs** *filter-sterilize the air that has passed over the work space in order to protect personnel from exposure to agents in the cabinet. They also filter-sterilize the air before it flows over the work space, thus protecting the work space from outside contamination.* This class of safety cabinet thus protects both the person and the cultures with which the person is working. There are other classes of safety cabinet that protect only the worker or only the culture, but these are less commonly used in classroom situations.

It is important to note that although BSCs superficially resemble chemical fume hoods, they do not protect people from chemical vapors unless they have been specially modified. Never assume that a BSC provides chemical protection in the same manner as a chemical fume hood.

BSCs may come equipped with germicidal UV lamps to help sterilize the cabinet surfaces after use. These lamps are not required, and the interior of the cabinet can be cleaned after use with chemical disinfectants. Germicidal UV lamps emit harmful levels of UV radiation. If a UV lamp is present in a BSC, it should only be turned on with the front sash of the cabinet completely closed and with no people in the same room.

FIGURE 1.7   Performing cell culture work in a Class II BSC.

D. DISINFECTION OF MATERIALS AND WORK SPACES

Standard microbiological practices require that work spaces are disinfected if there is a spill, at the completion of a task, and at the end of the day. **Disinfection** is *a physical or chemical process that removes many pathogens (although often not resistant spore forms of bacteria).* **Disinfectants** *are chemicals that are able to kill pathogens or other hazardous particles, such as viruses.* Routine disinfection of laboratory surfaces does not guarantee that those surfaces become **sterile**, *that is, free from absolutely all living materials.*

One of the most serious challenges to disinfection is that various biohazardous agents are resistant to various disinfectants. The relative susceptibility of biohazardous agents to disinfection, from most susceptible to most resistant, follows:

- Some viruses (e.g., human immunodeficiency virus [HIV] and herpes)
- Most bacteria
- Most fungi
- Small viruses, including poliovirus
- Bacterial spores (most resistant)

Other factors that affect the effectiveness of the disinfection process include the following:

- The **amount of time** during which the disinfectant and biohazard are in contact.
- The **extent of contamination**.
- The **texture of the surface** to be disinfected; smooth, unmarred surfaces are relatively easily decontaminated. Surfaces with scarring, pitting, or a rough texture provide crannies in which microbes can "hide" from the disinfectant.

Disinfectants are rated as "low," "medium," or "high" based upon their relative abilities to kill biohazardous agents.

- **70% alcohol and 10% bleach** are medium-strength disinfectants that are able to kill most bacteria and some viruses and fungi. These disinfectants are commonly used in teaching laboratories and elsewhere for routine work. Get in the habit of using these disinfectants on your work space before you leave.
- **Disinfectants containing phenol** kill most bacteria, fungi, and yeasts, and are used to disinfectant surfaces that have come in contact with bodily fluids. Note that the concentration of phenol in disinfectants is always very low because phenol is extremely hazardous to humans and is corrosive to equipment.

- **High concentrations of hydrogen peroxide or bleach** are used to inactivate bacterial spores, although they may not kill all spores. Bacterial spores are more easily killed using an autoclave (a sterilizing chamber that uses high-pressure steam), but some types of equipment and laboratory benches cannot be autoclaved.

E. DISPOSAL OF BIOHAZARDOUS WASTE

Biohazardous waste includes any bacterial, fungal, or viral cultures; any materials that have come in contact with these cultures; and human or animal cells, waste, or fluids. Biohazardous materials cannot be discarded in the regular laboratory trash receptacles or in the sewer. Biohazardous liquids should be autoclaved and then poured down the laboratory drain once they have been rendered harmless. Any contaminated "sharps," such as syringe needles and razor blades, must be placed in hard-sided biohazard receptacles designed for this purpose. Non-sharp, solid materials should be placed in leakproof plastic biohazard bags. These bags should be placed in a solid-sided receptacle to contain any leaks and be autoclaved prior to disposal. A significant percentage of laboratory-acquired infections occur in custodial and other support staff, and it is your responsibility to ensure proper disposal of biohazardous laboratory waste to protect them and the environment.

Laboratory Exercise 2: Production of Bioaerosols and Factors Affecting Aerosol Production

OVERVIEW

In this laboratory exercise, you will prepare to work with biological agents, including bacteria and cultured animal cells, by examining how and when bioaerosols are produced. You will perform these tasks:

- Intentionally create aerosols to see how they are produced.
- See how aerosols can scatter over a flat surface.
- Design an experiment to maximize aerosol production—so you will know what not to do in the future.

Because infectious particles are too small to see—and are dangerous—a fluorescent dye solution is used to simulate a biohazardous liquid.

Planning Your Work: Production of Bioaerosols
1. What data will you collect?
2. How will you analyze your data?

Safety Briefing

Do not look directly into the UV lamp's light. UV light can damage the retina of your eye. Wear UV-safe glasses when using UV lights.

LABORATORY PROCEDURE

Part A: Bioaerosol Production When Using Micropipettes

A.1.   Cover lab benches with light-colored absorbent paper.

A.2.   Using micropipettes, as directed by your instructor, practice pipetting any volume of the saturated fluorescent dye from the stock bottle into a microcentrifuge tube, a test tube, and a beaker.

A.3.   Using a handheld UV lamp, look for spills or sprays, indicating aerosol production. The dyes will fluoresce under the UV lamp, making small spills or sprays visible. Check hands, lab coats, and your work area for further contamination.

A.4.   Record your results on the worksheet at the end of this exercise. You may also be able to use photography for documentation.

Part B: Bioaerosol Production When Using Serological Pipettes

B.1.   Drop an empty but wet serological pipette (one that was used to pipette water) from a height of 12 inches above the lab bench surface. (You want a little residual liquid left in the pipette for this activity.)

B.2.   Circle the water droplets you can see on the lab bench paper.

B.3.   Drop an empty but wet serological pipette that was used to pipette the fluorescent dye from a height of 12 inches above the lab bench surface.

B.4.   Circle the dye droplets with a different colored pen or pencil.

B.5.   Compare the number and sizes of droplets produced by the water aerosol and the dye aerosol.

B.6.   Use the handheld UV lamp to search for droplets not visible to the naked eye.

B.7.   Record your results on the worksheet at the end of this exercise. You may also be able to use photography for documentation.

Part C: Designing an Experiment to Create Bioaerosols

C.1. Make a list of possible ways of creating aerosols in the laboratory. Possibilities include, for example, centrifuging, opening and closing microfuge tubes, and vigorous stirring.

C.2. Compare your list with at least two other groups.

C.3. Conduct an experiment to maximize the production of aerosols. Each group may want to try different methods and compare the results. Try to determine what activities create the most aerosols.

C.4. Record your observations and results on the worksheet at the end of this exercise.

C.5. Compare your results to those of the student results shown in Figure 1.8.

Bioaerosols Worksheet		
Observations Part A	**Observations Part B**	**Observations Part C**

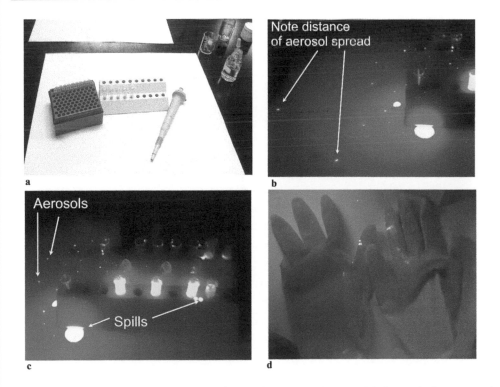

FIGURE 1.8   Creating aerosols. (a) The work space with normal lighting, before aerosol exercise. (b and c) Students intentionally created "spills" but did not realize until using the fluorescent light that smaller droplets had been released across their work surface. (d) Gloves after performing this exercise have small amounts of contamination.

LAB MEETING/DISCUSSION QUESTIONS

1. What did you learn in this activity concerning the ease with which bioaerosols can be created in the laboratory? What activities are most likely to generate aerosols?
2. What do the results of this laboratory suggest about the need for proper cleanup following handling of biohazardous liquids?
3. In what ways can you envision personnel (both laboratory and nonlaboratory) contracting laboratory-acquired infections?
4. Can you suggest ways to minimize bioaerosol production in any of the laboratory procedures investigated? For each task that you performed, are there alternative procedures that would reduce bioaerosol production?

UNIT DISCUSSION: SAFETY IN THE LABORATORY

Working safely in the laboratory is one of the overarching themes of good laboratory practices that we are exploring in this manual. Working safely also relates to other themes; for example, working safely requires preparation and planning. Jot down notes on this table or on another piece of paper that describe how the ideas and skills that you practiced in this unit relate to the overarching themes that weave throughout this manual.

Good Laboratory Practices
DOCUMENT
PREPARE
REDUCE VARIABILITY AND INCONSISTENCY
ANTICIPATE, PREVENT, CORRECT PROBLEMS

ANALYZE, INTERPRET RESULTS

VERIFY PROCESSES, METHODS, AND RESULTS

SAFETY CONTRACT

This copy is for you to keep for reference.

1. **PREPARE for every laboratory exercise before you come to the laboratory.** Follow all safety instructions in the exercise and those provided by the instructor.
2. **WEAR EYE PROTECTION at all times; do not wait for an instructor to remind you to do so.**
3. **WEAR PROPER ATTIRE**:

   * Wear closed-toe shoes.
   * Wear long slacks.
   * Wear laboratory coat.
   * Shorts are always a poor idea, and dresses and skirts should be avoided.
   * Sometimes gloves are also required.
   * Tie back long hair.

4. **LOCATE EYEWASH STATIONS, FIRE EXTINGUISHERS, AND SAFETY SHOWERS.**
5. **IF ANYONE'S EYES COME IN CONTACT WITH HAZARDOUS MATERIALS, RINSE THEM FOR 15 MINUTES AND NOTIFY AN INSTRUCTOR.** A second person should assist the injured person immediately.
6. **LEAVE COATS, EXTRA BOOKS, AND PERSONAL ITEMS IN YOUR LOCKER.** Avoid tripping over these items and taking pathogenic, toxic, radioactive, and other hazardous materials home to your families.
7. **DO NOT EAT, DRINK, SMOKE, CHEW GUM, OR APPLY COSMETICS IN THE LABORATORY.** Never use a laboratory microwave, laboratory, or freezer for food. Never eat or drink from laboratory glassware.
8. **SORT YOUR WASTE:**
   * **Dispose of biological materials in the special receptacles for biological hazards.** Do not throw other materials into these bags.

Container for sharp
waste only

Container for biohazardous
waste only

- **Dispose of glass and other sharp materials in the "sharps" disposal boxes.** Do not throw materials other than "sharps" into the "sharps" receptacles. Never use cracked or broken glassware.
- **Be sure you know where to dispose of hazardous chemicals.**
  - Do not wash solvents or hazardous materials down the drains.
  - Do not combine chemicals for disposal unless you know they are not reactive with one another.

9. **WASH YOUR HANDS WHEN LEAVING THE LABORATORY, even if you wore gloves.**
10. **BE VERY CAREFUL WITH BUNSEN BURNERS.** Do not leave a burner lit unless you are standing next to it because someone else might be injured if they do not realize it is on.
11. **BE FAMILIAR WITH SAFETY DATA SHEETS (SDSs).** These documents are valuable sources of information regarding hazards of a given chemical. Know where they are located.
12. **NEVER "MOUTH PIPETTE."** Use pipette aids.
13. **NEVER WORK ALONE.** As a student you need a staff person available. Later, make sure there is a colleague around whenever you are working.
14. **USE CHEMICAL AND BIOLOGICAL HOODS, AS REQUIRED.** Use a chemical fume hood for volatile and hazardous chemicals. Use a biological safety cabinet for particulate and biological hazards.
15. **DO NOT WEAR HEADPHONES IN THE LABORATORY.**

Note: Consult a physician if you are pregnant, plan to become pregnant, or have any other medical condition that might render you susceptible to exposure to the chemicals used in the laboratory.

SAFETY CONTRACT *This copy is to be initialed and turned in to your instructor. Your initials mean that you agree to follow these rules.*

Name (Please print) _____

1. **Prepare for every laboratory exercise before you come to the laboratory.** Follow all safety instructions in the exercise and those provided by the instructor.

   Initials _____
   Date _____

2. **Wear eye protection at all times; do not wait for an instructor to remind you to do so.**

   Initials _____
   Date _____

3. **Locate eyewash stations, fire extinguishers, and safety showers.**

   Initials _____
   Date _____

4. **If anyone's eyes come in contact with hazardous materials, rinse them for 15 minutes and notify an instructor.** A second person should assist the injured person immediately.

   Initials _____
   Date _____

5. **Wear proper attire.** Wear closed-toe shoes, long pants, and a laboratory coat. Shorts are always a poor idea, and dresses and skirts should be avoided. Sometimes gloves are also required. Tie back long hair.

   Initials _____
   Date _____

6. **Leave coats, extra books, and personal items in your locker.**

   Initials _____
   Date _____

7. **Do not eat, drink, smoke, chew gum, or apply cosmetics in the laboratory.** Never use a laboratory microwave, laboratory, or freezer for food. Never eat or drink from laboratory glassware.

   Initials _____
   Date _____

8. **Dispose of biological materials in the special waste receptacles for biological hazards.** Do not throw other materials into these bags.

   Initials _____
   Date _____

9. **Dispose of glass and other sharp materials in the "sharps" disposal boxes.** Place only "sharps" in the "sharps" receptacles. Never use cracked or broken glassware.

   Initials _____
   Date _____

10. **Consult your instructor regarding disposal of hazardous chemicals.** Do not wash solvents or hazardous materials down the drains. Do not combine chemicals for disposal unless you know they are not reactive with one another.

    Initials _____
    Date _____

11. **Be very careful with Bunsen burners.** Do not leave a burner lit unless you are standing next to it because someone else might be injured if they do not realize it is on.

    Initials _____
    Date _____

12. **Be familiar with SDSs.** Know where they are located.

    Initials _____
    Date _____

13. **Never "mouth pipette."** If you used this method in the past, learn to use pipette aids.

    Initials _____
    Date _____

14. **Do not work alone.** As a student you need a staff person available. Later, make sure there is a colleague around whenever you are working.

Initials _____
Date _____

15. **Use chemical and biological hoods, as required.** Use a chemical fume hood for volatile and hazardous chemicals. Use a biological safety cabinet for particulate and biological hazards.

Initials _____
Date _____

16. **Do not wear headphones in the laboratory.**

Initials _____
Date _____

17. **Wash your hands when leaving the laboratory, even if you wore gloves.**

Initials _____
Date _____

Consult a physician if you are pregnant, plan to become pregnant, or have any other medical condition that might render you susceptible to exposure to the chemicals used in the laboratory.

Documentation in the Laboratory

DOI: 10.1201/9781003360742-2

UNIT INTRODUCTION

The palest ink is better than the best memory.

—Chinese proverb

A. WHAT IS DOCUMENTATION?

Documentation is of critical importance in any system of good laboratory practices; it is also a difficult word to define. Documentation includes a wide variety of tangible items (e.g., forms, information recorded on paper, written instructions). Documentation also may be electronic, that is, recorded using a computer. The term "documentation," moreover, also refers to methods of verifying that tasks were performed in a certain way and that results and products are of good quality. **Documentation** *includes a system of records* where a **record** *is anything that provides permanent evidence of, or information about, past events.*

It is helpful to classify the many types of documents into three functional categories to better understand their roles:

- **Directive documents** *tell personnel how to do a task, identify a material, or define something.* **Protocols**, *which guide researchers in investigating a question,* are directive documents. Information is not added to protocols when work is performed.
- **Data collection documents** *facilitate the recording of data and provide evidence that a directive document has been properly followed.* Information is added to data collection documents during operations. Examples include forms, logbooks, and laboratory notebooks.
- **Commitment documents** *lay out the organization's goals, standards, and commitments.*

Table 2.1 summarizes types of documents common in laboratories.

TABLE 2.1   Examples of Documents That Are Common in Laboratories

Directive Documents
1. **Standard operating procedures (SOPs)** or simply "procedures," *detail what is to be done to complete a specific task and how to document that the task was done correctly.*
2. **Protocols,** like SOPs, *explain how to do a task.* The term "protocol," however, is often reserved for situations where an experiment is involved or a procedure is going to be performed only once.
3. **Numbering systems** are *used to identify and track materials, equipment, and products.*
4. **Labels** are *attached to solutions, products, or items to identify them.*
Data Collection Documents
5. **Laboratory notebooks** are *a chronological log of everything that an individual does in a laboratory.*
6. **Forms** are *filled out by an analyst to record information.* Forms are typically associated with SOPs or other documents and contain blanks.
7. **Reports** are *documents generated as a result of the execution of a protocol.* Reports summarize and interpret data that were previously collected.
8. **Equipment/instrument logbooks** *keep track of maintenance, calibration, and problems for a given instrument or piece of equipment.*
9. **Analytical laboratory documents** *record information regarding the testing of a sample.*
10. **Recordings from instruments.**
11. **Chain of custody forms** are *used to trace the movement of a sample throughout a facility and to keep samples and sample test results from being confused with one another.*
12. **Training reports** *document that individuals were properly trained to perform particular tasks.*

B. CONTROLLED DOCUMENTS

In a company or a regulated workplace, most documents are likely to be **controlled**, *that is, they are prepared and distributed using a formal process to ensure that each document is identified and accounted for.* Consider, for example, a laboratory where an analyst writes a **standard operating procedure (SOP)**, *a procedure to perform a laboratory test in a standardized step-by-step format.* The analyst's draft SOP is reviewed and possibly revised by a supervisor, other analysts, and a representative from the quality assurance department (the department charged with overseeing quality processes and documentation). Once the SOP is completed and approved, the original (master) SOP is signed, dated, and stored in a secure location. An analyst who needs to perform this laboratory test is issued a working copy of the approved SOP and receipt of the copy is recorded. If the SOP is revised, the new version is reviewed, approved, signed, and dated. The revision number is noted on the new version. The older version is removed from active use so that only copies of the most recent version are available to analysts. A controlled system ensures that everyone in the laboratory always uses the correct, most up-to-date version of every document. Controlling documents also helps to prevent unauthorized individuals from obtaining access to confidential information. In an academic research setting, documents are usually not formally controlled as they are in a company or regulated environment. In an academic environment, documents are typically maintained by individuals who share them as they, or their supervisors, see fit.

C. GOOD DOCUMENTATION PRACTICES

Documentation must be honest, secure, and verifiable. Standard practices have evolved in order to meet these documentation requirements. You may see these standard practices referred to as "good documentation practices" (GDPs). These practices are summarized in Table 2.2.

TABLE 2.2   Good Documentation Practices*

Rule 1: Records must be accurate, legible, and understandable. To accomplish this, perform these actions: • Write legibly. • Define abbreviations. • Paginate all forms and documents with the page number and the total number of pages (e.g., page 2 of 10). This helps ensure that no pages get lost or omitted over time. • Avoid ditto (") marks or arrows when recording repetitive information.
Rule 2: Records must record events with clear and verifiable dates. To accomplish this, perform these actions: • Date every record (e.g., every label, signature, and page of a laboratory notebook). • Never "backdate"; show the actual date. • Use the format that is required in your organization (e.g., April 15, 2020 or 4/15/2020). Note that conventions for dates vary in different countries.
Rule 3: Documents must be secure and safe from natural disaster, theft, and access by unauthorized individuals. To accomplish this, perform these actions: • Use filing cabinets and locks to protect paper documents. • Use software security and organization to safeguard electronic documents.

Rule 4: Records must be attributable to a particular individual.
 	
To accomplish this, perform these actions:
 	
• Sign every record to identify the person making the record and to attest to the truth of the data recorded.
  • A traditional signature is used with paper documents.
  • An electronic signature is used for computer records. This typically requires entering a unique user ID and password. Everything recorded onto the computer while that individual is logged on is attributed to that person. In situations requiring more assurance of an individual's identity, sophisticated authentication methods, such as voice recognition and retinal scans, can be used.
• Never sign a document for another person or log in as another person.
• Never sign before you have completed a task.

Rule 5: Documents must provide information for traceability. **Traceability** *means that all the materials used in an experiment or analysis, or used in making a product, and all associated documents can be identified.* To accomplish this, perform these actions: • Identify all chemicals, equipment, documents, samples, and other items with unique identification numbers, tags, or labels. • Always record complete information about all chemicals, equipment, documents, samples, and so on when the item is used.
Rule 6: Records should not be capable of being altered, either accidentally or intentionally. To accomplish this, perform these actions: • Enter data directly onto the correct form or into your laboratory notebook, never onto a piece of scrap paper or the back of your hand. • Always use permanent ink. • Cross out mistakes with a single line so that the original recording is visible. Never use whiteout; never write over data to obscure it. In many organizations, corrections must be initialed, dated, and briefly explained. • Draw a line through any unused space in a laboratory notebook so that no one can later add anything. If a whole page is supposed to be blank, label the top of the page as "intentionally left blank." • If you are filling out a form and a blank does not apply, write NA, "not applicable." Do not leave any fields empty as this can be interpreted as missing data or accidentally omitted data. • Electronic documents require a software method of tracking modifications to data to ensure that original recordings are not erased. If someone legitimately tries to add information to a record, the computer must be programmed to "know" that this act is legitimate and to show both the original record and the revision.

*These rules are not found in any official document or regulations. The rules, as they are stated here, are common practices compiled from a number of sources.

Classroom Activity 7: Being an Auditor

OVERVIEW

As you complete this Classroom Activity, you will perform the following tasks:

- Review basic principles of documentation.
- Experience the format and use of batch records in a production environment.
- Practice performing an audit, which is an important part of a quality program.

BACKGROUND

An **audit** in a biotechnology organization is *a process in which a person (or team) collects objective evidence to permit an informed judgment about whether or not an organization is complying with its quality system.* In this activity, you act as an auditor who is checking the documentation practices in a fictitious company, Cleanest Genes Corporation.

Your fictitious audit will have two parts. In the first part, you will walk around the company and observe the documentation practices of employees. In the second part of the audit process, you will review a single data collection document that was previously filled in by employees in order to see if the document was properly completed. The particular data collection document you will review is a small section of a batch record. A **batch record** is *a document used In production facilities to guide the production of a product.* A batch record accompanies a product as it is made. A batch record directs the operators in exactly how to make the product—and the operators must follow the batch record instructions just as they are written. The batch record not only guides the production operators in how to produce a product, it also provides blanks that are filled in as the operator performs tasks. A witness watches the production operator perform critical steps in the procedure and signs off that these steps were performed as directed. By filling in the batch record properly, production technicians demonstrate that they have produced a batch properly. If the batch record is not properly completed, then the quality of the product is questionable.

Each time a product is to be made, the operators are issued a fresh copy of the current version of the batch record by an authorized representative. A batch record is thus an example of a controlled document, as was discussed in the Unit Introduction.

CLASSROOM ACTIVITIES

Part A: Tour of the Facility

Scenario

BEING AN AUDITOR

You are a professional auditor hired by officials of the Cleanest Genes Corporation to help them make sure their documentation is in order. As you walk around their gleaming facility, you observe the following scenes pictured. For each scene pictured, what rule of documentation is violated, as described in Table 2.2 in the Unit Introduction?

SCENE 1

SCENE 2

SCENE 3

SCENE 4

Part B: Auditing a Specific Document

A brief portion of a batch record for making 6 molar (M) hydrochloric acid is shown in the following document, Batch Record Example. *(It is based on a batch record created by Michael Barazia, a quality professional.) A production technician filled out the batch record as a product was manufactured and a witness signed it. Apparently their minds were not on their work! The filled-in batch record has many errors, see* Filled-in Batch Record.

B.1.   Begin by simply finding these features of the batch record.

    B.1.1.   **Title or Subject.** The title of the batch record is on each page.

    B.1.2.   **ID Number.** Each batch record should contain a unique identification number.

    B.1.3.   **Page Number.** Each page of the batch record should be properly numbered.

    B.1.4.   **Hazard Communication.** This section warns the operator of any hazards associated with the procedure and any required safety precautions.

    B.1.5.   **Procedure.** The core of the batch record details what the operator will do in a step-by-step chronological manner. Every batch record has a procedure.

        B.1.5.1.   The operator writes information into blanks on the batch record while performing the procedure.

        B.1.5.2.   Another person witnesses and signs the "verified by" lines.

    Note that two sections, "Label Information" and "QA Review," were omitted here to save space.

B.2. See how many errors you can find in the way the employees filled out the batch record.

B.3. Check your answers with the answer key.

FIND THE MISTAKES ON THIS BATCH RECORD

Cleanest Genes Corporation	COMPOUNDING INSTRUCTIONS: 6 Molar Hydrochloric Acid (HCl)	page 1 of 2
LOT NO.	**EFFECTIVE DATE** **August 12, 2009**	Batch Record C89001

QA Issued By _____   Date _____

HAZARD COMMUNICATION:

Hydrochloric Acid, 6M

DANGER:	CORROSIVE. AVOID CONTACT WITH SKIN AND EYES. AVOID INHALATION OF VAPOR AND MIST. DO NOT MIX WITH CAUSTICS OR OTHER REACTIVES.

Water for Injection

WARNING	HOT. WFI IS MAINTAINED AT A TEMPERATURE OF 80°-95°C. TAKE NECESSARY PRECAUTIONS TO AVOID SKIN CONTACT AS BURNING MAY OCCUR.

PROCEDURE:

1. Into a clean, calibrated container, collect approximately 100 L of water for injection (WFI) and cover the container. Let cool to ambient temperature.

Vessel ID _____ *Wagner LC55*

Amt WFI Collected _____ *10L*

Collected By _____ *JDS*

Date/Time____ *7/25/00* ____ *1:50* _____AM/PM

Temperature at Collection_____

Verified By _____ *PRS*

2. *Slowly* add 200 L (236kg) of 36-38% [~12 M] HCl into the vessel containing the WFI and stir slowly.

Temperature of WFI_____ *82° F*   Ambient Temperature ____ *70° F*

Rec # HCl ___ *Good Chem 55L24*   HCl Added By _____ *JDS*

Balance ID _____ *Bettler 20000X*   Date/Time __ *7/25/00* ___ *2:32* ___AM/**PM**

Amt Weighed _____ *236 kg*   Addition Verified By ___ *PRS*

Weighed By _____ *JDS*   Volume_____ *298* _____L

Date/Time__ *7/25/00* ___ *1:47* ____AM/**PM**

Weight/Volume Verified By ____ *PRS*

3. QS with WFI to 400 L. Add WFI *slowly.*

* Amt WFI Added _____ *158* _____L

Added By _____ *JDS*

Date/Time__ *7/25/00* ___ *3:55* ___AM/**PM**

Final Vol in Container ____ *400* _____L

Verified By _____

(* Final Volume - Volume at end of **Step 2**.)

Continues…rest of document is not shown to save space

Batch Record Example

ANSWERS

Page 1 of 2

| Clean Gene Corporation | COMPOUNDING INSTRUCTIONS: |
| | 6 Molar Hydrochloric Acid (HCl) |

| LOT NO. | EFFECTIVE DATE |
| | August 12, 2009 |

Lot number not filled in

Does not state who issued this document

QA Issued By _____   Date _____

Does not state when this document was issued

HAZARD COMMUNICATION:

Hydrochloric Acid, 6M

DANGER: CORROSIVE. AVOID CONTACT WITH SKIN AND EYES. AVOID INHALATION OF VAPOR AND MIST. DO NOT MIX WITH CAUSTICS OR OTHER REACTIVES.

Water for Injection

WARNING: HOT. WFI IS MAINTAINED AT A TEMPERATURE OF 80°-95°C. TAKE NECESSARY PRECAUTIONS TO AVOID SKIN CONTACT AS BURNING MAY OCCUR.

PROCEDURE:

1. Into a clean, calibrated container, collect approximately 100 L of water for injection container. Let cool to ambient temperature.

   Units of volume not recorded

   Vessel ID _____ *Wagner 6253*
   Amt WFI Collected _____ *10L*

   AM/PM not circled

   Collected By _____ *JDS*
   Date/Time ___*7/25/00*___ ___*11:50*___ AM/PM

   Temperature not recorded

   Temperature at Collection _____
   Verified By _____ *MK*

2. *Slowly* add 200 L (236kg) of 36-38% [~12 M] HCl into the vessel containing the WFI and stir slowly.

   Temperature of WFI ___*55° F*___   *Temperature not at ambient*   Ambient Temperature ___*70° F*___
   Rec # HCl ___*Good Chem 55624*___   HCl Added By ___*JDS*___
   Balance ID _____ *Settle 10000K*   Date/Time ___*7/25/00*___ ___*2:32*___ AM/(PM)
   Amt Weighed _____ *236 kg*   Addition Verified By ___*MK*___
   Weighed By _____ *JDS*   Volume _____ *298*   L
   Date/Time ___*7/25/00*___ ___*1:47*___ AM/(PM)

   Time 1:47 pm before previous time of 2:32 pm

   Weight/Volume Verified By _____ *MK*

3. QS with WFI to 400 L. Add WFI *slowly*.

   Calculation not correct 400L - 298L=102L

   * Amt WFI Added _____ *158*   *Should be PM*
   Added By _____ *JDS*
   Date/Time ___*7/25/00*___ ___*3:55*___ (AM)/PM

   No final verification

   Final Vol in Container _____ *400*   L
   Verified By _____
   (* Final Volume - Volume at end of **Step 2**.)

Continues... rest of document is not shown to save space

Filled-in Batch Record.

Laboratory Exercise 3: Keeping a Laboratory Notebook

OVERVIEW

The purpose of this laboratory exercise is to introduce you to the skill of maintaining a laboratory notebook and to help you understand its place in a system of documentation. In this exercise, you will perform the following tasks:

- Set up your laboratory notebook for future use.
- Practice recording information regarding materials and equipment.
- Practice documenting your work as you conduct a simple procedure, making "slime," which is a combination of polyvinyl alcohol and Borax.
- Practice witnessing another person's laboratory notebook.

BACKGROUND

A. Laboratory Notebooks

Laboratory notebooks are a laboratory investigator's primary data collection document. **Laboratory notebooks** *are assigned to individuals and are a chronological log of everything that individual does and observes in the laboratory.* The primary user of a laboratory notebook is the investigator, who uses the notebook to track the progress of a project, archive the data generated in experiments, record observations, and record all the details that must be remembered. Investigators use their laboratory notebook when they analyze their experiments, write reports, plan their upcoming investigations, and troubleshoot problems.

A laboratory notebook is an important legal document that may be viewed by people in addition to the investigator. A laboratory notebook may provide evidence that the scientific community uses to assign credit for a research discovery. The notebook documents the honesty and integrity of data that are published in research journals and used in grant applications. Laboratory notebooks can be subpoenaed in litigations. They can be examined by auditors from the U.S. Food and Drug Administration, U.S. Environmental Protection Agency, and other regulatory agencies. Laboratory notebooks are of particular importance in patent law, where they document the invention of a new technique, item, or other innovation.

It is essential that investigators maintain a proper laboratory notebook. Recall that in the Documentation Unit Introduction we discussed "good documentation practice" rules. These rules apply to laboratory notebooks; see Table 2.3.

Students often wonder what to include in their laboratory notebooks. This depends on the situation, but here is some general guidance:

- **Identification.** Begin the laboratory notebook with identifying information: your name, the date of beginning the notebook, your company or academic institution, and the project/course name.
- **Table of contents.** Reserve the first few pages of the notebook for a table of contents that will include page numbers and descriptions with sufficient detail to allow easy searching of the notebook's contents.
- **Title and purpose.** Begin each experiment/assay/exercise with a descriptive title and its purpose or rationale. It is important to record the justification for pursuing a particular line of investigation.
- **Introductory information.** Consider including introductory information, such as a literature review and background information from colleagues.
- **Date.** It is common practice to sign and date the bottom of the page.
- **Materials and methods**
  - Include complete *identifying information for chemicals*, such as the manufacturer's name, catalog number, and lot number. In some companies and regulated workplaces, the date of expiration must also be noted and possibly the date a stock bottle is first opened.

TABLE 2.3   Applying Good Documentation Practices to Laboratory Notebooks

The same six rules of good documentation outlined in Table 2.2 apply to laboratory notebooks.

Rule 1: Records must be accurate, legible, and understandable.

- Note all problems and be honest; never try to obscure, erase, or ignore a mistake. However, avoid derogatory statements about your ideas or techniques.
- Remember that in a workplace, your supervisors, colleagues, patent attorneys, and regulatory agency inspectors may review your entries.

Rule 2: Records must record events with clear and verifiable dates.

- Use only a bound notebook, not a spiral notebook or ring binder from which pages can be removed or into which pages can be inserted.
- Make sure every page is numbered consecutively *before* using the notebook.
- Never rip out a page.
- Never skip a page to insert information later. Keep moving forward; see, for example, Figure 2.1b.
- Record data at the time they are observed.
- Date and sign every page.
- Cross out blank lines or unused portions of the page with a diagonal line so nothing may be added to the page at a later date. Do this every day.
- Securely attach and date printouts and other materials added to the notebook.

Rule 3: Documents must be secure and safe from natural disaster, theft, and access by unauthorized individuals.

- Avoid loaning your laboratory notebook to another student.
- In a workplace, do not remove the laboratory notebook from the facility.

Rule 4: Records must be attributable to a particular individual.

- Sign and date the bottom of every page.
- Never write in someone else's lab notebook or allow him or her to write in yours.

Rule 5: Documents must provide information for traceability.

- Record complete information about all chemicals, equipment, documents, and samples.
- Record identifying information for all major equipment.

Rule 6: Records should not be capable of being altered, either accidentally or intentionally.

- Enter data directly onto the correct form or into your laboratory notebook, never onto a piece of scrap paper. *Never* rewrite your laboratory notebook to make it "nicer"!
- Always use permanent ink.
- Cross out mistakes with a single line so that the original recording is visible. If your instructor/supervisor requires, initial, date, and explain the corrections.
- Draw a blank line through any unused space in a laboratory notebook to prevent later additions.

- Identify the *equipment* that is used. In a regulated environment, the date of last calibration may be required.
- Write the *methods* used completely and clearly enough that you or another individual could exactly repeat the work described. In cases where you are following a standard procedure exactly and where a copy of that procedure is always available, you probably do not need to record the procedural details in the lab notebook. You can reference the standard procedure. If you are not exactly following a procedure that will be readily available for the foreseeable future, then record all procedural information into the laboratory notebook.
- **Calculations.** Include equations and calculations with correct units and enough explanation so that another person can understand your reasoning.

- **Raw data** *are the first records of an original observation.* Raw data may be written into the notebook with pen by the operator, may be included as a paper output from an instrument, or increasingly, may be recorded into a computer medium. The researcher must save all raw data and ensure that the data are not altered or edited and that data are retrievable. Printouts from instruments, photos, and other paper data are generally signed, dated, and securely taped into the laboratory notebook with a permanent adhesive. It is common to sign or write across both the inserted paper and the page on which it is taped in order to authenticate the paper's placement on the page. An explanation of the data should be provided with it. If data cannot be affixed in the notebook, they may be titled, signed, explained, dated, and filed securely. The laboratory notebook should reference the data and storage location.
- **Observations.** Record observations (e.g., changes in pH, temperature, humidity readings, and instrument operational parameters). These observations may be critical in analyzing your results or in trouble-shooting problems later.
- **Sample information.** Whenever samples are tested, record complete information regarding their source, storage, identifying information, disposal, and so on.
- **Ideas.** Include ideas as well as experiments. Explain why each experiment is performed and at the end, summarize what the results show, avoiding negative comments about your work or ideas (even if something did not go as planned). Summarize the day's work and provide conclusions when appropriate. Documenting ideas is particularly important in cases where patents are potentially involved.
- **Sign and date.** Sign and date the bottom of every page; put a diagonal line through unused portions of the page.
- **Witness**. Laboratory notebooks should be witnessed whenever research might lead to a patent and in regulated workplaces. The witness, who is not one of the inventors, reads, signs, and dates each entry. The witness should be sure that he or she understands the entries because the witness is corroborating that the work described really happened. It is preferable that the witness is someone who actually observed the experiments, although this may not be possible in practice.

One of the challenges of keeping a good laboratory notebook is that it must clearly record what actually happened in the laboratory. If you outline in advance what you *intend* to do, use the present imperative or future tense. Clearly record what *actually occurred*, using the past tense.

Another challenge is that a laboratory notebook must be chronological—it must keep moving forward in time. A researcher might be working on more than one project, in which case it is simplest to keep a different laboratory notebook for each project. If only one laboratory notebook is used, it is not correct to leave empty pages or spaces on a page in order to fill them in later. Rather, if a researcher returns to a project after recording information about something else, it is correct to state at the top of the page that the work is "continued from page ___."

Figure 2.1 shows examples from a student laboratory notebook to illustrate these points.

B. Scientific Reports

A laboratory *notebook* is not a laboratory *report*. What is the difference?

- **Word processor.** A report is prepared using a computer word processor; it is not written in a laboratory notebook.
- **Format.** A final report is a formal discussion of laboratory work that is written in a particular format. A typical report format begins with an *Introduction*, followed by *Materials and Methods*, then *Results*, and finally a *Discussion*.

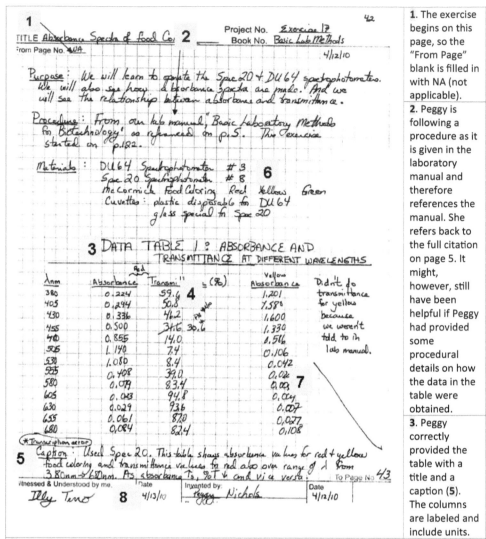

The figure shows a handwritten laboratory notebook page with the following content:

TITLE Absorbance Spectra of Food Coloring   Project No. Exercise 17
From Page No. NA   Book No. Basic Lab Methods   42   4/12/10

Purpose: We will learn to operate the Spec 20 & DU 64 spectrophotometers. We will also see how absorbance spectra are made. And we will see the relationship between absorbance and transmittance.

Procedure: From our lab manual, "Basic Laboratory Methods for Biotechnology" as referenced on p.5. This exercise started on p.182.

Materials: DU 64 Spectrophotometer #3
Spec 20 Spectrophotometer #8
McCormick Food Coloring Red Yellow Green
Cuvettes: plastic disposable for DU 64
glass special for Spec 20

DATA TABLE 1: ABSORBANCE AND TRANSMITTANCE AT DIFFERENT WAVELENGTHS

λ nm	Red Absorbance	Transmi^{11} T (%)	Yellow Absorbance	
380	0.224	59.6	1.201	Didn't do transmittance for yellow because we weren't told to in lab manual.
405	0.294	50.8	1.580	
430	0.336	46.2	1.600	
455	0.500	31.6 30.6	1.330	
480	0.855	14.0	1.586	
505	1.140	7.4	0.106	
530	1.080	8.4	0.042	
555	0.408	39.0	0.02	
580	0.079	83.4	0.003	
605	0.043	94.8	0.004	
630	0.029	93.6	0.007	
655	0.061	87.0	0.027	
680	0.084	82.4	0.108	

* Transcription error

Caption: Used Spec 20. This table shows absorbance values for red & yellow food coloring and transmittance values for red also over range of λ from 380 nm → 680 nm. As absorbance ↑s, %T ↓ and vice versa.   To Page No. 43

Witnessed & Understood by me. Illy Tino   Date 4/13/10   Invented by: Peggy Nichols   Date 4/12/10

1. The exercise begins on this page, so the "From Page" blank is filled in with NA (not applicable).

2. Peggy is following a procedure as it is given in the laboratory manual and therefore references the manual. She refers back to the full citation on page 5. It might, however, still have been helpful if Peggy had provided some procedural details on how the data in the table were obtained.

3. Peggy correctly provided the table with a title and a caption (**5**). The columns are labeled and include units.

4. Peggy made an error when recording this datum and, therefore, crossed out the incorrect value with a single line. She recorded the correct value alongside it, initialed it, and dated the correction. She also used a footnote to explain the error. Footnotes are required in regulated workplaces, but you may not see this practice in an academic laboratory.

6. When available, equipment serial numbers should be recorded.

7. Peggy placed a diagonal line across the part of the page that was not used in order to show that it is supposed to be blank.

8. The page is properly filled out at the bottom with Peggy's signature in the "inventor" blank and the date. Another student properly witnessed the page a day later.

FIGURE 2.1A   An annotated page from Peggy's laboratory notebook.

In contrast, a laboratory notebook is written as a chronological account of what you did, in the order that you did it. Some materials or methods may thus show up in the notebook *after* some results. Some results may show up in the notebook *before* a particular material is described. The key issue in a laboratory notebook is to make sure it is chronological. Never go back later and record information on previously used pages. Keep moving forward in the notebook.

- **Discussion.** While the laboratory notebook should include all the materials, methods, and results that will be included in the formal report, the laboratory notebook seldom includes the formal discussion required for a report. The laboratory notebook may or may not include all the information that goes into the introduction section of a formal report.

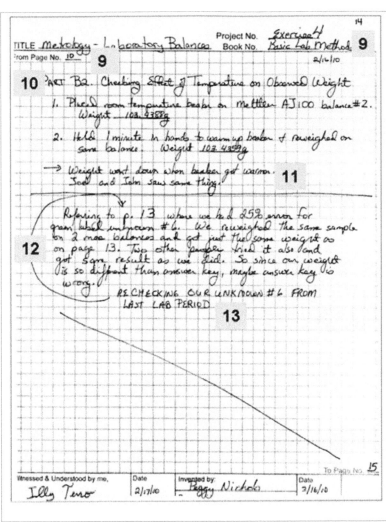

9. This exercise is continued from p. 10. Presumably Peggy was working on another activity on pp. 11-13. This is correct – never skip pages in order to come back to something later. You can always continue an earlier analysis on a later page as Peggy did here.

10. This title is sufficient to explain the two data points that are recorded. It is clear that the data are intended to determine whether an object's temperature can affect its weight.

11. Peggy includes useful observations and comments that will later help her interpret her results.

12. It is important to recognize problems and to try to determine their causes. In this case, Peggy's result deviates substantially from the answer key. She therefore retests the same sample using different equipment. She also asks colleagues to check her work. She consistently gets the same answer and hypothesizes that the answer key is incorrect.

13. It is important to clearly label each section. Peggy initially forgot to put a title on this analysis and so added a title later.

FIGURE 2.1B   Another annotated page from Peggy's laboratory notebook.

- **Appearance.** A report is written in paragraph form with correct grammar and careful attention to spelling. A laboratory notebook may or may not include text written in paragraph form. A laboratory notebook must be clear and legible, but it is not supposed to be a "thing of beauty."

C. Making Slime

You will prepare "slime" as part of this exercise. Slime is prepared by mixing together polyvinyl alcohol (PVA) with sodium borate (Borax). PVA consists of a chain of repeating subunits of vinyl alcohol that link to form chains when dissolved in water. If this solution is mixed with Borax, the chains of PVA form bridges from one to another to form a gel. The bridges are formed by relatively weak chemical bonds, so the resulting slime has a gel-like consistency. It is possible to manipulate the components of the gel to get slime with slightly different characteristics, as you will do in this exercise. (This activity is described on the website http://www.west.net/~science/slime.htm.)

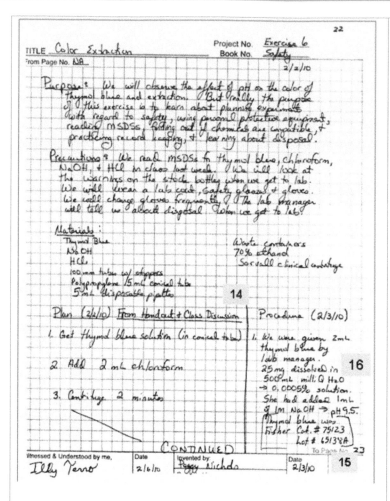

14. It is a challenge to distinguish between what one plans to do and what one actually does in the laboratory. Peggy has done some planning in advance. She read the chemical MSDSs and listed the required materials. She began to work out the procedure that she will follow. In this case, Peggy has successfully distinguished between what she planned and what she actually did. Observe her use of dates, the use of the imperative when writing her plans, and her use of past tense to indicate what she actually did.

15. Peggy has correctly recorded information for this chemical: manufacturer's name, chemical name, catalog number, and lot number. She also reports the concentration, the type of water used (MilliQ is highly purified water), and the pH of the solution.

16. The concentration of thymol blue is reported, but the concentration is incorrect—it should be 0.005%. Peggy should have shown her calculation. If she had, she might have seen the error. The witness also should have checked for this type of error so that Peggy could correct it.

FIGURE 2.1C   A third annotated page from Peggy's laboratory notebook.

Planning Your Work: Keeping a Laboratory Notebook

1. Purchase a laboratory notebook that meets your instructor's requirements.

2. Read the Documentation Unit Introduction.

3. Read this laboratory exercise.

   a. What will you record in terms of materials and methods when you are in the laboratory?

   b. What will you record in terms of observations, data, and results when you are in the laboratory?

   c. How will you analyze your results?

LABORATORY PROCEDURE

Part A: Preparing Your Laboratory Notebook

A.1.   Label the first page of the notebook or the cover with your name, school or company name, course/project name and number, and date.

A.2.   Number every page of the notebook sequentially if this has not already been done.

A.3.   Set aside the first three pages for a table of contents. Label each page "Table of Contents."

Part B: Making Slime

B.1.   Prepare the fourth page for this laboratory exercise. Record the following at the top of the page:
  - The title of the laboratory exercise
  - The date (optional because you will date the bottom)
  - Your name (optional because you will sign the bottom)
  - Partner's name (if applicable)

B.2.   Write a brief statement of the purpose of this exercise.

B.3.   Record the following information for each chemical provided. If any of this information is not available, write "not available":
  - Name of chemical
  - Company that manufactured it
  - Lot number
  - Catalog number (What is the difference between a catalog number and a lot number?)
  - Date the bottle was first opened
  - Expiration date

B.4.   Measure 25 mL of 4% polyvinyl alcohol (PVA) into a plastic disposable cup.

B.5.   Add 5 mL of 5% sodium borate (Borax) to the PVA solution.

B.6.   Add food coloring as desired.

B.7.   Stir slowly until the liquid has gelled.

B.8.   Experiment with slightly different proportions of the ingredients to see if you can improve the properties of your slime. Remember to record the following:
  - Amounts of each component combined.
  - Method of mixing.
  - Results. Try to be as objective as possible. For example, you might test and record whether your slime bounces, whether it sticks to surfaces, how much length it can achieve if stretched, and so on.

B.9.   When finished, sign and date the bottom of the page. Put a diagonal line through any unused portion of the page.

Part C: Witnessing a Laboratory Notebook

C.1.   Witnessing should be done regularly. It is not necessary to witness a lab notebook the same day an entry is completed, but it is much better to do so within a few days:

  C.1.1.   Exchange notebooks with another student. Read the other student's lab notebook and check it according to the following checklist.

  C.1.2.   If your witness finds errors, cross out mistakes with a single line so that the original recording is visible. If your instructor/supervisor requires, initial, date, and explain corrections.

C.2.   Make sure that your partner similarly corrects any errors in his or her notebook.

C.3.   Look at five more people's notebooks to get ideas for how to best keep a notebook.

LAB MEETING/DISCUSSION QUESTIONS

1.   What will you do to make your lab notebook better in the future?

2.   What did you see in other students' notebooks that was effective?

3.   Suppose that a company was making slime as its product for commercial sale. What properties could they test to be sure that it is "good"?

	Checklist for Your Laboratory Notebook and for Witnessing Another Person's Notebook
	Is the notebook legible?
	Is the notebook filled out in nonerasable ink?
	Are the investigator's name and date on every page?
	Is there a title at the top of every page?
	Is every page numbered?
	Are mistakes crossed out with a single line so they can be read?
	Is there a diagonal line drawn through every blank space?
	Is the table of contents up to date?
	Does every new experiment/exercise begin with a purpose/rationale?
	Is the notebook chronological; do the dates recorded always move forward and never backward?
	Are all the calculations correct?
	Is it clear what the author intended to do versus what the author actually did?
	For instruments, are the name, model, serial number,* and operating conditions recorded?
	For instruments, is the date of the last calibration recorded*?
	For chemicals, are the manufacturer, catalog, and lot number recorded?
	Is the expiration date for every chemical recorded*?
	Is the date noted when each chemical stock bottle was first opened*?
	Is the procedure written clearly and completely? Alternatively, is an SOP or manual referenced?
	Are all experimental/test conditions recorded?
	If applicable, are all relevant sample data recorded?
	Are all raw data present and properly recorded with units (if applicable)?
	These may not be required in a teaching or academic laboratory setting.

Classroom Activity 8: Writing and Following an SOP

OVERVIEW

One of the most important goals of documentation is to help ensure that personnel perform their activities properly to obtain good quality results or products. Written procedures are essential in this regard. The purpose of this activity is to introduce you to written procedures so that you can use them effectively, and when appropriate, write procedures for others to follow. You will practice these skills by writing a procedure to make a model of a molecule. Another student will follow your written instructions and will construct the molecule that you describe. You will also practice following a procedure written by a classmate.

BACKGROUND

Many organizations have documents that instruct personnel in how to perform particular tasks. In a research laboratory, people might call these documents "procedures" or "protocols." (Some people reserve the term "protocol" for a procedure that is nonroutine and/or is designed to answer a question.) People in companies often call their procedures **standard operating procedures (SOPs)**. Regardless of what they are called, written procedures are of critical importance because they help ensure that everyone in an organization performs tasks correctly and consistently every tlme.

Many organizations use templates so that within the organization, every procedure has the same format and includes the same sections and types of information. There can be slight differences between organizations in how SOPs are formatted and what information they include, but they all usually have these features:

- They provide a step-by step outline of how a task is to be performed.
- They indicate safety concerns where relevant.
- They indicate in whlch situations the task is to be performed.
- They specify who is qualified to do the task and who is responsible for the task.
- They specify how to document that the task was performed properly.

Figure 2.2 shows the parts of a typical SOP, and Figure 2.3 shows an excerpt from an SOP.

When performing a task according to an SOP, it is expected that you will not deviate from the procedure as written. You should, therefore, always consult an instructor/supervlsor if for some reason an SOP cannot be followed exactly.

It is not easy to write an effective SOP that everyone in an organization wlll be able to follow exactly. SOPs are often written by research and development scientists after they have investigated and come to deeply understand the system of interest. Entry-level employees usually do not write SOPs because they do not have enough experience. It is nonetheless good practice for you, as a student, to try writing this type of document in order to better understand its purpose and construction.

When writing your SOP, think about these items:

- What is the purpose of the procedure? What will be the outcome or product of the procedure?
- What resources and materials are required in order to achieve the desired result?
- What equipment is required?
- What steps are required to achieve the desired product?
- What tests ensure that the product of the procedure is good?
- Are there any potential problems that may occur, and if so, how can these problems be avoided?

Before class, choose a biological molecule that interests you, for example, a fragment of DNA, a hormone, a saturated fat. A molecule consisting of 20–40 atoms is about right. Find a diagram of that molecule and bring it with you to class.

```
Name of organization                                          page x of x
Title of SOP
Effective Date (the first date the SOP can be used by personnel)
SOP Identification #
Revision Number
Prepared by (author) _____ Date _____
Reviewed and approved by _____ Date _____
Reviewed and approved by _____ Date _____
```

1. Purpose: *What is the purpose of this procedure?*

2. Scope: *To whom in the organization does this SOP apply or, what equipment or situations are covered by this SOP?*

3. Definitions: *Define terms that may be unfamiliar to the user or that are specific to the organization.*

4. References: *Refer to any other documents that may be associated with this procedure. For example, does the procedure require that a form be filled out? Are there other procedures needed to use associated equipment?*

5. Materials, Reagents, Equipment: *What is required to perform the task described in this SOP?*

6. Responsibility: *Who does this work?*

7. Hazard Communication: *Are there safety concerns?*

8. Procedures: *This is usually the bulk of the SOP and consists of numbered steps guiding the user in how to perform a task.*

FIGURE 2.2   Common parts of an SOP. The top box (shown here in gray) contains information that is generally repeated at the top of every page. The eight sections below are typically present. If one or another section does not apply, include that section but write "NA."

```
Cleanest Genes, Inc.                                          page 1 of 3
Operation of Balance: BRP 88
Effective Date: 7/22/10
SOP Identification #  BRP8762
Revision Number 1.0
Prepared by (author) E.J.Reid Date July/01/10
Reviewed and approved by D. Buttersworth Date July/15/10
Reviewed and approved by G. Worthy  Date July/17/10
```

1. Purpose: Describes routine operation of Balance Type BRP 88.
2. Scope: The routine (daily) operation and use of Balance model BRP 88 by laboratory analysts.
3. Definitions: NA
4. References: Manufacturer's Bulletin No 2; Form: Balance Instrument record (QF 15.3.6.3)
5. Materials, Reagents, Equipment: NIST Traceable standard mass set (2000 g); part number SMS 100. Set must have been checked by metrology department within the last year.
6. Responsibility: Training level of "General Lab" personnel or greater is required for use of this equipment.
7. Hazard Communication: NA
8. Procedures:
8.1. Calibration and verification of balance
Frequency: Daily before use:
      8.1.1. Switch on the balance for at least 20 minutes to allow the internal components to come to working temperature.
      8.1.2. Check that the leveling bubble is centered.
           8.1.2.1. If the bubble is off center, adjust the leveling feet to bring back to center.
      8.1.3. Check that the balance is clean.
… and so on

FIGURE 2.3   Example of a portion of a SOP.

CLASSROOM ACTIVITIES

1. Choose a partner. Separate from your partner by going into a different class-room or space where you cannot see one another.
2. Construct a model of your chosen molecule using the toothpicks, marshmal-lows, markers, and other supplies provided. Learn as much as you can about how to best use your construction materials to make a molecular model. You should build a complete model that you think is correct.
3. Write a draft of an SOP to make your chosen molecule. Write this first-draft SOP in your laboratory notebook because you are exploring the best way to make the model. Include:
    3.1.   A list of materials required.
    3.2.   The procedure itself, written as a series of steps.
    3.3.   Sketches as needed.
    3.4.   Features that the completed model must have:
        3.4.1.   For example, does the technician who constructs the molec-ular model need to use a protractor to measure the angles between toothpicks? If so, what is the acceptable range for each angle? Is the height or length of the model important?
        3.4.2.   Does it matter whether the narrow ends of the toothpicks all face in one direction?
        3.4.3.   Whatever features you identify to be important should be clearly incorporated into the SOP.
4. Meanwhile, your partner is in a different room writing an SOP for a different molecular model. After both partners have completed their SOPs, your instruc-tor will exchange them so that you will have the SOP your partner wrote and your partner will have the SOP you wrote. You will not see the model your part-ner worked from, only the SOP that she or he wrote.
5. Assemble a model according to the SOP provided by your partner. Follow your partner's directions *exactly*. If you have any difficulties following the directions in the SOP your partner wrote, make careful notes on the procedure Itself, explaining the problems.
6. Once you have finished constructing a model according to the SOP your part-ner wrote, and your partner has constructed a model according to the SOP you wrote, get back together in the same room. Examine both models and compare them with the expected results. Note and *document (write down)* any differences between the expected models and the models that you and your partner actually made. Evaluate each model as to how successful it is.
7. Revise your SOP to improve it. Think about these points: How easily did your partner follow your procedure? Was your partner able to use your directions to produce an acceptable model? What specific problems did your partner encounter and how could your written instructions have prevented those problems? Did you use illustrations in your SOP, and if so, were they help-ful? Was your writing legible and your spelling proper? Ask your partner for suggestions.
8. SOPs are written on a word processor. If your teacher requires it, prepare your revised SOP on a word processor with all the components of a complete SOP. Figure 2.2 shows the parts of the SOP that you should include. Observe the text in the box that goes at the top of every page; it is a header. The other items go in the body of the SOP. The majority of the SOP will be the procedure, but do not forget the other information. Refer to the following checklist to make sure you have all the required parts.

LAB MEETING/DISCUSSION QUESTIONS

1. What is involved in writing an SOP? What do you need to think about to ensure that another person will be able to follow your directions?

2. Perhaps the model you constructed was close to, but not identical to, the model your partner intended you to make. Maybe it was a little lopsided or the glue looked a bit thick—is it acceptable anyway? Consider that what is "good enough" depends on the situation. In a company making a real product, it is essential to know how "good" the product must be to be "good enough." For example, we would not want an x-ray machine to be a "little lopsided" or the glue to be a "bit thick" on a medical device. How can you decide if your models are "good enough"? How do you think biotechnology companies decide how "good" is "good enough" when developing their products?

SOP Checklist			
	Header		**Body of SOP**
	Organization/company name		Purpose of SOP
	SOP title		Scope
	SOP author signature with date		Definitions (may say NA)
	Date SOP becomes effective		References (may say NA)
	Reviewers' signatures with date		Responsibilities
	Pagination		Safety statement present or noted as NA
	Revision number		Materials and equipment required
	Identification number		Procedure written in step-by-step format, clearly, with diagrams, as needed

UNIT DISCUSSION: DOCUMENTATION IN THE LABORATORY

1. What does it mean to say that "good" laboratory results are "trustworthy"?
2. How does a system of documentation help ensure that laboratory results are trustworthy?
3. Consider the overarching themes of good laboratory practices in the following table. Jot down notes on the table or on another piece of paper that describe how the ideas and skills that you practiced in this unit relate to these overarching themes.

Good Laboratory Practices
DOCUMENT
PREPARE
REDUCE VARIABILITY AND INCONSISTENCY
ANTICIPATE, PREVENT, CORRECT PROBLEMS
ANALYZE, INTERPRET RESULTS

VERIFY PROCESSES, METHODS, AND RESULTS
WORK SAFELY

Metrology in the Laboratory

UNIT

III

DOI: 10.1201/9781003360742-3

UNIT INTRODUCTION

The only man who behaved sensibly was my tailor; he took my measurement anew every time he saw me, while all the rest went on with their old measurements and expected them to fit me.

—George Bernard Shaw, playwright, essayist, winner of 1925 Nobel Prize for literature

A. MEASUREMENTS SHOULD BE "GOOD"

Making measurements is an integral part of everyday work in any biology laboratory. For example, preparing solutions to support the activity of cells and biological molecules requires measuring weights of chemicals, volumes of water, and the pH of the resulting mixture. There are countless measurements made in any laboratory, each of which must be "good."

Metrology is *the study of measurements, where* **measurements** *are numerical descriptions* (e.g., the girth of one's waist or the temperature of a solution). The goal of studying metrology is to ensure that laboratory measurements are "good." Determining what makes a measurement good, however, is not simple. Consider, for example, a man who weighs himself in the morning on a bathroom scale* that reads 165 pounds. Shortly after, he weighs himself at a fitness facility where the scale reads 166 pounds. At this point he might be somewhat uncertain as to his exact, correct weight; perhaps he weighs 165 pounds, perhaps 166. Perhaps his weight is between 166 and 165 pounds, or perhaps both scales are wrong. Despite this uncertainty, he is likely to conclude that he weighs about 165 pounds and that this measurement is "good enough."

Uncertainty of a pound or two in an adult's weight is seldom of great concern. A measurement must be much "better," however, in other situations. A one pound difference in the weight of a newborn infant could mean the difference between a healthy baby and one who is severely dehydrated. An error of 1 gram in a laboratory measurement could mean the difference between a successful experiment and a wasteful failure. A **"good" measurement** then might be defined as *one that can be trusted when making a decision in a given situation.* "Good," trustworthy measurements have three characteristics:

- Good measurements *are traceable to international standards.*
- Good measurements *achieve a required level of accuracy and precision.*
- Good measurements *include an indication of their uncertainty.*

B. GOOD MEASUREMENTS ARE TRACEABLE TO INTERNATIONAL STANDARDS

I. STANDARDS AND CALIBRATION

Consider an object that is said to be "100 cm in length." Length is the property described, 100 is the numerical value of that property, and centimeters are the measurement **units**. A **unit of measure** is *an exactly defined amount of a quantitative property.* Units must be defined clearly, and everyone who uses the unit must agree on its definition. Establishing units of measure, therefore, requires international cooperation.

Metrologists, *people who work with measurements*, devote much effort toward ensuring international consistency in measurement. One result of this effort is the International System of Units, SI; the SI definitions of units are used by laboratory professionals around the world.

Consider measuring length with a ruler; the ruler is an external authority. The ruler is satisfactory if it was marked by the manufacturer so that its lines are correct according to the

---

* We use the term "scale" for the everyday instruments used outside the laboratory and reserve the term "balance" to refer to the sensitive and accurate weighing instruments used in laboratories.

internationally accepted definition of a "meter." The ruler is a standard where a **standard** is *a physical embodiment of a unit.*

The unit of a kilogram formerly was defined by international treaty to have as much mass* as a special platinum-iridium bar located at the International Bureau of Weights and Measures near Paris. All other mass standards were defined by comparison to this special metal bar. Every country that signed the treaty received a national kilogram prototype whose mass was determined by comparison with the standard in France. The United States' standard, K20, is still housed at the National Institute of Standards and Technology (NIST). Since May 2019 the kilogram has been defined based on a physical constant, that is, Planck's constant, which relates the energy carried by a photon to its frequency. Laboratories, such as those at NIST, have instruments that can measure the mass of objects in terms of Planck's constant. Thus, each country can now transform the definition of a kilogram, based on Planck's constant, into a physical object that is the authority for mass in that country.

The standards at NIST are **primary standards**, *standards whose values may be accepted without further verification by the user. A primary standard is used to establish the value for a* **secondary standard**. Companies that manufacture **working standards**, *standards for use in individual laboratories*, use secondary standards as the basis for their products.

Calibration is *the process of adjusting a measuring system so that the values it gives are in accordance with an external standard(s).* In the example of length, the manufacturer calibrates a ruler by placing lines on it that are in accordance with the SI definition of meters. Glassware that is used to measure the volume of liquids is similarly calibrated by the manufacturer with lines that indicate volume.

Instruments in the laboratory are also calibrated. A user might calibrate an instrument by placing a standard in or on the instrument and turning a knob, keying in a value, or adjusting a dial until the instrument displays the value of the standard. The response of the instrument after calibration is in accordance with the standard. Instruments are subject to the effects of aging, and their response may be altered by changes in their environment. Laboratory equipment must, therefore, be periodically recalibrated by the user or a service technician to correct for this deviation. If an instrument is not properly calibrated, its measurements will deviate from their correct values. Improper calibration is a common cause of laboratory error, and maintaining instruments "in calibration" is critical.

Let's return to the person weighing himself at home and at a gym. If one of the scales was not recently calibrated, then there is no way to know if its results are correct. Indeed, if neither scale was recently calibrated, the man's true weight may not be in the range of 165–166 pounds at all.

II. TRACEABILITY

Suppose you use a balance to weigh an object. How can you be certain and demonstrate to others that the balance you used was indeed properly calibrated? The answer is that your balance calibration must be "traceable." **Traceability** *describes the chain of calibrations that establish the value of a standard or of a measurement.*

Consider the traceability of mass standards. A working mass standard used in an individual laboratory was calibrated according to a secondary standard, which in turn was calibrated against a national, primary standard. There is thus a "genealogy" for the standard that is used in a particular laboratory (see Figure 3.1). The purpose of tracing the genealogy of a standard is to ensure the trustworthiness of measurements or calibrations made with that standard. Manufacturers use the term "traceable to NIST" in their catalogs when they have documented the genealogy of standards used in manufacturing their product.

To summarize the relationship between standards, calibrations, and traceability, consider the example of making a measurement using a balance. The balance was *calibrated* by the

---

* "Weight" and "mass" are not synonyms. Weight measures the force of gravity on an object, while mass is the amount of material an object contains. Biologists customarily weigh samples and do not correct their readings to account for the difference between mass and weight; thus, in practice, they use the words interchangeably. This manual follows this common practice and does not distinguish between mass and weight.

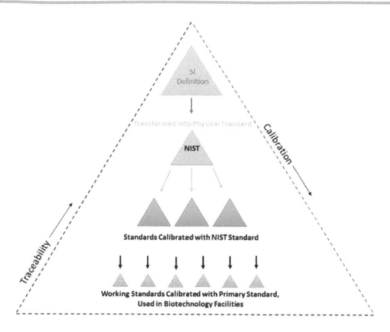

FIGURE 3.1 **Traceability of a mass standard showing its genealogy in the United States.**

user or a service technician so that its readings were correct according to the internationally accepted definition of a gram. To calibrate the balance, the technician used working mass *standards.* The person performing the calibration knows that the standards were correct embodiments of the unit called a "gram" because the comparisons of those standards to the primary, national standard were properly performed and documented. The working standards are, therefore, *traceable* to NIST. Thus, a "good" measurement is traceable to an accepted standard.

III. VERIFICATION

It is good laboratory practice to regularly check the performance of instruments to make sure they are functioning properly. **Verification** or **performance verification** is *the process of checking the performance of an instrument or system.* Verification includes checking whether the instrument is properly calibrated. Verification is performed in the laboratory of the user and is typically documented in a logbook or on a form. For example, in many laboratories, balance accuracy is checked regularly by weighing a standard and recording its weight. Based on repetition of the test and knowledge of the system, it is possible to establish a range within which values for the test should fall. If the value for the standard's weight does not fall within the desired range, then the balance is taken out of service and repaired. This check of the balance's performance verifies that the balance is properly calibrated but should not be mistakenly called "calibrating the balance." It is similarly good practice to periodically check the performance of micropipettes and other laboratory devices, thus helping ensure "good" measurements results.

C. "GOOD" MEASUREMENTS ACHIEVE REQUIRED ACCURACY AND PRECISION

I. ACCURACY AND PRECISION

A "good" measurement achieves a required level of accuracy and precision. **Precision** is *the consistency of a series of measurements or tests.* **Accuracy** is *how close an individual value is to the true or accepted value.* Accuracy and precision are not synonyms to scientists, although they are often used interchangeably in nonscientific English. It is possible for a series of measurements to be precise but not accurate. A series of measurements may

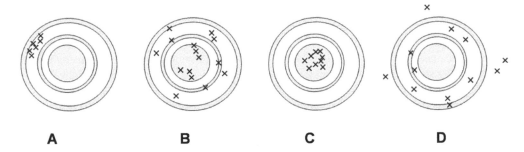

A   **B**   **C**   **D**

FIGURE 3.2   **Accuracy and precision.** Precision and accuracy are illustrated using the analogy of a target where the bull's-eye is the correct value. (a) Archer A is precise but not accurate. (b) Archer B is inconsistent. Archer B's results average the correct value although they are not precise. (c) Archer C is not very skilled; the results are neither accurate nor precise. (d) Archer D is both precise and accurate.

also average the correct answer yet lack precision. A series of measurements may be neither accurate nor precise, or may be both accurate and precise (see Figure 3.2). One of the main objectives of the exercises in this unit is to bring an awareness of accuracy and precision into your everyday work in the laboratory.

Experienced laboratory professionals know that measurements repeated in succession on the same day tend to be relatively consistent. In contrast, measurements performed on different days, by different people, using different materials and equipment tend to be more variable. There are different words for precision to make this distinction clear. **Repeatability** is *the precision of measurements made under uniform conditions*. **Reproducibility** is *the precision of measurements made under nonuniform conditions, such as in two different laboratories or by two different analysts*. Repeatability and reproducibility are, therefore, two practical extremes of precision. It is challenging, but important, to develop procedures that give results that are as reproducible as possible. Standard operating procedures (SOPs), previously explored in the Documentation Unit, are a key to ensuring reproducibility. You will, therefore, practice writing a measurement-related SOP in an exercise in this unit.

II. ERRORS REDUCE ACCURACY AND PRECISION IN MEASUREMENTS

Errors reduce the accuracy and precision of measurements. Obviously, as a laboratory professional, you will want to reduce errors in your own work. Metrologists classify measurement errors into types. One such classification scheme is as follows:

1. **Gross errors** are *caused by blunders*. If a person were to accidentally misread the numbers on a pipette, it would be a gross error. Gross errors cause measurements to be less accurate and/or less precise.
2. **Systematic errors** are normally more subtle than gross errors. There are a number of causes of systematic errors, including the use of equipment that is improperly calibrated, not well maintained, or operated improperly; solutions that have degraded; and environmental fluctuations.

    An important feature of systematic error is that it results in **bias**, that is, *measurements that are consistently either too high or too low*. Thus, by definition, systematic errors reduce accuracy. If, for example, a 500 mL volumetric flask has its 500 mL mark set slightly too low, then it will consistently deliver volumes that are a little less than they should be. Systematic error is not detected by repeating measurements because every time the measurement is repeated, it will incorporate the same error and tend to be too high or too low. It is the job of the laboratory professional to minimize systematic errors in measurement by being knowledgeable and attentive to potential problems and properly maintaining and calibrating equipment.
3. **Random errors** are *difficult or impossible to find and eliminate. Random errors cause measurement values that are sometimes too high and sometimes too low*. Random

error leads to a loss of precision because it leads to inconsistency. Consider a 500 mL flask that is perfectly marked at exactly 500 mL. Although the flask is marked correctly, every time it is used there will be a tiny variation in how much liquid is delivered because of imperceptible variations in the environment and the person using the flask. This slight variability is random error and will result in volumes being delivered that are sometimes a bit more than 500 mL and sometimes a bit less. If the flask is used many times, the average volume delivered will be 500 mL. It is good laboratory practice to repeat critical measurements, assays, and tests to determine the impact of random error. As measurements are repeated, you can see their variability.

D. "GOOD" MEASUREMENTS HAVE AN INDICATION OF UNCERTAINTY

Good measurements should have an indication of their uncertainty. The term "uncertainty" is used often in the literature of metrology. To some extent, the meaning of "uncertainty" is familiar. We mentioned uncertainty earlier in relation to the person who weighs himself twice and gets two different results. You would know from experience that, for a variety of reasons, the man's weight may or may not be exactly as shown on the scales. There is uncertainty because error exists. Even when people are very careful, random errors and sometimes systematic errors persist.

Metrologists try to estimate the effect of errors so they can know how much confidence to place in a measurement. Estimating the uncertainty in a measurement is complex. It requires identifying potential sources of error, estimating the magnitude of each error, and combining all the effects of error into a single value. **Uncertainty** is *an estimate of the inaccuracy of a measurement caused by all the errors present.* It is outside the scope of this manual to go into the evaluation of uncertainty in detail. However, one aspect of uncertainty is relevant to everyone in the laboratory: recording measurement values with the correct number of significant figures. This is, therefore, the topic of Laboratory Exercise 4.

E. PUTTING IT ALL TOGETHER IN THE LABORATORY

This Unit Introduction has established basic principles that relate to making good measurements. We will apply these ideas in the laboratory exercises in this unit (see Figure 3.3).

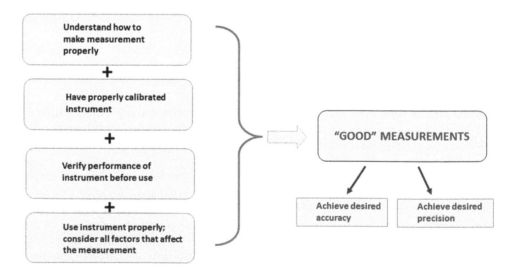

FIGURE 3.3 Making "good" measurements. Good measurements achieve the accuracy and precision required in a particular situation. Making good measurements requires that measuring instruments are properly maintained and calibrated, that their performance is verified and documented, and that the person making the measurement does so knowledgeably and correctly.

Many properties might be measured in a laboratory setting, e.g., length, density, chemical composition. This chapter cannot cover every type of measurement so it will focus on the measurement of three properties of great importance to biologists:

- **Weight** (Classroom Activity 9 and Laboratory Exercises 5 and 6)
- **Liquid volume** (Laboratory Exercises 7 and 8)
- **pH** (Laboratory Exercise 9)

Although we cover only these three types of measurements, the principles that you will explore are broadly applicable to all of the many measurements you will make throughout your career in the laboratory.

**Laboratory Exercise 4: Recording Measurements with
the Correct Number of Significant Figures**

OVERVIEW

The purpose of this short laboratory exercise is to practice recording values from laboratory instruments with the correct number of significant figures. Your instructor will provide you with the following:

- Samples whose lengths you will measure.
- Water samples whose temperatures you will measure.
- Samples whose weights you will measure.
- Water samples whose volumes you will measure.

BACKGROUND

Recall from the Unit Introduction that "good" measurements should have an indication of their uncertainty. One of the causes of measurement uncertainty is that measuring devices cannot measure any property with an infinite "exactness." Consider, for example, the rulers in Figure 3.4. In Figure 3.4a, the ruler's gradations divide each centimeter in half. We know the arrow is somewhat longer than 4 cm, and it is reasonable to estimate the tenth place and report the arrow's length as "4.3 cm." Information will be lost if we report the length as simply "4 cm" because it is possible to tell that the arrow is somewhat longer than 4 cm. It would be unreasonable to report that the arrow is "4.35 cm" because the ruler gradations give no way to read the hundredths place. When measurements are recorded, it is customary to record all the digits that are certain plus one that is estimated. In this example, the "4" is certain and the "3" is estimated, so the measurement is said to have two significant figures. Thus, by reporting the measurement to be 4.3, we are telling the reader something about our certainty in the length measurement value. We are sure about the "4" and not as sure about the "3."

The subdivisions in the second ruler (see Figure 3.4b) are finer and divide each centimeter into tenths. With the second ruler, we can reliably say the arrow is "4.3 cm," and in fact, it is reasonable to estimate that the arrow is about "4.35 cm." The "4.3" is certain, the "5" is estimated, and so there are three significant figures in the measurement. Thus, the arrow is the same length in both Figure 3.4a and 3.4b, yet the length *recorded* is different because the rulers are differently subdivided. The fineness of the measuring instrument, in this case the rulers, determines the certainty in the length of the arrow. The measurement certainty is indicated by the number of significant figures used in recording the measurement. It is your responsibility in the laboratory to always record measurements with the correct number of significant figures.

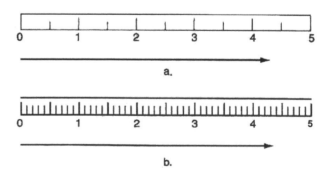

a.

b.

FIGURE 3.4   Measurements of length with two rulers. (a) A ruler where each centimeter is subdivided in half. (b) A ruler where each centimeter is subdivided into tenths. The first ruler provides measurements with two significant figures, while the second ruler provides measurements with three significant figures. (Not to scale.)

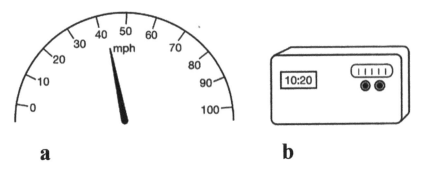

FIGURE 3.5   An analog readout compared to a digital one. (a) Analog. (b) Digital.

Rulers provide an **analog readout**, that is, *a smoothly changing reading* (see Figure 3.5a), as compared to a **digital display** that is *discontinuous* (see Figure 3.5b). Most modern electronic instruments display results with a digital display; there is no meter to read. With any instrument having a digital display, the last place is assumed to have been estimated by the instrument.

The concept of significant figures applies to all measuring devices. Suppose a particular laboratory balance can reliably weigh an object as light as 0.00001 g. On this balance, sample Q is found to weigh 0.12300 g. It would not be correct to report that sample Q weighs 0.123 g because information about the certainty of the measurement is lost if the last zeros are discarded. In contrast, if sample Q is weighed on a balance that only reads three places past the decimal point, it is correct to report its weight as 0.123 g. It would not be correct to record the weight as 0.1230 g because the balance could not read the fourth place past the decimal point. The first balance gives more certainty about the sample weight. The difference in measurement certainty between the two balances is shown by the number of significant figures used: 0.12300 g has five significant figures, while 0.123 g has only three.

You can see in these examples that a basic principle in recording measurements from instruments is to report as much information as is reliable plus one last figure that is estimated and might vary if the measurement were repeated. The number of figures reported by following this principle is the number of significant figures for the measurement. A **significant figure** is *a digit that is a reliable indicator of value*. The number of significant figures reported in a measurement is an indication of the certainty of a measurement.

Suppose that one report states that the number of COVID-19 cases last year was 156,375,000 while another reports it as 156,400,000. Let's assume that both reports are correct, but the second report rounded the number to make it easier to read. The first number, 156,375,000, is a more exact figure that allows the reader to be more certain about the actual number of cases than does the second number. We say that the number 156,375,000 has more significant figures (six) than the number 156,400,000 (which has four significant figures).

This COVID-19 case example can be used to illustrate an important point regarding zeros. The zeros in both reports are essential; without them the number would be reported as a paltry 1,564 or 156,375. The zeros are "placeholders," but they are not correct indicators of value. Perhaps the true number of cases was really 156,375,178 or 156,375,091, or other possibilities. Sometimes zeros are placeholders and are not reliable indicators of value. In contrast, in the report that there are "30 students in the laboratory," the zero is a reliable indicator of value. The zero shows that there are 30 students, not 29 or 31. Zeros that are placeholders are not called significant figures, while zeros that indicate value are significant.

Suppose the number given in a report is 45,000. The three zeros in this number each may be placeholders or they may be indicators of value. There are various ways to tell the reader whether the zeros at the end of a number are significant. One method is to use scientific notation. For example, given the number 45,000, if none of the zeros are

TABLE 3.1   Rules to Record Measurements with the Correct Number of Significant Figures

1. The number of significant figures is related to the certainty of a measurement.
2. When reporting a measurement, record as many digits as are certain plus one digit that is estimated. When reading a meter or ruler, estimate the last place. When reading an electronic digital display, assume the instrument estimated the last place.
3. All nonzero digits in a number are significant. For example, all the digits in the number 98.34 are significant; this number has four significant figures. A reader will assume that the 98.3 is certain and the 4 is estimated.
4. All zeros between two nonzero digits are significant. For example, in the number 100.4, the zeros are reliable indicators of value and not just placeholders.
5. Zero digits to the right of a nonzero digit but to the left of an assumed decimal point may or may not be significant. In the number 156,400,000, the decimal point is assumed to be after the last zero. In this case, the zeros are ambiguous and may or may not be reliable indicators of value. Methods of clarifying ambiguous zeros are discussed in the text.
6. All zeros to the right of a decimal point or to the right of a nonzero digit before a decimal place are significant. The following numbers all have five significant figures: 3.4000, 0.34000 (the zero to the left of the decimal point only calls attention to the decimal point), and 340.00.
7. All the digits to the left of a nonzero digit and to the right of a decimal point are not significant unless there is a significant digit to their left. 0.0098 has two significant figures because the two zeros before the 98 are placeholders. In contrast, the number 0.4098 has four significant figures.

significant, then the number could be reported as 4.5×10^4 with two significant figures. If there are three significant figures the number could be reported as 4.50×10^4. If all the zeros in the number 45,000 are significant, this can be shown by placing a decimal point after the number. The numbers 45,000. and 4.5000×10^4 both have five significant figures.

Table 3.1 summarizes rules regarding how to record measurements with the accepted number of significant figures. These rules also show how to decide when a zero is significant and when it is a placeholder.

LABORATORY PROCEDURE

1.   Temperature
    1.1.   Record the temperature measured by each thermometer provided.
    1.2.   Is each display analog or digital?
    1.3.   How many significant figures are in each reading?
2.   Weight
    2.1.   Record the weight of the object on each balance provided.
    2.2.   Is each display analog or digital?
    2.3.   How many significant figures are in each reading?
3.   Volume
    3.1.   Record the volume of water in each graduated cylinder provided.
    3.2.   Is each display analog or digital?
    3.3.   How many significant figures are in each reading?
4.   Length
    4.1.   Record the length of each sample provided.
    4.2.   Is each display analog or digital?
    4.3.   How many significant figures are in each reading?

EXAMPLES

Example 1. How long is this ant's body? Is this an analog or digital reading? How many significant figures are there?

The answer is <u>0.67</u> cm. It is also correct to say 0.65, 0.66, or 0.68 cm because the last digit is estimated. This is an analog display. There are <u>2</u> significant figures.

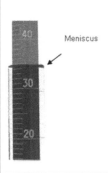

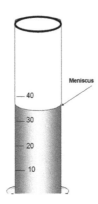

Example 2. This is a photograph of a graduated cylinder with liquid. What is the volume of the liquid? How many significant figures are there? **Answer** <u>34.9 mL</u> with <u>3</u> significant figures.	**Example 3.** What is the volume in this simulated graduated cylinder? How many significant figures are there? Read from the bottom of the meniscus. **Answer** <u>34 mL</u>, maybe <u>33</u>. This is also an analog reading. There are two significant figures. (A real graduated cylinder would be calibrated to have more than two significant figures.)	**Example 4.** What is the reading on this digital display? **Answer** <u>2.30</u>, not 2.3!

LAB MEETING/DISCUSSION QUESTIONS

1. Figure 3.6 is an example of the data recorded by a class for this laboratory exercise. For temperature, three thermometers were placed in a single beaker of water equilibrated to room temperature. The thermometers were labeled "A," "B," and "C." Each student recorded the temperature of each thermometer, one student at a time.

    a. Did the three thermometers provide the same value for room temperature?

    b. What *was* the temperature in this room at that time? How certain are you of the value? Can you provide a range that you believe contains the actual temperature of the room?

    c. Should all 11 students have recorded measurement values with the same number of significant figures? Did all students record values with the same number of significant figures?

2. Refer to Figure 3.6. For length, two objects were measured with a ruler. The objects were labeled "1" and "2." Each student recorded the length of each object using the same ruler.

    a. How long is each object?

    b. How many significant figures are in these measurements? According to the usual significant figure conventions, all measurements should be recorded with all values that are certain, plus one more place that is estimated.

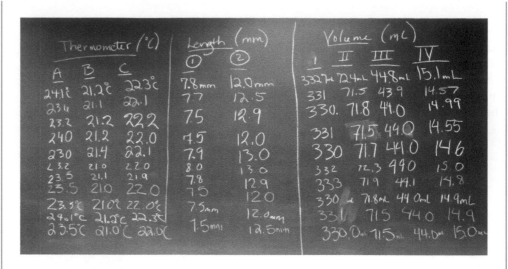

FIGURE 3.6   An example of class data.

3. Refer to Figure 3.6. For volume, liquids were placed into four graduated cylinders labeled "I," "II," "III," and "IV." Each student read the volume in each graduated cylinder. There is a problem for columns I and IV; the students did not record their data with the same number of significant figures. Given that you cannot see the graduated cylinders, do you have any way to know which students read the values with the correct number of significant figures and which students were wrong?

4. Put the values for your class on the blackboard. In reading analog measurements, your values should all have the same number of significant figures, but the last digit might vary from person to person because it is estimated. What about your readings from instruments with a digital readout? Does your class agree on all the readings? If not, why?

5. Explain in your own words how significant figures relate to the uncertainty of a measurement.

Classroom Activity 9: Constructing a Simple Balance

OVERVIEW

The purpose of this activity is to introduce you to the most basic principles of weight measurement. To do so, you will construct a simple balance. As you construct and operate your balance, you will learn the following:

- The basic principles of weighing.
- The effect of a measuring instrument's sensitivity on the results.
- Design and operation considerations that govern weight measurements in the laboratory.

BACKGROUND

Weight is *the force of gravity on a sample*; weighing a sample involves comparing the pull of gravity on the sample with the pull of gravity on standard(s) of established mass. **Balances** are *laboratory instruments used to determine the weight of substances*. The weighing method illustrated in this activity dates back to antiquity (see Figure 3.7a). The object to be weighed and the standard(s) are placed on pans attached to opposite ends of a **lever** or **beam**. When the beam is exactly balanced, gravity is pulling equally on the sample and the standard; they are the same weight. Thus, instruments that weigh objects in the laboratory have come to be called "balances." Mechanical devices that balance objects across a beam are still used in the laboratory, although electronic balances are most commonly used when high sensitivity is required.

Sensitivity is *the smallest value of weight that will cause a change in the response of the balance*. Sensitivity determines the number of places to the right of the decimal point that the balance can read accurately. Manufacturers express balance sensitivity by **readability**, which is *the smallest division of the scale dial or the digital readout*. For example, a balance that reads to four places to the right of the decimal point (0.0001 g) has a readability of 0.1 mg.

In order to achieve high sensitivity, the manufacturers of mechanical analytical balances orient the beam on a fine knife edge (see Figure 3.7b). To operate these mechanical balances, it is therefore important to place the balance in an **arrested position**, *where the beam is locked so it will not move*, before standards and samples are added to or removed from the weighing pan(s). If the beam is not arrested, it can actually fall from its position on the knife edge when samples are added or removed.

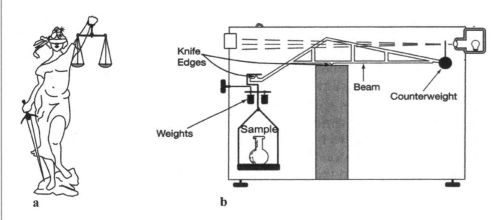

FIGURE 3.7   Mechanical balances. (a) Balances with a beam and two pans have been used since antiquity. In Greek mythology, Themis is the goddess of justice. She is usually represented with a double-pan balance in her hand. (b) This mechanical analytical balance shows the beam balanced on a knife edge. The weighing pan is also balanced on a knife edge. While you are unlikely to see this style of instrument in a modern laboratory, it illustrates basic principles for designing a weight measuring instrument.

CLASSROOM ACTIVITY

Use building blocks and rulers to construct a simple weighing device. Use your weighing instrument to determine the weight of a quarter in units of M&M's candies. Try to achieve as much accuracy and sensitivity as you can.

LAB MEETING/DISCUSSION QUESTIONS

1. Compare your class results and determine how much a quarter weighs in M&M's units. How much agreement is there in the class results?
2. What does it mean to "weigh" something?
3. How can you maximize the accuracy of your balance?
4. How can you maximize its precision?
5. What does the term "sensitivity" mean with regard to weighing? Which group constructed a balance with the most sensitivity? What are the disadvantages to a balance that is very sensitive?
6. Older style mechanical analytical balances almost always have an arrested position where the beam is locked so it will not move. Standards and samples are added to the beam while the balance is in the arrested mode. Based on your observations, why is this true?
7. Do you have any balances in your laboratory that are mechanical and have a beam? Do you have any electronic balances in your laboratory?
8. How many significant figures do your measurements have?
9. What factors would a user of your balance need to consider in order to obtain the best possible measurements?

Laboratory Exercise 5: Weight Measurements 1: Good Weighing Practices

OVERVIEW

The measurement of weight is among the most common tasks performed in a biology laboratory. There are two key factors to consider in order to avoid systematic error when weighing. First, your technique in operating the balance must be good. Second, the balance must be properly maintained and calibrated. This first laboratory exercise on weight measurements focuses on good techniques for operating different types of balances. The second weight measurement exercise involves verifying that a balance is calibrated and is performing properly.

Modern laboratory balances are easy to operate and appear to be deceptively simple. They are, in fact, complex, expensive, and carefully constructed instruments that are capable of achieving excellent accuracy and precision—if operated knowledgeably. In this exercise you will perform the following tasks:

* Learn how to operate laboratory balances.
* Explore factors that affect the accuracy of weight measurements.
* Analyze the accuracy of your measurements using percent error.
* Write an SOP to operate a balance.

BACKGROUND

A. Electronic Laboratory Balances

In Classroom Activity 9, you constructed a balance with a beam, across which the samples and standards were balanced. Modern laboratory balances are often electronic and do not have a beam. Electronic balances do not need a beam because an electromagnetic force counterbalances the sample, rather than standard weights. The magnitude of the force required to counterbalance the sample is related to the weight of the sample. In electronic balances, the comparison between the sample and a standard is made sequentially. The balance is first calibrated using a standard. The standard is removed and the sample is weighed afterward. Electronic balances do not use knife edges, and it is not necessary to arrest the balance before placing objects on the weighing pan.

B. Range, Capacity, and Sensitivity of a Balance

Balances vary in these characteristics:

1.   **Range.** *The span from the lightest to the heaviest weight the balance is able to measure.*
2.   **Capacity.** *The heaviest sample that the balance can weigh.*
3.   **Sensitivity and readability. Sensitivity** is *the smallest weight that will cause a detectable change in the response of the balance.* You probably observed in Classroom Activity 9 that the more sensitive a balance, the more difficult it is to design, construct, and operate. Sensitivity determines the number of places to the right of the decimal point that the balance can read accurately. Manufacturers express balance sensitivity as readability. **Readability** is *the value of the smallest unit of weight that can be read; it is the smallest division of the scale dial or the digital readout.* A balance that reads weight to four places to the right of the decimal point has a readability of 0.1 mg and is 10 times more sensitive than one with a readability of 1 mg. **Analytical balances** are *those that are designed to optimize sensitivity and can weigh samples to the nearest 0.1 mg or better.*

Range, capacity, and sensitivity are interrelated. A very sensitive balance will not be able to weigh materials in the kilogram range, and a balance intended for heavier samples will not detect a weight change of a milligram. Always select the right balance for your application.

C. Good Weighing Practices

Modern analytical balances are capable of consistently and accurately weighing samples in the microgram range. To get good quality results, especially with samples of very low weight, it is important that the operator uses good practices, as shown in Box 3.1.

Figure 3.8 illustrates a modern electronic analytical balance. Figures 3.9 to 3.11 illustrate the use of laboratory balances.

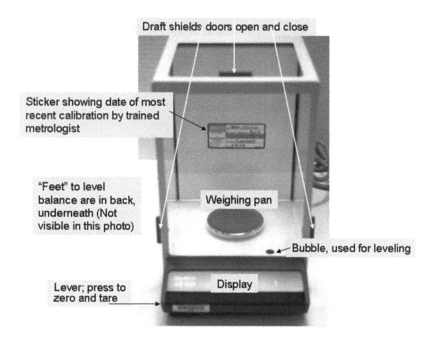

Draft shields doors open and close

Sticker showing date of most recent calibration by trained metrologist

"Feet" to level balance are in back, underneath (Not visible in this photo)

Weighing pan

Bubble, used for leveling

Lever; press to zero and tare

Display

FIGURE 3.8 A typical analytical balance. These balances are easy to operate and appear relatively simple—but they are, in fact, complex instruments that must be used properly.

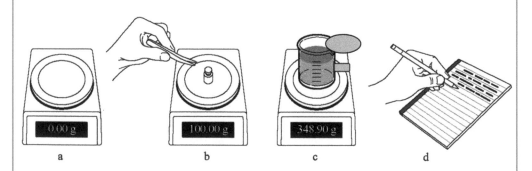

0.00 g

100.00 g

348.90 g

a   b   c   d

FIGURE 3.9 Using a top-loading balance. This balance has a capacity of 3000 g and a readability of 0.01 g. It is not suitable for weighing samples in the milligram range, but it is appropriate for samples weighing hundreds of grams. (a) The balance is first adjusted to zero with the weighing pan clean and empty. (b) A 100 g standard weight is tested to verify that the balance is operating properly. This verification is not necessary every time the balance is used, but it was done this time in accordance with the laboratory's quality control procedures. (c) The sample is placed on the balance and its weight is read. (d) All steps are documented.

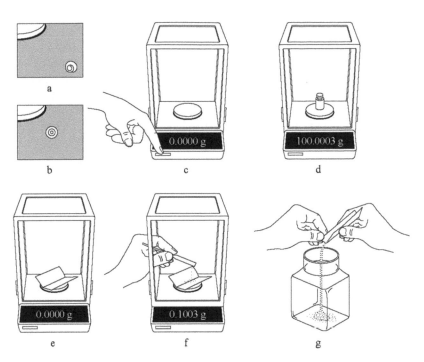

FIGURE 3.10   Using an analytical balance to weigh out a chemical: Strategy 1. In this example, the analyst needs to prepare a solution that contains 100 mg of NaCl and 100 mL of water. An analytical balance is required. (a and b) The balance is leveled so that the bubble is centered using the adjustable feet. (c) The balance is set to 0.0000 with the pan clean and empty. (d) The performance of the balance is verified and is found to be acceptable based on the laboratory's quality procedure. (e) The analyst folds a piece of weighing paper and places it on the balance's pan. (Alternatively, a small plastic weigh boat would work.) The analyst closes the draft shield and tares the balance to subtract out the weight of the paper by pressing the front lever (see Figure 3.8). f. The analyst carefully adds NaCl until the display shows 0.1000 g (= 100 mg). The figure in the last decimal place is likely to fluctuate, which is acceptable. (g) The analyst carefully pours the chemical into a container to which the water will be added.

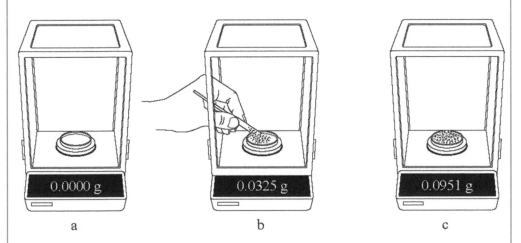

FIGURE 3.11   Using an analytical balance to weigh out a chemical: Strategy 2. The analyst now needs to prepare a solution that contains 100 mg of bovine serum albumin (BSA) and 100 mL of water. The analyst would like to weigh out exactly 0.1000 g BSA, however, it is difficult to handle and the analyst is not able to obtain exactly 0.1000 g. (a) She uses a metal weigh boat to reduce static and tares the balance. (b and c) She weighs out 0.0951 g of BSA. Later she adds only 95.1 mL of water to make her solution. This approach is correct because she adjusts the volume of water to compensate for the lesser amount of BSA weighed out on the balance.

D. Reporting the Accuracy of a Measurement

Recall that accuracy is the closeness of agreement between a measurement or test result and the true value or the accepted reference value. Although we generally do not know the "true" value for a measurement, we often have a standard or other basis by which to evaluate our answer. We can then express accuracy as **absolute error** using a simple equation as shown in Formula 3.1. Alternatively, we can express our answer using as **percent** or **relative error** using the equation in Formula 3.2. You will use these mathematical expressions of accuracy in this laboratory exercise.

Formula 3.1

Absolute Error

$$\text{Absolute error} = \text{Average measured value} - \text{True value}$$

where
error is *an expression of accuracy* and
true value may be *the value of an accepted reference material.*

Formula 3.2

Percent Error

$$\text{Percent error} = \frac{\text{Average measured value} - \text{True value}}{\text{True value}} \times 100\%$$

Planning Your Work: Weight Measurements: Good Weighing Practices

Part A: Operation of Balances; Factors Affecting Accuracy

1. What is the purpose of this section of the exercise?
2. What will you record in your laboratory notebook?
3. What numerical data will you collect?
4. How will you analyze your numerical data?
5. What will you have learned after you analyze your numerical data?
6. Prepare a table in which to record the data you collect in steps A.5 and A.6.
   - The table should have a descriptive title.
   - Every column should be clearly labeled.

Write the table in your laboratory notebook before coming to class, if directed to do so by your instructor. Alternatively, your instructor may have you prepare the table on a separate sheet of paper and copy it into your laboratory notebook when you come to lab, before making any measurements. A third possibility is to prepare your table using a word processor, print it out, and bring the page to lab. In this case, record your data into the table and then securely tape the completed table into your laboratory notebook. *A separate paper, like this one, used to record data and information* is called a "**form.**" Remember that whichever method you use, the values you record are your raw data, and their veracity must be protected.

Part B: Writing a Procedure to Weigh a Sample by Using a Specific Balance

1. What information will you need to record in your laboratory notebook in preparation for writing an SOP?
2. What is the value of this SOP in the future?

Box 3.1   Good Weighing Practices

- Choose a balance with the right capacity and sensitivity for the sample. Don't exceed the capacity of a balance.
- *Gently* place items onto the weighing pan to avoid breaking the delicate apparatus underneath.
- Check that analytical balances are level before use.
- Avoid vibration, drafts, and jostling, especially with analytical balances.
  - Analytical balances are extremely sensitive to vibration.

- Analytical balances have draft shields with doors that should be closed when taking a measurement. Draft shields are less common on less-sensitive balances.
- Temperature changes will affect the measured weight of a sample. A temperature change of 1.5°C will affect a weight reading at the fourth decimal place.
  - Keep the sample, balance, and surroundings at the same temperature.
  - Avoid touching samples and containers; use tongs if possible.
  - Do not locate balances near windows, radiators, or air conditioners.
  - Allow a balance time to warm up before use or leave it in "standby" mode.
- Reduce the effects of static charge. If a sample is charged with static electricity, a field builds up between the sample and balance. The force of this field can simulate a change in weight that may extend into the gram range (see Figure 3.12). You can recognize this effect by a weight display that rapidly fluctuates. In extreme cases, a charged sample can fly out of its container. Methods to reduce the effects of static charge include the following:
  - Use commercial products designed to avoid static charge.
  - Maintain room humidity between 40% and 60%.
  - Use metal weighing containers; avoid plastic ones.
  - Clean the walls of the chamber with a glass cleaning solution.
- Some samples gain moisture from the air and others lose volatile components to the air.
  - Work quickly but carefully to minimize such changes in the sample's composition.
  - Maintain the humidity in a weighing room between 40% and 60% if possible.
- Place the sample near the center of the weighing pan.
- Do not touch chemicals, magnetic stir bars, or the insides of beakers with your fingers to avoid contaminating them.
- Do not return unused chemicals to their storage bottles.
- Keep balances and weighing rooms clean.
  - Some chemicals are toxic.
  - Chemicals left on balances can cause corrosion.
  - Materials from one person's work may contaminate someone else's.

FIGURE 3.12 The effect of static charge. (a) A rectangular sheet of plastic was suspended on two blocks so that it was close to but not touching the weighing pan of an analytical balance. The balance was zeroed. (b) The plastic sheet was removed and rubbed briefly with a cloth to establish a static charge. It was then replaced on the blocks. As you can see, even though nothing was placed on or removed from the weighing pan, the display now reads –2.1430 g. The static charge on the plastic sheet therefore disturbed the reading on the balance by more than 2 grams!

LABORATORY PROCEDURE

Work individually or in teams, as directed by your instructor.

Part A: Operation of Balances; Factors Affecting Accuracy

A.1. Learn how to operate a particular balance based on its instruction manual. Then, teach other students how to use that balance.

A.2. Learn about all the balances in your laboratory; record the following information for each balance:

   A.2.1. Manufacturer and model number

   A.2.2. Serial number (if present)

   A.2.3. Range and capacity

   A.2.4. Readability

   A.2.5. How many figures to the right of the decimal point should you record when weighing items with that particular balance? (Recall that for a balance with a digital readout, you should record exactly what is displayed by the balance. For a balance with an analog readout, you should record all the values of which you are certain, plus one that is estimated.)

A.3. Which (if any) balances in the laboratory have a digital readout? When would you use this balance(s)? Which (if any) balance(s) in the laboratory have an analog readout? When would you use this balance(s)?

A.4. Most modern electronic balances can be **tared**, meaning that *the weight of the container is automatically subtracted from the total weight of the container plus the sample.* With a balance that can be tared, the empty weighing container is first placed on the balance, and a knob, dial, or lever is used to set the balance display to zero. The sample is then added to the container, and the balance displays only the weight of the sample; the weight of the container has been subtracted automatically. Which of your balances can be tared?

A.5. Changes in temperature can affect the accuracy of weight measurements. Convince yourself of this:

   A.5.1. Weigh a 50 mL beaker. (Use an analytical balance; do not exceed its capacity.)

   A.5.2. Hold the beaker in your hands for a minute or two to allow it to warm up. (Make sure your hands are warm first.) What do you predict will happen to the weight of the beaker now that it is warmer?

   A.5.3. Weigh and record the beaker's weight again; use the same balance. Can you detect the effect of the temperature change on weight? If so, did the weight go up or down? Is this what you predicted? How will you control for the effect of temperature in the future?

A.6. Check your skills by measuring the weights of "unknown" samples provided.

   A.6.1. Review Box 3.1 and pay attention to factors that will affect the measured weights of your samples.

   A.6.2. Choose the most appropriate balance in your laboratory for each unknown and weigh each one. Record your results in tabular form with the correct number of significant figures. Be sure you have labeled the table of data with a title at the top and headings for each column.

   A.6.3. Check your answers with the key provided by the instructor. Record the values from the key in your table.

   A.6.4. Data analysis

      A.6.4.1. Make two assumptions when analyzing your data for the unknowns. Assume that the teacher's answer key is correct. (Is this a reasonable assumption?) Assume also that your balance was properly calibrated. (Is this a reasonable assumption?) Calculate and record the *absolute and percent error* for each of your measurements of the unknowns. Record your results in your table.

      A.6.4.2. Write a few sentences below your table explaining the data you collected, why you collected it, and how you analyzed it.

Part B: Writing an SOP to Weigh a Sample Using a Specific Balance

Everyone must use laboratory instruments properly and consistently to ensure the accuracy and consistency of the team's work. Providing SOPs is an excellent way to guide everyone in the operation of instruments.

- B.1. Review the section on SOPs in Unit II.
- B.2. Learn how to use a particular analytical balance. Take notes in your laboratory notebook. Use sketches where appropriate.
- B.3. Write an SOP for this balance. Your final version must be written using a word processor. Your SOP must include directions to perform the following tasks:
    - B.3.1. Level the balance.
    - B.3.2. Zero the balance.
    - B.3.3. Use a standard weight to verify whether the balance is working properly. Decide and define for the analyst how accurately the balance must read the standard's weight to be considered acceptable. For example, if you use a 100 g standard, is it acceptable for the balance to read 100.0003 g for the standard? Is 99.9998 g acceptable? (There is no simple answer; choose a range that you can justify.) Tell the analyst what to do if the balance is "out of spec," that is, does not achieve the desired level of accuracy. (Normally the balance would be taken out of service, checked, and repaired, if necessary.)
    - B.3.4. Tare the balance (if it can be tared).
    - B.3.5. Weigh a sample.
    - B.3.6. Clean the balance when done.
    - B.3.7. Document that all steps were performed properly.
- B.4. Check that your SOP has all the required parts according to the SOP Checklist in Unit II.

LAB MEETING/DISCUSSION QUESTIONS

Put class data on the blackboard.

1. What do the terms "accuracy" and "precision" mean in the context of weight measurements?
2. What was the effect of temperature?
    a. In Part A, were you able to detect the change in weight of the beaker as its temperature changed? If not, speculate as to why not. If so, explain the phenomenon you observed.
    b. Do the results of everyone in class agree?
    c. If your class results demonstrate an effect of temperature on weight, what does this mean in terms of getting accurate and precise weight measurements?
    d. Figure 3.13 shows the results for a class. Everyone in this class agreed that temperature affected the weight of an object—but some found that the weight increased with temperature, and others saw the opposite result. Do you think the effect of temperature should always be the same? How do the results from your class compare to those in Figure 3.13?
    e. If your class finds an effect of temperature, is this effect always going to be important? Are there situations where you would not worry about temperature?
3. What factors in addition to temperature will affect the accuracy of your weight measurements?

FIGURE 3.13   Example of class results for the effect of increasing temperature on the weight of an object. Four groups observed a decrease in weight with increased temperature, but two groups saw the opposite effect.

4.   What did you find out about the accuracy of your weight measurements?
    a.   What percent errors did you get for the unknowns? What percent error did your classmates obtain?
    b.   Decide what percent error is low enough to be acceptable for your weight measurements; that is, how close is close enough? On what basis can you make this decision? Will your decision depend on the weight of the samples being measured? Are your own measurement values "close enough"?
5.   In Classroom Activity 9, you designed and built a simple balance that compared the force of gravity on a coin to the force of gravity on your M&M's standards. How does the design of the balance(s) you used for the current laboratory exercise compare to your M&M's balance?

Laboratory Exercise 6: Weight Measurements 2: Performance Verification

OVERVIEW

Performance verification of a balance *checks that the instrument is operating properly.* You will verify the performance of a balance by performing the following tasks:

- Checking a balance's calibration.
- Checking a balance's linearity.
- Checking a balance's precision.
- Verifying that instruments are working properly is a routine part of laboratory work.

BACKGROUND

A. Calibration and Calibration Error

As discussed in the Unit Introduction, measuring instruments, including balances, are calibrated using traceable standards to bring their readings into accordance with internationally accepted values. You will calibrate a balance using traceable standards. Be sure to handle these standards properly; never touch them with your bare fingers or drop them.

It is the responsibility of the user to make sure that a balance is calibrated. Balances should be calibrated in the location in which they are used because the gravitational pull of the Earth varies depending on location. Modern electronic balances are usually readily calibrated by the user. In contrast, older mechanical balances are complex to calibrate and must be periodically calibrated by a trained technician.

Balance calibration involves setting two points on the balance scale to known values. The first point is zero. The zero point is set by making sure the weighing pan is clean and empty and then bringing the balance readout to zero. Mechanical and older electronic balances are zeroed by turning a knob. Newer instruments are zeroed by pressing a "tare" or zero button or lever. The balance should be readjusted to zero each time a substance is weighed.

The second calibration point is near the capacity of the balance. A standard whose weight is nearly equal to the full scale of the balance is gently placed on the pan and the balance's readout is adjusted to the value of the standard. This procedure is called a "two-point calibration."

Balances vary from one another in the details of how they are calibrated. Consult the manufacturer's instructions to determine how to calibrate a particular instrument.

Calibration error is *when there is an error (usually small) in the upper calibration of the balance.* Thus, if the heaviest object that a balance can weigh is 100 g, it would generally be calibrated at 0 and 100 g. In this case, calibration error occurs if the 100 g standard was not correctly manufactured, if it was damaged, or if something else went wrong with the calibration. If the upper limit of the balance is not adjusted exactly, then all subsequent measurements from that balance will be biased, that is, they will be too high or too low.

Calibration error is detected by weighing a standard whose weight is about that of the highest capacity of the balance. A standard other than the one used for calibration is required for this purpose. The weight result for the standard should be correct, otherwise the balance needs attention. It is common practice to perform this test routinely, perhaps each morning, and record the results in a logbook.

B. Linearity and Linearity Error

Linearity refers to *the ability of an instrument to provide identical accuracy throughout its range.* Balances are expected to have a linear response. For example, if a balance exhibits perfect linearity, and if its range is from 0.0000 g to 100.0000 g, then it will provide exactly half the response at 50.0000 g as it does at 100.0000 g (see Figure 3.14a). In the presence of linearity error, the balance's response at 50.0000 g will deviate from this expected relationship. In the presence of linearity error, an instrument will be less accurate in one part

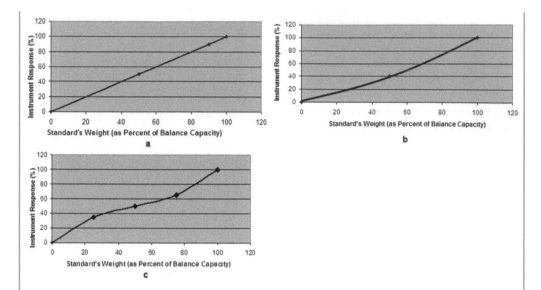

FIGURE 3.14   Linearity. Standards are weighed on three balances. (a) A balance with perfect linearity; the instrument's response is consistent across its entire weighing range. (b) "Bow-shaped" nonlinearity. The instrument is accurate at 0 and at full capacity, but the instrument is not as accurate at its midpoint. (c) "S-shaped" nonlinearity. The instrument is accurate at 0, full capacity, and the midpoint, but does not have a linear response at 25% and 75% of capacity.

of its range than in other parts. Checking for calibration error will not reveal if there is linearity error.

Most balances that exhibit nonlinearity have a nonlinearity curve that is bow shaped, as in Figure 3.14b. Observe in this figure that the maximum linearity error occurs at 50% of capacity, the midpoint of the balance. An intuitive way to test the linearity of a balance is to weigh two stable objects one at a time; each object should weigh approximately one-half the capacity of the balance. Then, both objects are weighed together. The sum of the two separate readings should equal the reading obtained when both objects are weighed together. This simple test indicates whether the balance is responding in a linear fashion in its midrange.

Occasionally a balance may have an "S-shaped" pattern of nonlinearity (see Figure 3.14c). To detect this type of nonlinearity, it is necessary to check the balance's linearity at 25% of its capacity, 50% capacity, and 75% capacity, as is described in Box 3.2.

Achieving good linearity depends largely on the design of the balance and is, therefore, the responsibility of the manufacturer. Linearity can, however, be checked periodically by the user to detect if something has gone amiss over time. The linearity of a balance is typically not checked as often as are calibration error and precision.

C. Precision

Precision is another aspect of balance performance that can be checked by the user. Balances usually provide excellent precision, that is, very consistent results over time—assuming that the good weighing practices outlined in the last laboratory exercise are followed.

Precision is easily evaluated by weighing the same object multiple times and calculating the standard deviation of the resulting data. **Standard deviation** is *a statistical tool that is used to describe how much the data deviate from the mean.* The standard deviation of repeated weighing results should be within the range specified by the laboratory's quality requirements. If it is not, the balance must be taken out of service and repaired. For more information about standard deviation and its calculation, see Appendix 5 in this manual.

Box 3.2   Procedure for Checking Balance Linearity

Notes:

a. Linearity tests do not require standard calibration weights. Any four objects that weigh about one-fourth the capacity of the balance can be used. For example, if a balance has a capacity of 100 g, then four items with weights of about 25 g are needed (see Figure 3.15).

b. Each of the weighings should be repeated, ideally 10 times, and the results averaged.

Test 1: Check the Balance Response at Its Midpoint

1. Select four items whose total weight is about the capacity of the balance; label them A, B, C, D.

2. Weigh all four pieces together. Record this weighing as the "full-scale value."

3. Weigh pieces A and B together and record their weights.

4. Weigh and record the weight of pieces C and D together.

5. Add the two values from step 3 and step 4 together. This is the "weight sum."

6. If the balance is exactly linear at its midpoint, then the weight sum will equal the full-scale value. If the two are not equal, divide the difference by two because two measurements were made. The result is the linearity error at the midpoint.

Test 2: Check the Balance Response at 25% of Capacity

1. Weigh all four pieces individually. Add their weights together and call this the "weight sum."

2. If the balance is linear at 25% of its capacity, then the weight sum should equal the full-scale value as determined previously. If the two differ, divide the difference by 4 to get the linearity error at 25% of capacity.

Test 3: Check the Balance Response at 75% of Capacity

1. Divide the pieces into the following groups:
   Group 1: pieces A, B, C
   Group 2: pieces A, B, D
   Group 3: pieces A, C, D
   Group 4: pieces B, C, D

2. Weigh all four groups, one at a time. Add their weights together and call the result the "weight sum."

3. Multiply the full-scale value by three and call it "75% full-scale value."

4. If the balance is linear at the 75% point, this weight sum will equal the 75% full-scale value. If the two differ, divide the difference by 4. This is the linearity error at 75% of capacity.

(Procedure from Jerry Weil, "Assuring Balance Accuracy," Product Note, Cahn Instruments Inc., 1991.)

FIGURE 3.15   Metal blocks. These blocks each weigh a little less than one-fourth the capacity of this analytical balance. They were manufactured by students in a machine shop in our college. They are not standards and should not be confused with standards.

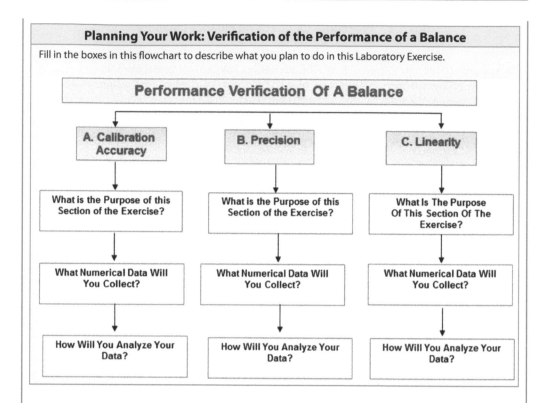

Planning Your Work: Verification of the Performance of a Balance

Fill in the boxes in this flowchart to describe what you plan to do in this Laboratory Exercise.

LABORATORY PROCEDURE

The parts of this exercise can be performed in any order. Work individually or in teams, depending on the availability of balances and standards.

1. If possible, calibrate one of the balances in your laboratory. (Older models must be calibrated by a service technician, so it is possible that no balances in your laboratory can be calibrated by the user.) Record the procedure that you followed. *Remember not to touch the standards with your bare fingers.*

2. Check the calibration accuracy of at least one balance. *Remember not to touch the standards with your bare fingers.*
   2.1. Weigh a standard whose weight is close to the upper capacity of the balance. (It may or may not be the same balance that you calibrate.) Use a standard other than the one you used for calibration. Record the result.
   2.2. Repeat by weighing the same standard three times.
   2.3. Average the values.

3. Check the precision of one of the balances by weighing the same object 10 times. Express the precision using standard deviation and record the result.

4. Check the linearity of a balance according to the directions in Box 3.2. In order to make this exercise faster, repeat each weighing only 3 times, rather than 10 times, as is suggested in the procedure. Describe and record the results.

5. **Data analysis**
   5.1. Calculate and record the percent error for your balance's weight measurements based on the average weight value for your standard.
   5.2. Record the standard deviation (precision) of your balance by calculating the standard deviation of repeated measurements.
   5.3. Calculate and record the linearity of your balance.

LAB MEETING/DISCUSSION QUESTIONS

1.   Fill in a "Class Results Table" on the blackboard.
2.   Discuss the results of the class. Here are some questions to consider:
     - How did your results compare to the rest of the class?
     - When and why would one perform these tests of a balance?
     - Who should perform these tests?
     - Is there a difference between the performance of different balances in terms of their precision, accuracy, and linearity? If so, why do you think the balances vary?
     - To what extent do you think your results were affected by your operation of the balances and to what extent were they affected by the balances themselves?
     - Are your laboratory's balances "good enough"? Justify your answer.
     - If your class decides that the balances are not functioning well enough, what should be done?
     - Did your class achieve consistent results from person to person? If not, how could the consistency of the class results be improved?

Example Format for a "Class Results" Table

Name(s)	Balance	Calibration Error (expressed as % error)	Precision (expressed as SD in units of g)	Linearity Error at 25% (in units of g)	Linearity Error at 50% (midpoint) (in units of g)	Linearity Error at 75% (in units of g)

3.   Explain linearity and its importance in your own words.
4.   Imagine that an auditor or inspector comes to your laboratory and asks, "How do you know that your weight measurements are correct"? How could you answer? (Hint: Use the term "traceable mass standard" in your answer.)
5.   The *U.S. Code of Federal Regulations* (CFR) outlines the following requirement for pharmaceutical companies:

The calibration of instruments, apparatus, gauges, and recording devices at suitable intervals in accordance with an established written program containing specific directions, schedules, limits for accuracy and precision, and provisions for remedial action in the event accuracy and/or precision limits are not met.

(21 CFR 211.160)

Suggest a program for ensuring that the balances in your classroom are compliant with this regulation.

Laboratory Exercise 7: Volume Measurements 1: Proper Use of Volume-Measuring Devices

OVERVIEW

This exercise briefly introduces the use of a variety of volume-measuring devices, but the primary tasks you will perform involve micropipettes. **Micropipettes** are *devices that are used to accurately pipette small volumes in the microliter range.* As we saw with balances, your technique in operating micropipettes must be good in order to avoid introducing systematic error. Also, micropipettes must be properly maintained and calibrated so that they provide accurate and precise measurements. This first volume laboratory exercise will provide you with practice operating micropipettes so that your technique is good. In the second volume exercise, you will verify the performance of a micropipette to make sure it is properly calibrated and functioning.

BACKGROUND

A. Volume and Its Measurement

Volume is *the amount of space a substance occupies.* Biologists commonly use units of **liters (L), milliliters (mL),** and **microliters (µL)** to express volume.

Various devices are used to measure the volume of liquids, depending on the volume being measured and the accuracy required (see Figure 3.16). Biologists use glass and plastic measuring vessels, such as **graduated cylinders** and **volumetric flasks**, when measuring larger volumes (i.e., greater than 10 or 25 mL) of liquid. **Measuring pipettes** are usually preferred for volumes in the 1–25 mL range. Various **micropipetting devices** are commonly used to measure volumes in the microliter range.

Beakers and Erlenmeyer flasks often have lines indicating volumes, but these lines are approximate; therefore, never use a beaker or flask as a measuring device (see Figure 3.17).

a   b   c

FIGURE 3.16 Pick the right glassware for your application; glassware varies depending on the volumes for which it is used and the accuracy with which the manufacturer applies its calibration lines. (a) Biologists commonly use graduated cylinders to measure volumes in the range of 10–2000 mL. Select the smallest-volume graduated cylinder that will contain the volume you want. For example, if you want to measure 38 mL, it is more accurate to use a 50 mL graduated cylinder than a 100 or 500 mL cylinder. Look at the calibration lines on the cylinders to see why this is so. (b) Volumetric flasks are calibrated with more accuracy than graduated cylinders but have the disadvantage that each one only measures a single volume. To use a volumetric flask, fill it exactly to the line. c. Measuring pipettes like these are commonly used for volumes in the range of 1–25 mL.

FIGURE 3.17   Some glassware is used only to contain liquids, not to measure volume. Although these (a) beakers and (b) flasks have calibration lines that appear to be like those on graduated cylinders, manufacturers do not apply the calibration lines on beakers and flasks with as much accuracy as is required on graduated cylinders and volumetric flasks. Therefore, do not use beakers and flasks as measuring devices. (c) Pasteur pipettes are convenient to dispense liquids but not to measure their volume.

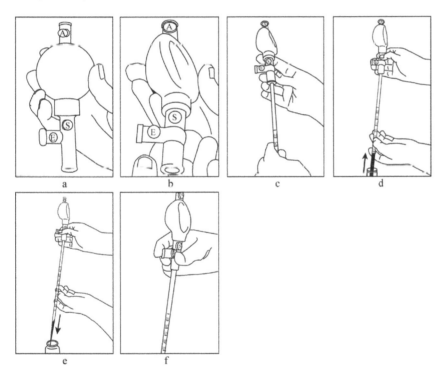

FIGURE 3.18   Pipette aids: Using a triple-valve bulb. A variety of pipette aids are used with measuring pipettes. This bulb style is relatively inexpensive and can be used to accurately dispense a measured volume. (a) The bulb has three "buttons" labeled "A," "S," and "E." (b) Begin by forcing air out of the bulb by pushing button A and squeezing the bulb. (c) Place the bulb on top of the pipette. (d) Aspirate the desired volume of liquid by pushing the button S. (e) Dispense (expel) the liquid by pushing button E. (f) Blow out the last drop by placing your finger on the side of the bulb, as shown, and squeezing.

B. Using Measuring Pipettes

Measuring pipettes are calibrated with a series of lines to allow the measurement of more than one volume. They come in different sizes, commonly, 1 mL, 5 mL, 10 mL, and 25 mL. These pipettes may be made of disposable plastic or glass, or they may be reusable and made from glass. Disposable pipettes can be purchased already sterilized, which makes them ideal for cell culture and microbial applications.

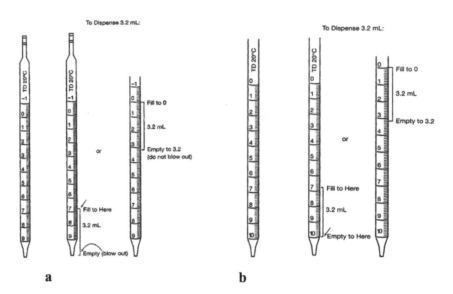

To Dispense 3.2 mL:

Fill to 0

3.2 mL

Empty to 3.2
(do not blow out)

Fill to Here

3.2 mL

Empty (blow out)

a

To Dispense 3.2 mL:

Fill to 0

3.2 mL

Empty to 3.2

Fill to Here

3.2 mL

Empty to Here

b

FIGURE 3.19   **Using measuring pipettes.** (a) Serological pipettes are calibrated so that the tips include the last milliliter of solution. The bands at the top indicate that these pipettes are to be "blown out." Note that these pipettes have a scale that extends above zero to expand the calibrated capacity of the pipette. Using a serological pipette to dispense 3.2 mL is illustrated. (b) Mohr pipettes are calibrated so that the tip does not include the last milliliter of solution. Using a Mohr pipette to dispense 3.2 mL is illustrated.

Some pipettes, those usually termed **serological**, are *calibrated so that the last drop needs to be "blown out" to deliver the full volume desired.* Manufacturers place one wide or two narrow bands at the top of a pipette to indicate that it is calibrated to be "blown out." In the past, these pipettes were literally blown out with one's mouth. Mouth pipetting, however, can be very dangerous and is now prohibited. Rather, we use pipette aids that fit over the top of the pipette (see Figure 3.18) to take up and dispense liquids.

Mohr pipettes *are calibrated by the manufacturers in such a way that the liquid in the tip is not part of the measurement and so the pipette does not need to be "blown out."* Figure 3.19 shows how to read serological and Mohr measuring pipettes.

Advice from the senior technician

Be Organized When Dispensing Multiple Liquids into Multiple Tubes—It Is Easy to Get Confused

- Plan exactly what will go into each tube.
- Obtain all materials required and arrange them for easy access.
- Label all tubes carefully before the addition of liquids.
- Place tubes to be filled in the front of the rack; move each one back after a sample is added to it.
- Check off components that have been added.
- If you are pipetting a single substance into more than one tube, you can use the same tip or pipette over and over if you are careful not to contaminate it with liquids already in the tube. Change the tip or pipette each time a new material is pipetted and whenever a tip touches components previously added to a tube.
- After adding all the components to a tube, be sure the liquids are well mixed, are not adhering to the tube walls, and are in the bottom of the tubes.
- Watch the sample as it comes into and leaves the tip or pipette to be sure the right volume is dispensed and the tip or pipette drains properly.
- Remember to have a proper waste container for used tips or pipettes.

C. Using Micropipettes

Micropipettes are among the most common instruments used by molecular biologists. It is essential that they be used properly; otherwise, any work done in a molecular biology laboratory is suspect. Richard Curtis and George Rodrigues published a study in which they evaluated the micropipetting accuracy and precision of students who had experience pipetting and also of experienced technicians ("Pipet Performance Verification: An Important Part of Method Validation," *American Laboratory*, February 2004: 12–17.) The results of their study were alarming: Many students and professional technicians pipetted incorrectly. Some of the individuals tested had percent inaccuracies as high as 35%. The imprecision of their volume measurements (reported as a coefficient of variation) was as high as 68% and almost always exceeded the manufacturers' specifications for the pipettes. In other words, many of the students and technicians had very poor pipetting skills. You will therefore practice micropipetting. In this exercise, you will simulate setting up a common molecular biology reaction, the polymerase chain reaction (PCR). The success of the PCR reaction depends on properly dispensing a series of liquid reagents, so it is a good example of a situation where your pipetting must be accurate and precise.

It is also important to remember that micropipettes are expensive and fragile. *Always use proper technique to avoid damage to them.* Figures 3.20 to 3.22 illustrate the use of micropipettes. The senior technician provides advice on how to avoid error when pipetting and how to avoid damaging micropipettes. Micropipette tips are identified and explained in Figure 3.23. Review all this information before coming to lab and refer back to it as you practice during the laboratory exercise.

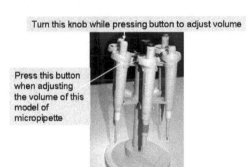

a   b

FIGURE 3.20   A brand of adjustable micropipettes. Biotechnologists commonly use micropipettes adjustable to different volumes. Each model can be used in a certain range of volumes. a. Micropipettes stored upright, according to the manufacturer's recommendation. To adjust the volume dispensed by this brand, the top knob is turned while the button on the side is depressed. (The device will be damaged if the button is not depressed.) (b) This micropipette can dispense volumes in a range from 0.5 to 10 μL; it is currently adjusted to dispense 2.50 μL. On this model, the line on the display represents a decimal point.

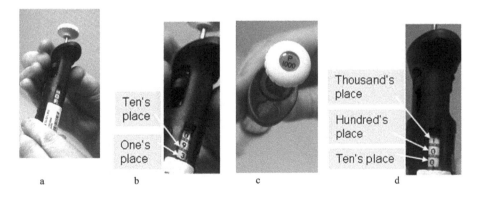

a   b   c   d

FIGURE 3.21   Another brand of adjustable micropipettes. (a) With this brand, volume is adjusted by rotating a knurled knob, as shown under the analyst's thumb and finger. This particular model is adjustable from 0 to 100 μL. The device is currently set to 093 μL (= 93 μL). (b) Here is the same brand and model as shown in Figure 3.21a indicating how the display is to be read. (c) This is the same brand as shown in 3.21a and 3.21b, but it is a model that is adjustable to volumes from 100 to 1000 μL. d. This micropipette is presently adjusted to dispense 1000 μL. The display appears to read 100, but on this particular model, the display is multiplied by 10.

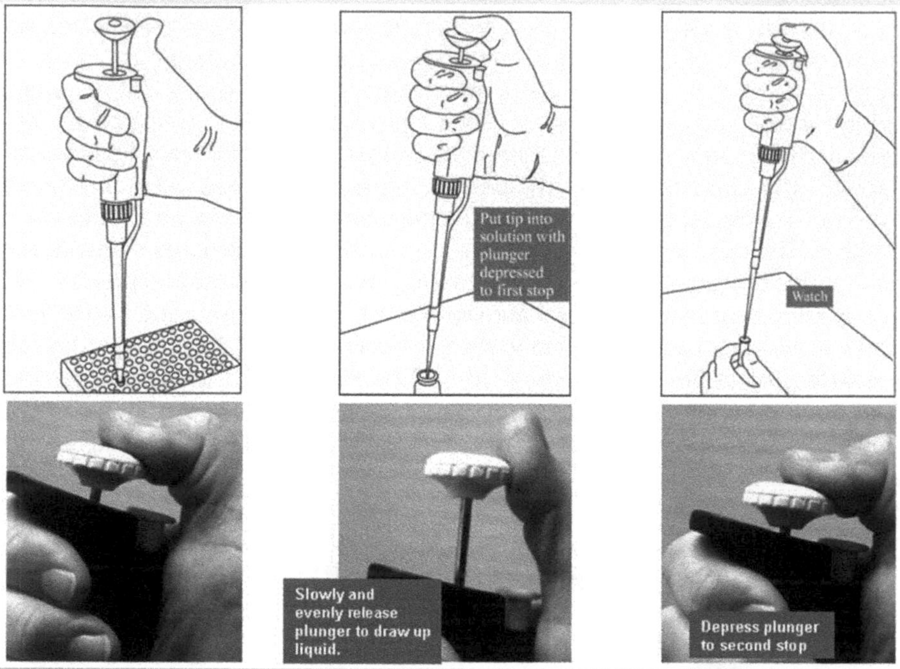

PREPARE	ASPIRATE LIQUID	EXPEL

1. Adjust micropipette to desired volume.
 	
2. Attach disposable tip. Press firmly to ensure airtight seal.
 	
3. Hold micropipette vertically; depress plunger to first stop.
	4. Immerse tip 2-6 mm into liquid.
	
	5. Allow plunger to slowly return to un-depressed position as liquid is drawn into the tip. Wait 2 seconds to allow complete filling.
	
	6. Pull micropipette straight out of container.
	
	7. Look at the tip. There should not be an air bubble at the end of the tip.If there is, perform these steps:
	
	  7.1. Try again, holding the device more vertically.
	
	  7.2. Consider that you may have withdrawn tip from liquid too quickly.
	
	  7.3 Make sure tip is firmly attached and is the proper type.
		8. Touch tip to side of receiving tube; liquid is "attracted" to side of tube by capillary action.
		
		9. Depress plunger to the first and then the second stop. Depressing to second stop blows sample from tip.
		
		10. Watch as sample is expelled into the container.
		
		11. Remove tip from vessel carefully, with plunger still fully depressed.
		
		12. Eject tip using the third stop, tip ejector button, or other mechanism.

FIGURE 3.22   **Using a micropipette.**

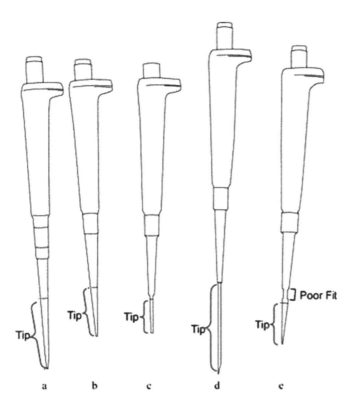

Tip   Tip   Tip   Tip   Tip   Poor Fit

a   b   c   d   e

FIGURE 3.23   Micropipette tips. Biotechnologists commonly use micropipettes that require disposable tips. Tips prevent liquid from entering and contaminating the pipette shaft. There are various sizes and styles of micropipettes that require different types of tips. Tips are sometimes color coded to match the micropipette. For example, 100–1000 µL micropipettes often use blue tips and may have a blue top. Not all manufacturers, however, use color coding. (a) A micropipette with a range from 100 to 1000 µL requires a larger tip. (b) A micropipette with a range from 10 to100 µL uses a medium size tip. (c) A micropipette with a range from 1 to 10 µL uses a small-size tip. (d) Long, thin tips are used for loading samples into the wells of electrophoresis gels. The wells are small and can be punctured; a fine tip easily fits in the wells and is unlikely to puncture the gel. (e) A mistake. This tip does not fit the micropipette. It can be "smashed" onto the end of the pipette, but it will not seal properly and will not deliver the proper volume.

Advice from the senior technician

Avoid Damaging Our Micropipettes

- Never drop a micropipette.
- Never contaminate the interior of a micropipette, for example, by inverting it and allowing liquid to flow into the shaft.
- Never rotate the volume of an adjustable micropipette beyond the upper and lower range of the instrument.
- Never pass a micropipette through a flame.
- Never use a micropipette without a tip, thus allowing liquid to contaminate the shaft assembly.
- Never allow the plunger to snap up when liquid is being aspirated.
- Store micropipettes vertically in a proper stand. It is best to store them adjusted at their highest volume to avoid tension on their internal springs.
- Be sure the plunger moves smoothly; if it does not, the interior of the device may be contaminated and should be checked by your instructor or laboratory manager.

Planning Your Work: Proper Use of Volume-Measuring Devices

1. Consider the laboratory procedure that follows and decide what should be recorded:

    a. What will you record in your laboratory notebook for Part A?

    b. What will you record in your laboratory notebook for Part B?

    c. What will you record in your laboratory notebook for Part C?

2. What numerical data will you collect and how will you analyze it?

3. What is the purpose of this Laboratory Exercise?

LABORATORY PROCEDURE

Part A: Familiarization with Volume-Measuring Devices and Measuring Pipettes

A.1.   Be sure you can use all types of pipette aids present in your laboratory.

A.2.   Dispense the following volumes of water into a container; select the best measuring device depending on the volume:

    A.2.1.   5.0 mL

    A.2.2.   3.7 mL

    A.2.3.   2.3 mL

    A.2.4.   17 mL

    A.2.5.   135 mL

Select the Best Device to Measure a Specific Volume

You might use either a 10 mL graduated cylinder or a 10 mL measuring pipette to measure 10 mL. If, however, you want to measure 3.7 mL, a measuring pipette is better. Do you see why? Compare the calibration lines on this measuring pipette and graduated cylinder.

A.3.   Check your results by comparing your dispensed volumes to those of another student or by consulting your instructor.

Scenario

PREPARING TO AMPLIFY DNA USING THE POLYMERASE CHAIN REACTION (PCR) METHOD

In this exercise, you will simulate preparing buffer and then samples for PCR, which is an important method that allows molecular biologists to amplify a specific sequence of DNA many thousands of times. Preparing samples for PCR requires pipetting small volumes of specific reagents into tubes. PCR will not work properly if this pipetting is not done accurately and consistently. You will actually use food coloring (which is easy to see) to simulate PCR reagents.

Part B: Practice Operating Micropipettes

B.1.   Learn about the micropipettes available in your laboratory.

    B.1.1.   Find each model of micropipette in your laboratory. Record the manufacturer of each and the range of volumes it can dispense.

    B.1.2.   Find the right size of disposable tip for each model of micropipette available in your laboratory. Check with your instructor to be sure you are correct—the wrong size of tip will leak or deliver the wrong volume.

B.1.3.   Practice operating a micropipette with the guidance of your instructor. Have the instructor watch you to make sure you are not making any mistakes that might damage the device or cause error in your measurements.

B.2.   Simulate mixing PCR buffer (practice using a larger-volume, 1000 µL device).

B.2.1.   Label two microcentrifuge tubes as "PCR Buffer A" and "PCR Buffer B" using a permanent lab marker.

B.2.2.   Table 3.2 shows the volume of each solution to add to each tube.

B.2.2.1.   Adjust the micropipette to 100 µL and add simulated 1 M Tris buffer, pH 8.3, to Tube A. *Remember not to let the plunger "snap."*

B.2.2.2.   Add the same amount of the simulated 1 M Tris buffer to Tube B. It is acceptable to use the same tip to add this solution to both tubes.

B.2.2.3.   Place a fresh tip on the micropipette and adjust it to the correct volume (500 µL). Add simulated 1 M potassium chloride (KCl) to a dry spot on the side of Tubes A and B.

B.2.2.4.   With a clean tip, dispense the correct amount of water (400 µL) to Tubes A and B.

B.2.3.   Close the tubes securely; use a microcentrifuge to bring the liquid to the bottom of the tubes.

B.2.3.1.   Place tubes directly across from one another in a microcentrifuge. This assures that the microcentrifuge is balanced, see Figure 3.24.

B.2.3.2.   A brief spin of about 5 seconds will centrifuge the liquid to the bottom of the tubes.

B.2.4.   A total of 1000 µL should have been added to each tube. Check your accuracy:

B.2.4.1.   Set a micropipette to 1000 µL. Withdraw the solution mixture from one of the two tubes. If you pipetted accurately, then the tip will be exactly filled.

B.2.4.2.   If there is an air space at the bottom of the tube, then your final volume is less than 1000 µL. You can determine the actual volume in the tip by rotating the volume adjustment to expel the air until the tip is just filled. Read and record the resulting volume—that is how much you dispensed.

B.2.4.3.   If the tip is filled and some solution remains in the tube, then your final volume is more than 1000 µL. You can measure the remaining amount with a lower-volume micropipette. Record the value.

B.2.4.4.   If your measurement is more than 100 µL (10%) off, repeat the exercise. This means you are aiming for a volume of 1000 µL ± 100 µL or a range from 900 µL to 1100 µL. (You may want to aim for a better percent accuracy, e.g., 5%.)

B.2.4.5.   Repeat your check with the other tube.

TABLE 3.2   Making Simulated PCR Buffer

Tube	Simulated 1 M Tris Buffer	Simulated 1 M KCl Solution	Water
A	100 µL	500 µL	400 µL
B	100 µL	500 µL	400 µL

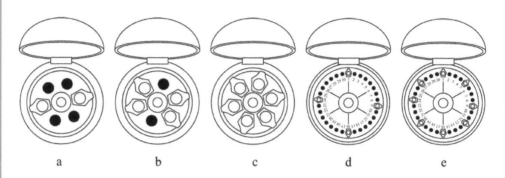

a   b   c   d   e

FIGURE 3.24   Examples of balanced microcentrifuges. A microcentrifuge is *a small centrifuge* for tubes that hold small volumes, on the order of 0.5–2 mL. The configurations of these tubes are all balanced. Running any centrifuge when it is unbalanced can cause damage to the centrifuge, damage to the samples, and possibly danger to people in the vicinity. (a) Two tubes are shown in a microcentrifuge that can hold six tubes. Four spaces are empty. (b) Four tubes are shown in a microcentrifuge that can hold six. (c) All spots are filled with tubes. (d and e) This microcentrifuge holds a maximum of 30 tubes.

Centrifuges Must Be Operated Properly

- If you ever hear a "rough" noise after beginning a centrifugation run, turn off the centrifuge immediately and check its balance.

- Never operate a centrifuge unless you know how to use it properly—they can be dangerous.

B.3.   Simulate preparing DNA samples (practice using a smaller-volume device, preferably a 10 or 20 µL model).

    B.3.1.   Label four tubes as "C," "D," "E," and "F," using a permanent marker.

    B.3.2.   Add simulated DNA sample, PCR mix, magnesium chloride ($MgCl_2$), and water to tubes C, D, E, and F according to the volumes in Table 3.3. Remember to change the tip each time you change solutions.

TABLE 3.3   Preparing Simulated Samples

Tube	Simulated DNA Sample	Simulated PCR Mix	Simulated $MgCl_2$	Water
C (negative control, no $MgCl_2$)	5 µL	4 µL	0 µL	1 µL
D	5 µL	4 µL	1 µL	0 µL
E	5 µL	4 µL	1 µL	0 µL
F (negative control, no DNA)	0 µL	4 µL	1 µL	5 µL

B.3.3.   Check and record your results, as you did in step B.2.4., but this time your total volume should be 10 μL. Repeat the activity if your percent error for any tube is greater than 10%.

Part C: Check Your Skills

C.1.   Check your skills by measuring the volumes of "unknown" samples provided. Pay attention to factors that will affect the measured volumes of your samples.

C.1.1.   Choose the most appropriate micropipette for each unknown provided.

C.1.2.   Record your results in tabular form in your laboratory notebook with the correct number of significant figures. Label the table of data with a title and caption.

C.2.   Check your answers with the key provided by the instructor. Record the values from the key in your table in your laboratory notebook.

C.3.   ⊞**Data analysis**

C.3.1.   Make two assumptions when analyzing these data for the unknowns. First, assume that the teacher's answer key is correct. (Is this a reasonable assumption?) Second, assume that the micropipettes you used were properly calibrated. (Is this a reasonable assumption?) Calculate and record the absolute and percent error for each of your measurements of the unknowns. Record your results in your table in your laboratory notebook.

LAB MEETING/DISCUSSION QUESTIONS

1.   What factors affect the accuracy of volumes dispensed by a micropipette?
2.   We use percent error as an indication of accuracy.
    a.   What percent errors did you get for the unknowns?
    b.   Decide as a class what percent error is low enough to be acceptable for your volume measurements, that is, how close is close enough? On what basis can you make this decision? Will your decision depend on the volume of the samples being measured? Are your measurement values "close enough"?

Laboratory Exercise 8: Volume Measurements 2:
Performance Verification of a Micropipette

OVERVIEW

This exercise continues to explore the use of micropipettes to get "good" measurements of small volumes. **Performance verification** of a micropipette *checks that the instrument is operating properly.* In this exercise, you will verify the performance of a micropipette by checking its accuracy and precision.

BACKGROUND

Micropipettes are calibrated by the manufacturer when they are made but can become less accurate as they are used (or abused). The performance of micropipettes should, therefore, be verified periodically. Users in some laboratories check their own micropipettes; some organizations regularly send their instruments to the manufacturer to be inspected and repaired; and some facilities have in-house technicians for this purpose. Laboratories that are compliant with current Good Manufacturing Practices (CGMP) or the standards of the International Organization for Standardization (ISO), or that meet other laboratory requirements have written policies for micropipette performance evaluation and routine maintenance.

Micropipette performance is most often evaluated using a **gravimetric**, *relating to weight*, procedure. Gravimetric methods take advantage of the high accuracy and precision attainable with modern balances. Briefly, during gravimetric testing, a desired volume of purified water is placed into or is dispensed from the device to be checked. This *desired volume* is termed the **nominal volume**. The volume measurement is then checked by weighing the liquid on a high-quality, well-maintained analytical balance. A calculation is performed to convert the weight of the water to a volume. If the volume-measuring device was properly operating, then the nominal volume will be identical to the volume calculated from the weight of the water. This gives an indication of the accuracy of the device being tested.

Micropipette precision is evaluated by adjusting the device to one volume and repeatedly dispensing water. The volumes of water are then weighed, and the weight measurements are converted to volumes. If the precision of the micropipette is perfect (which is seldom the case), then the measurement values will all be the same. If not, the standard deviation of the repeated measurements is calculated.

To determine if a micropipette is performing with acceptable accuracy and precision, the values obtained can be compared with the manufacturer's specifications (see Table 3.4) or with the requirements of the laboratory's quality control procedure. If a micropipette is performing outside of its specifications for accuracy, it may be possible to recalibrate it by using directions from the manufacturer. In other cases, the instrument will need to be repaired by the manufacturer or an in-house specialist.

Table 3.4 gives the specifications for six models of volume-measuring devices, as established by their manufacturer, Gilson, Inc. These specifications state how accurately and precisely these devices dispense the volumes for which they are set (assuming proper calibration, maintenance, and operation). The specifications are shown in tabular format:

- The first column in the table is the model of the measuring device.
- The second column indicates the *volume that the device is set to dispense*, the **nominal volume** (i.e., the "true volume").
- The third column is the permissible systematic error, which is a quantitative evaluation of the accuracy of the device. The systematic error is expressed here as the **absolute error**, *the difference between the dispensed volume and the nominal volume*. It is generally determined based on the average of 10 measurements. Observe that the lower the value specified for systematic error, the better the accuracy and, therefore, the performance of the device.

- The fourth column is the **random error**, *which is a quantitative evaluation of the precision of the device.* The random error is expressed as the standard deviation of a series of measurements, usually 10. Observe that the lower the value specified for standard deviation, the better the precision and performance of the device.
- The fifth and sixth columns are from the ISO document **ISO 8655**-1:2002, which specifies general requirements for piston-operated volumetric apparatus.

LABORATORY PROCEDURE

Scenario
VERIFYING MICROPIPETTE PERFORMANCE IN A RESEARCH LABORATORY

You are a research scientist in a molecular biology laboratory. Your ability to routinely pipette very small volumes of reagents is critical to the success of your research. You have, therefore, developed the strategy of verifying the performance of your micropipettes on a routine basis. This is the procedure you follow.

- Work individually or with a partner. You will need an analytical balance for each person or pair.
- *Note:* This procedure simplifies the procedure for checking a micropipette by assuming that 1 mL of water weighs exactly 1 g, which is not true. Consult the bibliography for a reference to the ASTM standard that covers the procedure in more detail.
    1. Label five microcentrifuge tubes with the numbers 1 to 5. Weigh each of them. Record the weights in tabular format.
    2. Pipette purified water.
        2.1. Obtain a micropipette of any size and any brand.
        2.2. Choose a volume that is in the midrange of the micropipette. Adjust the micropipette to the chosen volume. This volume is the "nominal volume."
        2.3. Pipette purified water into each of the five microcentrifuge tubes:
            2.3.1. You may use the same tip all five times or change the tip each time. Whichever method you choose, be consistent. If you use the same tip each time, pre-wet it before pipetting the first time.
            2.3.2. Record the brand and serial number of your micropipette. If there is no serial number, then label it with one.
            2.3.3. Be sure to record the details of your procedure, including whether or not you changed tips each time.
    3. Weigh each of the five tubes. By subtraction, determine the weight of water in each tube. Record the results in your table.
    4. **Data analysis**
        4.1. Convert the weight of the water to dispensed volume, using the conversion factor that 1 mL water = 1 g.
        4.2. Calculate the standard deviation (repeatability) of the micropipette based on these five volumes. Record the calculations on your form. (See Appendix 5 for information about calculating standard deviation.)
        4.3. Determine how close the average weight of the water is to the expected weight. For example, if you adjusted the micropipette to 5 μL, you would expect the average weight to be 5 mg. The expected value is the "true value" in this case.

      4.3.1.   Determine the accuracy of the micropipette by calculating the absolute error of the average weight. Record the results and calculations.

5.   If you have time, repeat the verification procedure with a second volume that is at the bottom of the micropipette's range.

6.   Compare your results to those in Table 3.4. Does your micropipette meet these specifications? If so, it passes; if not, it fails.

Note: Two instruments are involved here: the balance and the micropipette. A third factor is also involved: your technique. All three of these factors will affect the results. If your balance was recently calibrated and maintained, then assume that it gives accurate results. Make sure that your technique is correct so that you are actually measuring the repeatability of the micropipettes—not your operating technique.

TABLE 3.4   **Example of Catalog Micropipette Specifications**

Model	Volume (µL)	Gilson Maximum Permissible Systematic Error (µL)	Gilson Maximum Permissible Random Error (µL)	ISO 8655 Maximum Permissible Systematic Error (µL)	ISO 8655 Maximum Permissible Random Error (µL)
A-2	0.2	±0.024	≤0.012	±0.08	≤0.04
	0.5	±0.025	≤0.012	±0.08	≤0.04
	2	±0.030	≤0.014	±0.08	≤0.04
A-10	1	±0.025	≤0.012	±0.12	≤0.08
	5	±0.075	≤0.030	±0.12	≤0.08
	10	±0.100	≤0.040	±0.12	≤0.08
A-20	2	±0.10	≤0.03	±0.20	≤0.10
	5	±0.10	≤0.04	±0.20	≤0.10
	10	±0.10	≤0.05	±0.20	≤0.10
	20	±0.20	≤0.06	±0.20	≤0.10
A-100	20	±0.35	≤0.10	±0.80	≤0.30
	50	±0.40	≤0.12	±0.80	≤0.30
	100	±0.80	≤0.15	±0.80	≤0.30
A-200	20	±0.50	≤0.20	±1.60	≤0.60
	50	±0.50	≤0.20	±1.60	≤0.60
	100	±0.80	≤0.25	±1.60	≤0.60
	200	±1.60	≤0.30	±1.60	≤0.60
A-1000	200	±3	≤0.6	±8	≤3
	500	±4	≤1	±8	≤3
	1000	±8	≤1.5	±8	≤3

Source: Reproduced courtesy of Gilson, Inc.

Note: These specifications include values for accuracy and precision for six models of micropipette, each at more than one volume.

LAB MEETING/DISCUSSION QUESTIONS

1.   Fill in a "Class Results Table" on the blackboard.
      Example Format for a "Class Results" Table

Name(s)	Device	Nominal Volume	Precision (expressed as SD)	Accuracy (absolute)	Pass/Fail?

2.   Examine the specifications for micropipettes in Table 3.4.
      a.   The third column specifies the accuracy of each type of micropipette when it leaves the factory. The lower the number, the lower the error and, therefore, the more accurate the device is. For each type of micropipette, the manufacturer has three or four values in this column. Why? Are micropipettes equally accurate across their entire range? If not, in what part of its range is each type of micropipette most accurate?
      b.   How many people in your class found that his or her micropipettes met the specifications in Table 3.4?
      c.   How did your results compare to the rest of the class?
      d.   If someone found that their micropipette did not meet the specifications in the table, does that mean there is something wrong with the micropipette?
3.   The CFR outlines the following requirements for pharmaceutical companies:

The calibration of instruments, apparatus, gauges, and recording devices at suitable intervals in accordance with an established written program containing specific directions, schedules, limits for accuracy and precision, and provisions for remedial action in the event accuracy and/or precision limits are not met.

(21 CFR 211.160)

Consider these requirements and suggest a program for ensuring that the micropipettes in your classroom laboratory are compliant with cGMP.

Laboratory Exercise 9: Measuring pH with Accuracy and Precision

OVERVIEW

Recall from chemistry that **pH** describes *the concentration of hydrogen ions in water*:

$$pH = -\log\left[H^+\right]$$

where concentration is expressed in moles per liter.

pH is an important measurement because biological systems are sensitive to the pH of their medium. You must measure pH correctly to ensure that your biological solutions provide consistent, suitable conditions for your laboratory work. In this laboratory exercise, you will, therefore, perform the following tasks:

- Learn to measure pH.
- Explore the factors that influence the accuracy and precision of pH measurements.

BACKGROUND

A. The Basic Design of a pH Meter System

A pH meter/electrode measuring system is the most common laboratory method of measuring the pH of an aqueous solution. A pH meter consists of (1) a **voltmeter** that *measures voltage*, (2) two **electrodes** connected to one another through the meter that *respond to differing H^+ concentrations*, and (3) the **sample** whose *pH is being measured*. When the two electrodes are immersed in a sample, they develop an electrical potential (voltage) that is measured by the voltmeter. The magnitude of the measured voltage depends on the hydrogen ion concentration in the solution; therefore the voltage reading can be converted by the instrument to a pH value (see Figure 3.25).

B. Electrodes

The two electrodes in the pH measuring system are different from one another. The **pH measuring electrode** *has a thin, fragile glass bulb at its tip that is sensitive to the concentration of H^+ ions in the surrounding medium*. An electrical potential develops between the

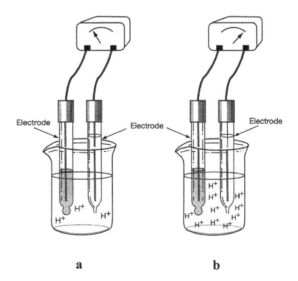

a   b

FIGURE 3.25   Overview of a pH Measuring System. Two electrodes are immersed in the sample and are connected to one another through a meter. (a) The concentration of hydrogen ions in the solution is relatively low. (b) A higher concentration of hydrogen ions leads to a change in the voltage difference between the two electrodes.

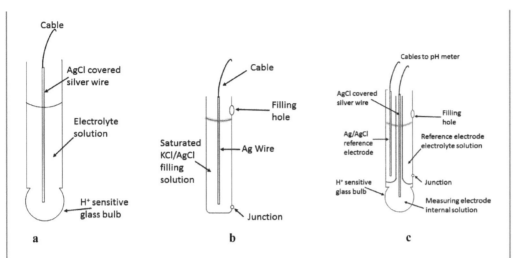

FIGURE 3.26   pH electrodes. (a) Measuring electrode with pH sensitive glass bulb. (b) A silver/silver chloride reference electrode. (c) A combination electrode.

inner and outer surfaces of the glass bulb when the measuring electrode is immersed in a sample solution. It is this potential whose magnitude varies depending on the concentration of hydrogen ions in the solution (see Figure 3.26a). The pH measuring electrode must be paired with a **reference electrode** *that has a stable, constant voltage*. The meter measures the potential between the measuring electrode and the reference electrode and converts it to a pH value. You will see, when you begin to operate a pH meter that you can have the meter display either voltage (in units of millivolts) or pH.*

The most common type of reference electrode is the **Ag/AgCl electrode**, *which contains a strip of silver coated with silver chloride* (see Figure 3.26b). The strip is immersed in an electrolyte solution of KCl and silver chloride.

The pH meters in biology laboratories usually are equipped with convenient, compact **combination electrodes** *made by combining the reference and pH-measuring electrodes into one housing* (see Figure 3.26c). pH meters can be made still more compact by combining the electrodes and the meter into a single housing.

Combination electrodes and compact meters work well for much routine work, but they are not suitable for measuring the pH of certain "difficult" samples, which include the following:

- Samples with very high ionic strength (e.g., a high concentration of salt)
- Very pure water
- Samples with very high pH
- Samples with very low pH
- Samples with a high protein concentration
- Samples that are slurries or sludges or that contain particulates

The end of the reference electrode that is submerged in the sample has a *porous plug*, called a **junction**. The electrode contains filling solution that ionizes when dissolved, and these ions flow slowly out of the reference electrode, through the junction, into the sample. Many electrodes, therefore, have a **filling hole** that *is used to replenish the electrolyte* (see Figure 3.27). Other reference electrodes are manufactured with a gelled filling solution that cannot be replenished. Gel-filled electrodes are discarded when the filling solution is depleted. With any reference electrode, pH cannot be properly measured if the electrolyte is depleted or if the junction is blocked so that electrolyte does not contact the sample.

---

* Some newer pH meters do not display the voltage, but older meters and more sophisticated ones usually do.

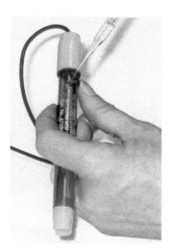

FIGURE 3.27   A combination electrode that requires periodic filling with filling solution. Filling solution (available from the electrode manufacturer) is being added. The filling hole must be closed when the electrode is not in use, and it must be open whenever the electrode is used to determine the pH of a sample. If you have a gel-filled combination electrode, there will be no hole for adding filling solution. This electrode is surrounded by a plastic shield to help protect the fragile glass bulb.

Electrodes must be selected, used, and maintained properly. If you need to purchase pH electrodes for your laboratory, keep in mind that opening a catalog and selecting a general-purpose, combination electrode may or may not be the best strategy. The best choice of electrodes depends on the type of samples whose pH you will be measuring. Manufacturers can help with this selection because they make different styles of electrodes for various purposes.

Read the manufacturer's instructions for storing and maintaining your electrodes because different types of electrodes have different requirements. Problems with pH measurements are often caused by electrode problems—never take their performance for granted.

C. Calibrating a pH Meter

pH meter systems need to be calibrated by the user every day or every time the system is used because the response of electrodes declines over time. **Calibration** *tells the meter how to translate the voltage difference between the measuring and reference electrodes into units of pH.* Calibration is accomplished using standards of known pH.

There is an inverse linear relationship between the voltage measured by the system and the pH of the sample. Therefore, as with balances, two standards are used to calibrate a pH meter; two points form a line. The first standard is a buffer that is normally pH 7.00. The second standard is typically pH 4.00, 10.00, or 12.00, although other standardization buffers are available. If the sample is acidic, an acidic calibration buffer is used as the second buffer, and if the sample is basic, a basic buffer is chosen.

The general method of calibration is to first immerse the electrodes in pH 7.00 buffer. Using the appropriate knob, dial, or button, the meter is set to display 7.00. Internally, the meter adjusts itself so that a pH of 7.00 corresponds to zero millivolts (mV) (see Figure 3.28a). The adjustment for pH 7.00 buffer may be called "set," "zero offset," or "standardize" because the meter "sets to" 0 mV. The electrodes are then placed in the second buffer and using the appropriate knob, dial, or button, the meter is adjusted to display the pH of the buffer. This second adjustment may be called "slope," "calibrate," or "gain" because internally the meter uses the reading to establish a calibration line with a particular slope (see Figure 3.28b). During calibration, the meter constructs a line that can be used to convert any voltage to its corresponding pH. The response of the

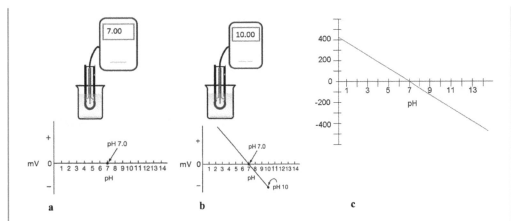

FIGURE 3.28   Conventional two-point pH meter calibration. (a) Electrodes are placed in pH 7.0 standard, and the instrument is "told this is pH 7.0." The meter adjusts itself so that 0 mV corresponds to pH 7.0. (b) The electrodes are rinsed, blotted dry, and placed in a second standard buffer, usually either 4.0 or 10.0. The meter is "told" the pH of the second standard, and the meter internally constructs a linear plot of pH versus millivolts. (c) Ideally, the slope of the resulting calibration line is -59.16 mV/pH unit at 25°C.

electrodes ideally should be linear throughout the range of the meter. In practice, if a meter is calibrated with pH 7 and 10 buffers, there might be some inaccuracy at acidic pHs. Conversely, if it is calibrated with pH 7 and 4 buffers, there might be some inaccuracy at basic pHs.

The slope of the calibration line measures the response of the electrodes to pH. The steeper the slope, the more sensitive the electrodes are to hydrogen ions. Ideally, the slope should be close to –60 mV for each pH unit when the temperature is 25°C. This means that if the pH changes by one pH unit, the electrodes will change their response by 60 mV (see Figure 3.28c). As electrodes age or become dirty, they are less able to generate a potential, so the slope of the line declines. Eventually, as the electrodes age, the system is no longer able to adjust to the value of both buffers. Some pH meters, therefore, display the slope of the calibration line as an indication of the condition of the electrodes; a new, properly functioning electrode has a calibration slope of close to –60 mV/pH unit. The slope is often displayed by the meter as a percent of theoretical value. For example, a 97% slope is equivalent to a slope of –58.2 mV/pH unit.

D. The Effect of Temperature on pH

Temperature has two important effects on pH:

1. The *measuring electrode's response to pH* (the potential that develops) is affected by the temperature.
2. The *pH of the solution* that is being measured may increase or decrease as its temperature changes.

Most pH meters can compensate for temperature-dependent changes in the electrode response. It is possible to measure the temperature of a solution with a thermometer and, using the proper knob or dial, "tell" the pH meter the solution's temperature. Alternatively, there are *electronic temperature probes*, **automatic temperature compensating (ATC) probes**, that can be placed in the sample alongside the pH electrodes. The probes are connected to the pH meter and automatically measure the sample temperature and report it to the meter. The meter automatically compensates for the sample's temperature. Compact devices may even have a temperature probe built into the electrode housing (see Figure 3.29).

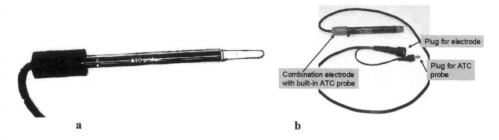

FIGURE 3.29 **Temperature and pH.** (a) An ATC probe. (b) A combination electrode with a built-in temperature probe.

In contrast, meters cannot compensate for temperature-dependent changes that occur in the *pH of the sample*. For example, a solution might be pH 7.6 when it is at 37°C, yet its pH might be 7.3 at room temperature. This is a chemical characteristic of the sample, not the pH meter. It is, therefore, best to measure the pH of a solution when the solution is at the temperature at which it will be used.

Box 3.3 summarizes all these factors and outlines the steps in operating a typical pH meter.

Box 3.3 A Conventional* Method for Measuring the pH of a Solution

Step 1. Allow the meter to warm up.

Step 2. Open the filling hole; be certain filling solution is nearly to the filling hole level (refillable electrodes only).

Step 3. If the meter has a "standby" mode, use it when the electrodes are not immersed. Use the "pH" mode to read the pH of a sample or standard.

Step 4. Calibrate the system each day or before use. Enter "calibration mode."

a. Adjust the meter temperature setting to room temperature or use an ATC probe.

b. Obtain two standard buffers: pH 7.00 at room temperature and an acidic standard if the sample is acidic and a basic standard if the sample is basic.

c. Rinse the electrodes with distilled water and blot dry. Do not wipe the electrodes as this may create a static charge leading to an erroneous reading.

d. Immerse the electrodes in pH 7.00 calibration buffer. Be certain that the junction is immersed and that the level of sample is below the level of the filling solution. Allow the reading to stabilize.

e. Adjust the meter to read 7.00.

f. Remove the electrodes, rinse with distilled water, and blot dry. Alternatively, rinse the electrodes with the next solution and do not dry.

g. Place the electrodes in the second standardization buffer, set the meter to read pH, and allow the reading to stabilize. Adjust the meter to the pH of the second buffer with the proper method of adjustment. Remove, rinse, and blot the electrodes.

h. With an older pH meter, recheck the pH 7.00 buffer as in step d and readjust as necessary. Recheck the second buffer and readjust the meter as necessary. Readjust as needed up to three times. If the readings are not ±0.05 pH units of what they should be after three adjustments, the electrode probably needs cleaning. Exit "calibration mode."

Step 5. Optional: Perform quality control checks at this point.

---

* Modern microprocessor-controlled pH meters sometimes allow the user to calibrate with more than two standards and/or to use calibration standards other than pH 7.0, 4.0, and 10.0. These features may improve the accuracy attained by the meter, but it is best to check in your own lab. For example, we found that using three standards with our pH meter caused results to be slightly less accurate than using two standards.

a. Perform a linearity check. Take the reading of a third calibration buffer. For example, if the meter was calibrated with pH 7.00 and pH 10.00 buffers, check a pH 4.00 calibration buffer. Immerse the electrodes in the third buffer, allow the reading to stabilize, and record the value. Do not adjust the meter to this third calibration buffer. If the reading is not within the proper range, as defined by the laboratory's quality control procedures, the electrodes require maintenance.

b. Test the pH of a control buffer whose pH is known and that has a pH close to that of the sample. It is common to set the maximum allowable error of the control buffer to be ±0.10 pH units. Do not adjust the meter to the pH of the control buffer; the purpose of this buffer is to check the accuracy of the system. If the pH reading of the control buffer is not within the required tolerance, the electrodes require maintenance.

Step 6. Set the meter to the temperature of the sample or place an ATC probe in the sample.

Step 7. Place the electrodes in the sample.

Step 8. Allow the pH reading to stabilize. Generally, the reading should stabilize within a minute. Initially the pH reading changes rapidly; then it slowly stabilizes. One of the causes of error in using pH meters is not allowing the reading to stabilize. Note the following points:

- If you wait too long, the pH of some samples will change because of exposure to air, chemical reactions, or other factors.

- Solutions of high or low ionic strength and nonaqueous solutions may require longer to stabilize.

- Many new pH meters have an "autoread" feature that automatically determines when the electrode has stabilized. For example, the meter may lock on the reading if it changes less than 0.004 pH units over 10 seconds. The autoread feature helps standardize readings made by different operators in a laboratory. It also ensures that impatient operators will wait long enough.

Step 9. Record all relevant information, including the temperature of the sample.

Step 10. Remove the electrodes from the sample, rinse them, and store them properly. Close the filling hole (refillable electrodes only).

Planning Your Work: pH Meters

Fill in the boxes in this flowchart to prepare for each section of the upcoming laboratory exercise.

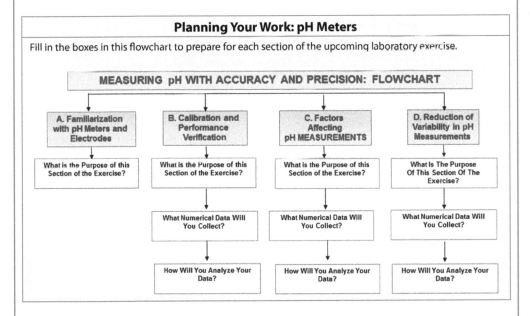

LABORATORY PROCEDURE

Perform these exercises individually or in pairs. If you work with a partner, make certain that both you and your partner(s) have the opportunity to operate each instrument.

Part A: Familiarization with pH Meters and Electrodes

Become familiar with the pH meters in your laboratory. If there is more than one type, study each.

    A.1.   Attach electrodes. Note how the electrodes plug into the meters (see Figure 3.30). Unplug a combination electrode and plug it back into the meter.

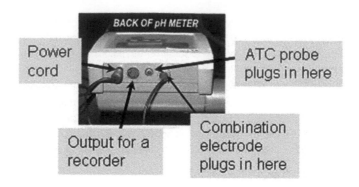

FIGURE 3.30   **Back of one model of pH meter.**

A.2.   Consider the temperature probes.
  A.2.1.   Which meters have a place to plug in an ATC probe?
  A.2.2.   Which meters have ATC probes?
  A.2.3.   What is the purpose of an ATC probe?
  A.2.4.   Do any of your electrodes have built-in temperature sensors?
A.3.   What type(s) of electrodes do you have in your laboratory?
  A.3.1.   Are they combination or separate?
  A.3.2.   Are they covered with a plastic sheath to protect the fragile glass bulb or not?
  A.3.3.   Are your electrodes designed to be compatible with Tris buffer? (Ask your instructor or check the instructions with the electrode; you probably cannot tell just by looking.)
  A.3.4.   Do the electrodes have filling holes or are they gel filled? If they can be refilled, make sure the electrodes have filling solution nearly to the level of the filling hole.
  A.3.5.   How are the electrodes supposed to be stored? Check their instructions.
A.4.   Find the junction of each electrode, if possible. (Often the plastic housing of the electrode makes the junction difficult to find.) The junction must be submerged when you measure the pH of a sample.
A.5.   Practice operating a pH meter.
  A.5.1.   Use a pH meter that was previously calibrated by someone experienced.
  A.5.2.   Follow the instructions with the meter and measure the pH of a sample provided by the instructor. Be sure to rinse the electrodes and blot them dry, as described in Box 3.3.
  A.5.3.   If there is more than one type of meter in your laboratory, learn how to operate each one.
B. **Calibration and Performance Verification**
  B.1.   Calibrate each type of pH meter in the laboratory.
    B.1.1.   Follow the directions that come with the pH meter.
    B.1.2.   Refer to Box 3.3 and perform step 5, the optional quality control checks, even if these are not in the instructions that come with the pH meter. Record the results of the quality control tests.
  B.2.   Determine the relationship between pH and millivolt reading. (This requires a calibrated meter that will display millivolts.)
    B.2.1.   Obtain four pH standards, ideally not those used to calibrate the instrument.
    B.2.2.   Using a calibrated pH meter, place the electrodes in each standard, one at a time, and read off the millivolts. Record your data in your laboratory notebook in a clearly labeled tabular format.

B.3.   ⊞ **Data analysis**

    B.3.1.   Interpret the results of your quality control checks from B.1.

        B.3.1.1.   Did your pH meter respond in a linear fashion? If not, how far off was it?

        B.3.1.2.   How accurate was your pH meter when you checked it in part B.1?

    B.3.2.   Interpret your results from B.2:

        B.3.2.1.   Graph the data from B.2 with pH on the X-axis and millivolts on the Y-axis. (Refer to Figure 3.28c.)

        B.3.2.2.   Your points should form a nearly straight line; if so, draw a straight line through the points.

    If the points do not form a nearly straight line, consult your instructor.

        B.3.2.3.   Based on the slope of your line, calculate:

$$\left(\text{Change in mV}\right)/\text{pH unit}$$

        B.3.2.4.   Calculate the percent efficiency for your electrodes where:

$$\%\,\text{Efficiency} = \frac{\left(\text{Change in mV}\right)/\text{pH unit}}{-60\,\text{mV}\,/\,\text{pH unit}}\left(100\%\right)$$

        B.3.2.5.   If your percent efficiency is less than 85%, consult your instructor.

C. Factors Affecting pH Measurements

    C.1.   Experiment with stabilization. (This requires a calibrated pH meter with the autoread function, if present, disabled.)

        C.1.1.   Obtain a solution of Tris buffer at a relatively high pH, such as 8 or 9. Also obtain Tris buffer at a lower pH, such as 4 or 5.

        C.1.2.   Place the pH electrodes in the higher pH Tris solution and immediately record the pH reading. Continue recording the pH every 2 seconds for a minute. Let another minute pass and then take another reading to see if the pH has changed.

        C.1.3.   Next, using the same meter, place the electrodes in the lower pH Tris solution and immediately record the pH reading and continue recording it every 2 seconds for a minute. Let another minute pass and then take another reading to see if the pH has changed.

        C.1.4.   ⊞ Data analysis

            C.1.4.1.   Graph the pH results from both solutions on a single graph with time on the X-axis and pH on the Y-axis. Explain the significance of your graph. Compare the stabilization of readings using the two solutions. What does this mean in terms of getting accurate pH readings?

    C.2.   Check the pH of water:

        C.2.1.   Get samples of tap water and purified water.

        C.2.2.   One at a time, check the pH of these samples for 3 to 6 minutes. Record the pH every 20 seconds.

        C.2.3.   ⊞ **Data analysis**

            C.2.3.1.   Graph the results for both samples.

            C.2.3.2.   What is the pH of water? Is it 7, as you might predict?

            C.2.3.3.   Is tap water different than purified water? If so, why do you suppose this is the case?

            C.2.3.4.   Which has a more stable measured pH? Do the two eventually have the same measured pH?

            C.2.3.5.   What does this tell us about measuring the pH of very pure samples?

C.3. Check the effect of temperature on the pH of Tris buffer.
- C.3.1. Remove a bottle of cold Tris solution (pH 7–8) from the refrigerator or ice bath; place about 200–300 mL in a small beaker.
- C.3.2. Place the Tris on a stir-plate heater with a stir bar. Turn on the stirrer and the heat to a low setting so that the Tris slowly heats.
- C.3.3. Using a pH meter with an ATC probe, or an electrode with a built in temperature probe, check the temperature every few minutes and then check the pH. Record the pH and the temperature. Continue until the temperature is around 60°C. (Alternatively, if no automatic temperature probes are available, check the temperature of the Tris with a thermometer and manually adjust the pH meter to the temperature before taking each pH reading.)
- C.3.4. **Data analysis**
  - C.3.4.1. Plot temperature versus pH on a piece of graph paper.
  - C.3.4.2. Connect the points into a best fit line.
  - C.3.4.3. Calculate the slope of the line on your graph. This is the change in pH per degree of temperature change. For Tris buffer, the published value for the slope is –0.028 pH units/1°C. This means a 10 degree increase in temperature should cause the pH to decrease by 0.28 pH units. How close is your value to the published value?

C.4. Check the effect of dilution on buffers.
- C.4.1. Obtain about 200–300 mL of 1 M Tris buffer, pH between 7 and 9; and 1 M phosphate buffer, pH between 7 and 9.
- C.4.2. Check the pH of the 1 M Tris buffer.
- C.4.3. Dilute the Tris buffer to 0.1 M by combining 20 mL of buffer with 180 mL of purified water. Check the pH of the dilution.
- C.4.4. Dilute the Tris buffer to 0.01 M by combining 20 mL of the 0.1 M buffer with 180 mL of purified water. Check the pH of this dilution.
- C.4.5. Repeat steps C.4.2 to C.4.4 with the phosphate buffer.
- C.4.6. **Data analysis**
  - C.4.6.1. Calculate the change in pH with each tenfold dilution for each buffer.
  - C.4.6.2. Summarize in your own words the results of this section.

D. Reduction of Variability in pH Measurements

Scenario

QUALITY PROBLEM IN THE MEDIA DEPARTMENT

Cleanest Genes Corp. has a department, media preparation, whose responsibility is to make all the buffers needed by all scientists in the research and development department of the company. A problem has occurred in the media preparation section. The pH of the buffers appears to be inconsistent from day to day. The quality control (QC) department (your class) has been called in to investigate. Your instructor will provide you with two bottles of buffer solutions made by the media preparation department with their recorded pH. Remove about 200–300 mL from each bottle for each group. Determine the exact pH of each solution. Each group should check both solutions by using all the different meters available in your classroom. Have different technicians measure the pH. Keep good records of your investigation.

Each team should document their investigation, prepare a report of their activities and findings, and make recommendations based on their findings.

LAB MEETING/DISCUSSION QUESTIONS

Place class data on the blackboard.

1. Discuss your results for Part B.
    a. Did any pH meters fail the quality control checks? Explain. If so, what did you do?
    b. Record the class values for percent efficiency on the board and discuss them. Do any of the electrodes appear to have an efficiency of less than 85%? If so, what should you do?
2. Discuss stabilization of pH measurements. What did you learn about the stabilization of pH readings in part C.1? Based on your observations, how long should you wait after placing the pH electrodes in your sample before reading the pH value? Based on your results, if you wait too little or too long, how will that affect your readings?
3. Discuss the pH of water. What was your experience measuring the pH of tap water and purified water? Was either pH 7? What does this tell you about trying to determine the pH of a sample that is relatively pure water?
4. Discuss the effects of temperature on pH measurements.
    a. What was the effect of its temperature on the pH of Tris buffer? How will knowing this affect your pH measurements of buffers in the future?
    b. Compare the slopes obtained for the entire class. They ought to be about the same because Tris has a consistent pH response to temperature. If the class results vary, why? How could your class reduce the variability in your results?
5. Discuss the effect of dilution on the pH of buffers.
    a. What is the effect of dilution on Tris buffer?
    b. Compare the class results. They ought to be about the same. If they are not, why is there variation? How could you reduce the variability?
    c. What Is the effect of dilution on phosphate buffer?
    d. Compare the class results. They ought to be about the same. If they are not, why is there variation? How could you reduce the variability?
6. How will knowing about the factors that affect pH measurements affect your methods of measuring pH in the future?
7. Place all class results for Part D on the board.
    a. Calculate the range and the standard deviation of the results.
    b. Is there variability in your results; if so, why? How can this variability be reduced?
    c. How many significant figures are in your results?
    d. Do different pH meters/electrodes give different measurement values?
    e. What range should a quality control procedure specify for the pH of this buffer?
8. Compare and contrast pH measurements and weight measurements in terms of the following:
    a. Whether or not they exhibit linearity.
    b. The accuracy that is achievable.
    c. The variability (precision) that you observed.
9. Study the class results shown in Figure 3.31.
    a. Find the results in Figure 3.31 for the dilution of Tris buffer. Did these students see a consistent effect of dilution? Did Tris consistently become more acidic or more basic with dilution? How do their results compare with yours? Should there be a consistent effect if the procedure Is properly performed?

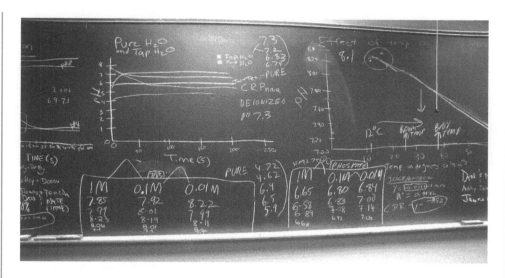

FIGURE 3.31   **Example of class results.**

b.   Find the results in Figure 3.31 for the dilution of phosphate buffer. Did these students see a consistent effect of dilution? Did phosphate consistently become more acidic or more basic with dilution? How do their results compare with yours? Should there be a consistent effect if the procedure is properly performed?

c.   Do you think that dilution affects the pH of buffers? How confident do you feel in your answer?

d.   Find results in Figure 3.31 that show the slope of the line on the graph of temperature versus pH of Tris. How do their results compare with your results? How do their results compare with the published value? What does this tell you about working with Tris buffer? (If you cannot read their results in the photo, Andrew and Jose got a slope value of −0.0247, and the group labeled "C.P.R." got a value of −0.0152.)

e.   Did you have an easy time understanding the presentation of data in Figure 3.31? How might this class have labeled their results so they would be more easily interpreted by people not in their classroom?

10.   The *U.S. CFR* outlines the following requirement for pharmaceutical companies:

The calibration of instruments, apparatus, gauges, and recording devices at suitable intervals in accordance with an established written program containing specific directions, schedules, limits for accuracy and precision, and provisions for remedial action in the event accuracy and/or precision limits are not met.

(21 CFR 211.160)

Suggest a program for ensuring that the pH meters in your classroom are compliant with this regulation. Consider all the sources of variability in pH measurements that you discovered.

UNIT DISCUSSION: METROLOGY IN THE LABORATORY

1. Electronic measuring instruments generate an electrical signal that is related to the magnitude of the property being measured. Show this relationship in two graphs: one for balances and one for pH meters.
2. How will you ensure that all your measurements in the laboratory are as accurate as possible? Provide general guidelines along with specific examples from measuring weight, volume, and pH.
3. How will you ensure that all your measurements in the laboratory are as precise as possible? Provide general guidelines along with specific examples from measuring weight, volume, and pH.
4. Show how each box in this flowchart relates to specific tasks you performed in this unit. For example, one of the boxes is about calibration. You calibrated instruments in Laboratory Exercises 6 and 9.

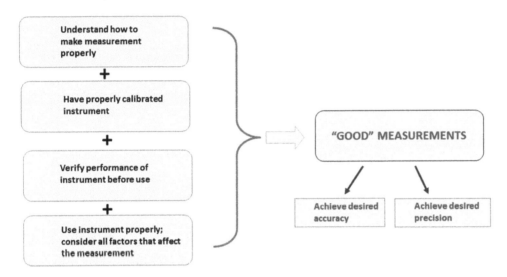

5. Consider the overarching themes of good laboratory practices in the table that follows. Jot down notes on the table or on another piece of paper that describe how the ideas and skills that you practiced in this unit relate to these overarching themes.

Good laboratory practices
DOCUMENT
PREPARE
REDUCE VARIABILITY AND INCONSISTENCY

ANTICIPATE, PREVENT, CORRECT PROBLEMS
ANALYZE, INTERPRET RESULTS
VERIFY PROCESSES, METHODS, AND RESULTS

Spectrophotometry and the Measurement of Light

DOI: 10.1201/9781003360742-4

UNIT INTRODUCTION

All the fifty years of conscious brooding have brought me no closer to answer the question, "What are light quanta?" Of course today every rascal thinks he knows the answer, but he is deluding himself.

—Albert Einstein, as quoted at http://www.todayinsci.com/ QuotationsCategories/P_Cat/Photon-Quotations.htm

A. COLOR AND THE WAVELENGTH OF LIGHT

Spectrophotometers *measure the interactions of light with substances of interest.* We therefore begin this unit with a brief introduction to the nature of light.

Light is *energy moving through space.* The motion of light is often compared to that of waves rippling on the surface of a pond, with crests and troughs. The distance from one crest to the next is called the wavelength (λ) of the light. Our eyes and brain perceive light to be of different colors depending on its wavelength (see Table 4.1). Light of a wavelength of 515 nm, for example, is perceived to be green. Light whose wavelength is slightly shorter than that of violet light is called ultraviolet (UV). Light whose wavelength is slightly longer than that of red light is called infrared (IR). UV and IR light are not visible to our eyes but can be detected by some laboratory instruments.

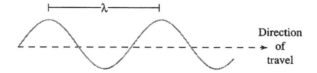

Substances can interact with light. Consider a liquid solution, such as a test tube of food coloring (see Figure 4.1). The red dye in the food coloring is the *compound of interest to us*, the **analyte**. The red dye is dissolved in water. The red food coloring solution appears to be red because the dye absorbs light energy of greenish wavelengths and allows other wavelengths of light to pass through. In general, if the components of a solution absorb visible light energy, the solution will appear to be a particular color, depending on which wavelengths of light are absorbed and which pass through (see Table 4.2). You can see from Table 4.2 that if, for example, the components of a solution absorb light energy of 450 nm, which is blue light, then the solution will appear to be orange. A liquid that does not absorb visible light energy will appear to be clear.

TABLE 4.1   Wavelength of Vis Light vs. Color

λ of Light Waves (in nm)	Color
380–430	Violet
430–475	Blue
475–495	Greenish blue
495–505	Bluish green
505–555	Green
555–575	Yellowish green
575–600	Yellow
600–650	Orange
650–780	Red

Note: Because color is subjective, these wavelengths are approximations.

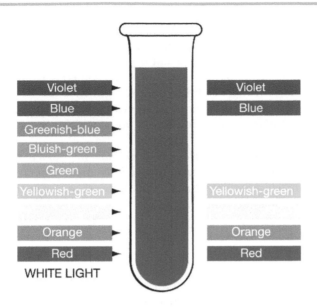

FIGURE 4.1   Absorption of light by a tube of red food coloring. The light that shines on the sample is white and consists of a mixture of light of many wavelengths. As the light passes through the food coloring, the dye absorbs light energy that is greenish blue to green. When this light is removed from the solution, we perceive the solution to be red.

TABLE 4.2   Absorbance of Light of Particular Wavelengths and the Color of a Liquid

Light Absorbed by Solution		Color of the Solution
λ (in nm)	Color	
380–430	Violet	Yellow
430–475	Blue	Orange
475–495	Greenish-blue	Red-orange
495–505	Bluish green	Orange-red
505–555	Green	Red
555–575	Yellowish green	Violet-red
575–600	Yellow	Violet
600–650	Orange	Blue
650-780	Red	Green

Note: Colors that appear alongside one another on this table are said to be "complementary" colors (e.g., green and red are complementary).

B. THE INTERACTION OF LIGHT WITH A SAMPLE PROVIDES INFORMATION

Laboratory analysts take advantage of the interaction of light with samples to obtain information about those samples. Consider again the test tube containing red food coloring in Figure 4.1. The more red dye present, the more light energy the solution will absorb. If we can measure how much light is absorbed by the solution, then we can obtain information about how much red dye is present. This is called **"quantitative information,"** which has to do with *the amount or concentration of a substance.* There are many situations in the laboratory where it is valuable to quantify the amount or concentration of an analyte in a sample. Biotechnologists, for example, might want to know how much DNA is present in a cellular extract, the percentage of a particular protein that is present in a sample, or the amount of active enzyme in a tube.

There is another kind of information that can be obtained by measuring the absorption of light by a sample. Refer again to Figure 4.1. Observe that red food coloring is strongly absorbing light in the bluish-greenish region, say around 500 nm. The food coloring transmits light of other wavelengths. This pattern of light absorption is a characteristic of the compound that is used to manufacture red food coloring. The compound that is used to make blue food coloring is different; it absorbs light in the "orangish" region of wavelengths and transmits other wavelengths. The compounds in blue and red food coloring thus interact differently with light, absorbing some wavelengths more than others. Different compounds have different patterns of interaction with light of different wavelengths. We can measure the pattern of interaction of a sample with light of different wavelengths to obtain qualitative information about the sample. **Qualitative information** *tells us about which compounds are present, or the features of those compounds.*

A spectrophotometer can be used to measure *how much* light of a particular wavelength is absorbed by a sample, thus providing quantitative information about that sample. It can also be used to measure *which wavelengths* of light are absorbed by a sample, thus providing qualitative information about the sample.

C. THE BASIC DESIGN OF A SPECTROPHOTOMETER

Spectrophotometers measure the interaction between light and a sample in order to obtain information about the sample. Figure 4.2 shows a simplified diagram of a conventional spectrophotometer. There is a bulb that emits light. The light beam consists of light of many wavelengths blended together. The beam of light from the bulb is directed into a **monochromator** ("wavelength selector") *where light of a single wavelength is selected.* This **monochromatic light** *of a specific wavelength* shines on a sample inside the sample chamber. The sample is generally a liquid. The liquid is contained in a small tube, called a **cuvette**. Some of the light's energy may be absorbed by the analyte in the sample, while the rest passes through unchanged. The spectrophotometer cannot actually "see" the absorbed light, but it can detect the *light that passes through without being absorbed*, the **transmitted light**. The spectrophotometer measures the amount of transmitted light with a detector that is sensitive to light.

Figure 4.3 diagrams a spectrophotometer in more detail than Figure 4.2. Beginning at the upper left of Figure 4.2, observe that this spectrophotometer has two light sources: one that emits visible light and one that emits UV light. Light from either bulb is focused into a narrower band by a mirror. Another mirror reflects the light beam so that it passes through a narrow entrance slit. *Light is "sharpened" as it passes through the* **entrance slit**. The sharpened light beam shines on a **diffraction grating**, whose purpose is *to separate the light into its constituent wavelengths.* For visible light, the result is a rainbow, much as you would see formed by a crystal. UV light is also separated into its constituent wavelengths, but our eyes cannot detect it. A single wavelength (or a narrow span of a few wavelengths) is selected by moving the exit slit until it lines up with the desired wavelength. Only light of the desired wavelength passes through the narrow slit. The now monochromatic light is directed into the sample, which may absorb some, or all, of the light.*

When you operate a spectrophotometer, you turn on one or both of the light bulbs—the visible bulb if you are working with visible wavelengths of light and the UV bulb for ultraviolet wavelengths. (Some spectrophotometer models do not have a UV light source.) You will also "tell" the spectrophotometer which wavelength(s) of light to select, depending on your application.

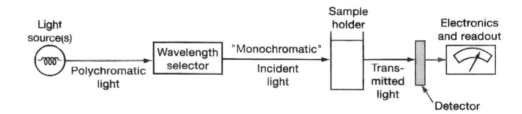

FIGURE 4.2   Simplified diagram of a conventional spectrophotometer.

---

* There are also spectrophotometers that have a different configuration. These instruments use CCD detectors, much like the sensors on a digital camera. In this type of spectrophotometer, there is no exit slit and no moving parts to select wavelength.

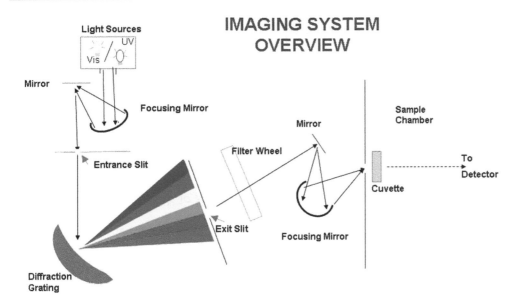

FIGURE 4.3   **A more detailed diagram of a conventional spectrophotometer.** See text for explanation.

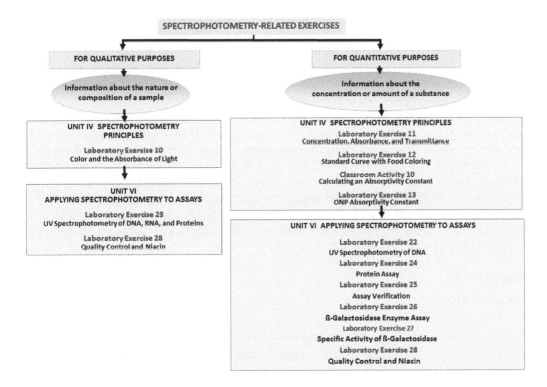

FIGURE 4.4   **Exercises that show how spectrophotometers are used to provide qualitative and quantitative information.**

D. THE PURPOSE OF THIS UNIT

The operation and function of spectrophotometers are the subjects of this unit. The first laboratory exercise uses colored dyes to explore qualitative spectrophotometry. You will then learn how spectrophotometers obtain quantitative information about a sample. There are more laboratory exercises relating to quantitative spectrophotometry because these applications are more common in biotechnology. In the upcoming unit on assays, you will apply what you have learned about spectrophotometers to answer questions about samples. Figure 4.4 provides a flowchart of the various spectrophotometry laboratory exercises in this manual.

Laboratory Exercise 10: Color and the Absorbance of Light

OVERVIEW

This is the first in a series of laboratory exercises dealing with UV/Vis (ultraviolet/visible) spectrophotometry. This exercise will introduce you to the operation of spectrophotometers and to concepts relating to color and light absorbance. As you explore spectrophotometers in this exercise you will perform the following tasks:

- Learn how to operate the spectrophotometers in your laboratory.
- Learn how to generate an absorbance spectrum to obtain qualitative information about a sample.
- Look at the relationship between light absorbance of specific wavelengths and the color of a sample.
- Learn about the relationship between transmittance and absorbance.

BACKGROUND

A. Spectrophotometers Measure Transmittance

When light shines on an analyte in a sample, the substance may absorb some or all of the energy of the light. Spectrophotometry is about the measurement of this absorption of light energy by an analyte. We obtain information about how much of the analyte is present and/or its nature by measuring this absorption of light. Once light is absorbed, however, the spectrophotometer cannot "see" it; therefore, the instrument cannot directly measure the absorption of light. The measurement of light absorption must be done in an indirect way. The spectrophotometer measures how much light is transmitted through the sample and how much light is transmitted through a comparable substance that has no analyte. It uses this information to calculate the amount of light that was absorbed by the analyte.

The substance that has no analyte is called the "blank." A **blank** is *a reference solution that contains no analyte but does contain the solvent and any reagents that are intentionally added to the sample.* There will always be some light absorbed by the blank even though it has no analyte. This is because the cuvette absorbs light, the solvent absorbs light (even if the solvent is just pure water), and compounds other than the analyte may absorb light. *The relationship between the light transmitted through the sample and the light transmitted through the blank* is called **transmittance (t)** (see Figure 4.5). The formula to calculate transmittance is shown in Formula 4.1a. **Percent transmittance (T)** is *the transmittance multiplied by 100%* (see Formula 4.1b).

Formula 4.1a

Transmittance

$$t = \frac{\text{Light transmitted through the sample}}{\text{Light transmitted through the blank}}$$

Formula 4.1b

Percent Transmittance

$$\%T = (t)(100\%)$$

Transmittance can range from 0 (no light passes through the sample relative to the blank) to 1 (the same amount of light passes through the sample and the blank). Percent transmittance ranges from 0% to 100%.

Spectrophotometers measure transmittance, that is, the ratio between the amount of light transmitted through a sample and the amount of light transmitted through a blank.

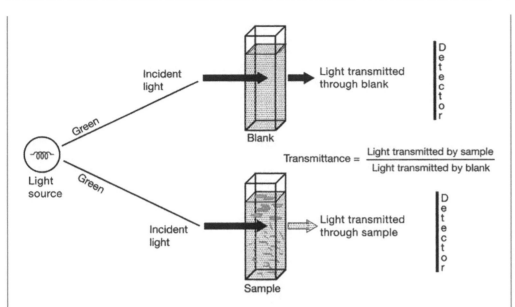

FIGURE 4.5   Transmittance. Two solutions are placed in a spectrophotometer (either sequentially or at the same time, depending on the instrument). One is a sample containing the analyte, in this case, red food coloring. The other is the blank, in this case, water. Green light is shined on the two solutions. The amount of light transmitted through each is measured by a detector and the transmittance is calculated.

Analysts, however, are generally interested in the amount of light *absorbed* by a sample. It is, therefore, customary to convert transmittance to a measure of light absorption. Early analysts devised a measure of absorption called **absorbance** (or **optical density, OD)**, *which is derived from the transmittance* (see Formula 4.2). Modern spectrophotometers can be adjusted to either display transmittance or to automatically convert the reading to an absorbance value.

To operate a spectrophotometer, the blank is first placed into the instrument, which is then adjusted to 0 absorbance or 100% transmittance. (Some instruments have two chambers: one for the sample and one for the blank. In these instruments, the blank is not removed to insert the sample.) The blank is removed and the sample is placed in the instrument, which displays the absorbance or transmittance of the sample relative to the blank. This means that the instrument must be "blanked" (set to 100% transmittance or 0 absorbance) at the beginning of every analytical procedure.

B. Absorbance Spectra

An **absorbance spectrum** is *a graph of the absorbance of an analyte (on the Y-axis) versus the wavelength of the light shined on the analyte (X-axis).* The resulting graph will have one or more peaks and valleys that are characteristic of that substance. The absorbance spectrum of a red food coloring sample is shown in Figure 4.6.

As discussed in the Unit Introduction, different analytes have a particular pattern of absorbance at different wavelengths. This pattern is the absorbance spectrum for that substance. Qualitative applications of spectrophotometry use the spectral features of a sample to obtain information about the sample's component(s). Sometimes it is possible to use an absorbance spectrum like a "fingerprint" to identify the components in an unknown substance. The infrared (IR) spectra of organic compounds form particularly distinctive "fingerprints." Organic chemists, therefore, frequently use IR spectra to identify compounds. The spectrophotometers routinely used in biological laboratories, however, less commonly use IR light.

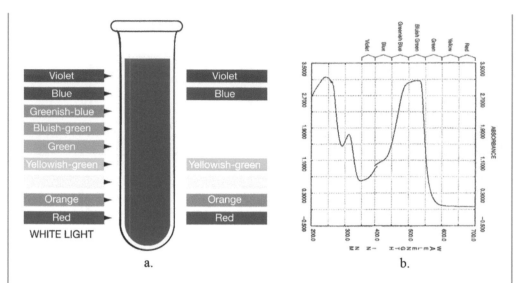

FIGURE 4.6   The absorption of light by red food coloring. (a) When white light is shined on a tube containing red food coloring, blue- to green-colored light is absorbed. (b) The absorbance spectrum for red food coloring, generated by a spectrophotometer. The X-axis is wavelength, which corresponds to the color of the incident light. The Y-axis is the amount of light absorbance. A peak of visible light absorbance occurs between 490 and 530 nm. (The large peak on the far left of the spectrum is in the UV region.)

Formula 4.2

Absorbance

$$A = -\log_{10}(\text{transmittance})$$

UV/Vis spectra are less complex and distinctive than IR spectra and so are used less commonly for identification of unknown substances. There are, however, situations where information about a sample can be obtained from its UV/Vis spectrum. The *U.S. Pharmacopeia*, for example, includes a number of tests of drug identity in which an absorbance spectrum from 200 to 400 nm is used to confirm the identity of a drug. UV/Vis identity tests are used because they are simple and rapid, and the equipment to prepare an absorbance spectrum is widely available. A problem with identifying substances using UV/Vis spectrophotometry is that some compounds whose structures are similar have spectra that cannot be distinguished. UV spectra are, therefore, frequently used in conjunction with other identification methods. An example of a test of identity from the *U.S. Pharmacopeia* is given in Laboratory Exercise 28. This test is used to confirm the identity and purity of niacin tablets.

C. Spectrophotometer Features

Spectrophotometers are available in a number of models from various manufacturers. Traditionally, teaching laboratories were equipped with a simpler style of spectrophotometer, sometimes called a "Spec 20." Although instruments with this basic design have been used successfully for years, these spectrophotometers have the disadvantage that the operator must manually reset the instrument to 100% transmittance (or 0 absorbance) with the blank each time the wavelength is changed. Manually selecting each wavelength and repeatedly "blanking" the instrument is tedious when an absorbance spectrum is being prepared. Manufacturers, therefore, developed instruments called **scanning spectrophotometers** that *are capable of automatically scanning through a range of wavelengths and constructing an absorbance spectrum*. This style of spectrophotometer is now common in both professional and teaching laboratories.

Simple Spec 20-style instruments have another limitation for biologists, which is that they can be used only in the visible range of light, not in the UV range. Most biological analytes, such as DNA, RNA, and most proteins, do not absorb any visible light; hence they appear clear to our eyes. Biological substances do, however, absorb ultraviolet light. It is, therefore, desirable to have a UV/Vis spectrophotometer that can operate in both a visible and UV mode.

D. Exploring Qualitative Spectrophotometry

The purpose of this laboratory exercise is to explore the relationships between color and the absorption and transmission of light. You may or may not be able to complete every part of this exercise depending on the types of spectrophotometers available. Your instructor will tell you which parts you can do.

Part A can only be performed with a spectrophotometer where the lamp stays on when the sample chamber is open. An older Spec 20-style instrument is perfect for this purpose. Part B involves creating absorbance spectra manually. This means that you adjust the wavelength, blank the instrument, insert your sample, and read the sample's absorbance and/or transmittance. You then repeat this process many times to slowly collect the data to create a spectrum. This is the only way to create an absorbance spectrum with an older Spec 20-style instrument or with any other model of nonscanning spectrophotometer. The instructions here are written for a Spec 20-style instrument.

A scanning spectrophotometer allows you to prepare an absorbance or transmittance spectrum without repeatedly inserting and removing the blank and sample. You insert the sample and the blank only once, and the instrument automatically takes care of changing the wavelength and collecting data at each wavelength. Part C is about creating absorbance spectra with a scanning spectrophotometer.

Advice from the senior technician

Use Cuvettes Properly

1. Remember that cuvettes are expensive and fragile laboratory items (except for disposable plastic ones).

2. Use quartz cuvettes for UV work; glass, plastic, or quartz are acceptable for work in the visible range. There are inexpensive plastic cuvettes that may be suitable for some UV work.

3. Be certain to only use clean cuvettes.

4. Use lens paper to clean the outside surface of quartz and glass cuvettes; these cuvettes are treated like microscope lenses. You can use ordinary laboratory tissue for plastic cuvettes.

5. Wash the insides of cuvettes immediately after use.
   - Do not allow samples to sit in a cuvette for a long period of time.
   - Do not clean them with brushes or abrasives.
   - Consult Figure 4.7 for an effective washing system that will not scratch the cuvettes.

6. Consider the use of disposable plastic cuvettes for colorimetric protein assays because dyes used to visualize proteins tend to stain cuvettes.

7. Make sure the cuvette is properly aligned in the spectrophotometer (see Figure 4.8).

8. Matched cuvettes are manufactured to absorb light identically so that one of the pairs can be used for the sample and the other for the blank.
   - If you do not have matched cuvettes, you will obtain the best accuracy by using the same cuvette for the blank and all the samples.
   - If you have matched cuvettes, keep them stored together.

9. Do not touch the base of a cuvette or the sides through which light is directed.

10. Do not scratch cuvettes; do not store them in wire racks or clean with brushes or abrasives.

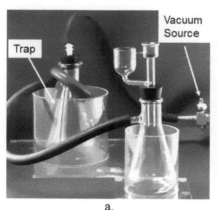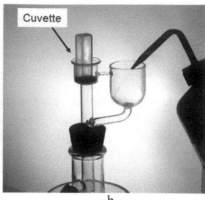

a.   b.

FIGURE 4.7   **Cleaning cuvettes.** (a) A trap is attached to the vacuum source with thick-walled vacuum tubing. (The trap keeps materials from contaminating the vacuum line.) The trap is connected to a flask with the glass cuvette washing apparatus. b. To use the apparatus, the vacuum is turned on, and the cuvette is placed in the apparatus as shown. Cuvette washing solution is applied, which vigorously sprays the cuvette. The recipe for cuvette washing solution is in Appendix 7.

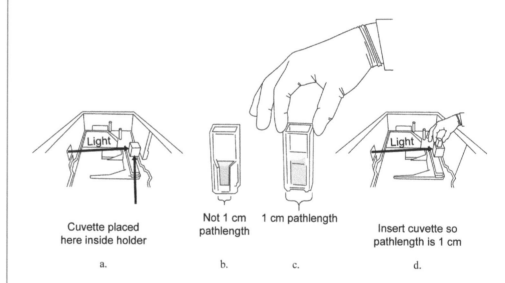

Cuvette placed   Not 1 cm   1 cm pathlength   Insert cuvette so
here inside holder   pathlength   pathlength is 1 cm

a.   b.   c.   d.

FIGURE 4.8   **Orient the cuvette in the spectrophotometer properly or results will be inaccurate.** (a) The sample chamber of a spectrophotometer. (b) This cuvette is designed to hold a small volume yet have the sample fill the cuvette high enough to be in the light path. The path that the light travels through the cuvette in this orientation is not 1 cm, as it normally should be. Measurements will be inaccurate if the cuvette is oriented this way when placed in the sample chamber. (c) The cuvette has the expected 1 cm path length when oriented properly. (d) Pay attention to cuvette orientation as you place it in the sample chamber.

Planning Your Work: Color and the Absorbance of Light

PART A: COLOR AND WAVELENGTH

1. What is the purpose of this section of the exercise?
2. Do you have the right kind of spectrophotometer to perform this activity? (Check with your instructor.)
3. If you perform this part of the exercise, what will you record in your laboratory notebook?

PARTS B AND C: USING A NONSCANNING SPECTROPHOTOMETER

1. What is the purpose of Parts B and C of this exercise?
2. Do you have a nonscanning spectrophotometer in your laboratory? Do you have a scanning spectrophotometer?
3. What data will you need to record in your laboratory notebook?
4. How will you analyze your data?
5. On the three blank graphs that follow, predict what you think the absorbance spectra of red, yellow, and green food coloring will look like. (Hint: See Figure 4.6b for the idea.) Label each one. When you go to the laboratory, do not allow your predictions to influence the results. If your predictions do not match your results, think about why this is the case. When predictions do not match results, alert scientists learn something they did not know before.

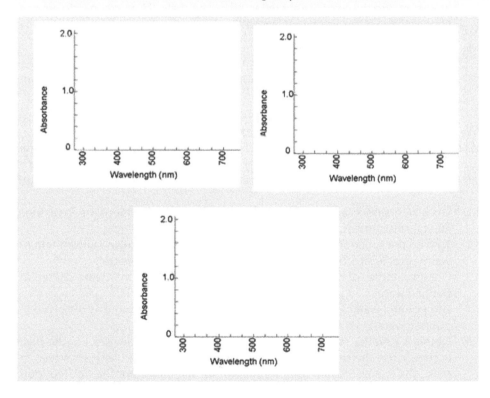

LABORATORY PROCEDURE

The instructions given in Part A and Part B assume that you are using a Spec 20-style of instrument. If you have a different type of instrument, obtain directions from your instructor as to its operation. Part A requires an instrument where the light remains on even when the sample chamber door is open. Skip Part A if you do not have this style of instrument.

PART A: COLOR AND WAVELENGTH

Table 4.1 in the Unit Introduction shows the relationship between the wavelength and the color of light. You can see this relationship for yourself using a Spec 20 by viewing the light exiting the monochromator. Adjust the instrument to a particular wavelength and look at the light emitted. You may find that the way you name colors is not exactly the same as it is in Table 4.1 because color naming is subjective.

A.1. Turn on the spectrophotometer and the visible light source and allow at least 15 minutes of warm-up to stabilize the source and detector.

A.2. Obtain a piece of white chalk that is about 15 to 20 mm tall. With a razor blade, bevel one end to make it smooth and reflective.

A.3. Place the chalk in a Spec 20 cuvette.

A.4. Place the cuvette into the spectrophotometer sample holder. When the spectrophotometer is on, the chalk will reflect the light exiting the monochromator so that your eyes can see it.

    A.4.1. You may need to adjust the transmittance/absorbance control knob to better see the light.

    A.4.2. It may be helpful to turn off the room lights.

    A.4.3. You may need to adjust the orientation of the chalk to optimize the reflection from it.

A.5. Dial in wavelengths at 10 nm increments from 380 to 780 nm. For each wavelength, record the color of light reflected by the chalk. (Many people do not see light until the wavelength is at least 400 nm.)

A.6. **Data analysis**

    A.6.1. Prepare a table like Table 4.1 in the Unit Introduction but with your own observations of color. Compare your table to the one in the Unit Introduction.

Part B: Preparing Absorbance Spectra Using a Nonscanning Spectrophotometer

Prepare absorbance spectra for red, green, and yellow food coloring. Also prepare a transmittance spectrum for red food coloring. Begin at 380 nm and increase the wavelength by 25 nm each time. Blank the instrument each time you change the wavelength.

B.1. Work with a nonscanning spectrophotometer; these directions are for a Spec 20-style instrument.

B.2. Turn on the spectrophotometer and the visible light source and allow them to warm up at least 15 minutes to stabilize the source and detector.

B.3. Obtain samples of red, green, and yellow food coloring that were diluted in purified water.

B.4. Use purified water as your blank (because the food coloring was dissolved in purified water).

B.5. Obtain a cuvette. You can use a plastic, glass, or quartz cuvette. For the best accuracy, use the same cuvette for every sample and the blank. However, to make the exercise faster, use one cuvette for the blank, another for red food coloring, another for green, and another for the yellow sample.

B.6. Set the desired wavelength with the wavelength control. Begin at 380 nm.

B.7. With a Spec 20-style instrument, perform the following activities:

    B.7.1. Set the display to "transmittance" by pressing the mode control button until "transmittance" is lit.

    B.7.2. With the sample chamber empty and the lid closed, set the display to 0% T with the zero control knob on the left front panel. (This step may not be required with other spectrophotometers. This step does not involve a blank. It is performed when the instrument is first turned on.)

B.8. Fill your clean cuvette with the blank solution.

    B.8.1. Remove liquid droplets, dust, and fingerprints from the cuvette.

        B.8.1.1. For glass and quartz cuvettes, always use lens paper. For plastic cuvettes, it is acceptable to use ordinary laboratory tissue.

    B.8.2. Place the cuvette into the sample compartment—align the guide mark of the cuvette with the guide mark at the front of the sample chamber. If there is no guide mark, consult your instructor.

    B.8.3. Close the lid.

B.9. Adjust to absorbance mode.

B.10. Adjust the display to "0.0 A" with the proper knob, dial, or keystroke; you have now "blanked" the instrument at this wavelength.

B.11. Remove the blank solution and rinse the cuvette (if using only a single cuvette).

B.12. Read the absorbance of red food coloring.

    B.12.1. Insert the red food coloring into the cuvette.

    B.12.2. Place the cuvette into the sample compartment, aligned as before.

    B.12.3. Close the lid. Read the absorbance.

B.13. Read the transmittance of the red food coloring sample by adjusting the mode to "transmittance." (You do not need to blank the instrument again because you have not changed the wavelength.)

B.14. Remove the red food coloring sample, rinse the cuvette (if using only one cuvette), and put the green food coloring sample into the same cuvette.

B.15. Adjust the instrument back to absorbance mode and read the absorbance of the green food coloring. (You do not need to blank the instrument at this point because the wavelength is the same as before.)

B.16. Remove the green food coloring sample and put in the yellow food coloring sample. Read the absorbance. (You do not need to blank the instrument at this point.)

B.17. Adjust the wavelength to 405 nm.

B.18. Put the blank solution into your cuvette and blank the instrument at the new wavelength.

B.19. Remove the blank. Insert red food coloring and read both its absorbance and transmittance.

B.20. Remove red food coloring and read the absorbances of green and yellow.

B.21. Repeat steps B.18–B.20, increasing the wavelength each time by 25 nm, until you reach 680 nm. Remember to blank the spectrophotometer each time you change the wavelength.

B.22. Choose any two colors of food coloring and combine them in any proportions that you like. Name the resulting color. Then, prepare an absorbance spectrum for this new sample, as you did for the other dyes.

B.23. **Data analysis**

    B.23.1. Plot the absorbance spectra for all three color dyes on a single piece of graph paper. (If possible, use colored pens or a computer graphing program with color.) Label the graph completely and tape it into your laboratory notebook. Alternatively, plot the graph directly in your laboratory notebook. Use most of a page so that the graph is easy to see and interpret.

    B.23.2. On another graph, plot both absorbance versus wavelength and transmittance versus wavelength for red food coloring.

    B.23.3. Prepare an absorbance spectrum for the sample that was a combination of two colors.

Part C: Preparing Absorbance Spectra Using a Scanning Spectrophotometer

Repeat Part B using a spectrophotometer that has the ability to automatically scan through wavelengths, if this type of instrument is available. Every model of scanning spectrophotometer is operated somewhat differently, so if you have this type of spectrophotometer, your instructor will direct you in its operation.

C.1. Prepare absorbance spectra for red, green, and yellow food coloring using a scanning spectrophotometer.

C.2. If you skipped Part B, prepare a spectrum of transmittance (Y-axis) versus wavelength (X-axis) for red food coloring. You do not have to repeat this spectrum if you have already done it.

C.3. If you skipped Part B, choose any two colors of food coloring and combine them in any proportions that you like. Prepare an absorbance spectrum for this new sample, as you did for the other dyes.

Lab Meeting/Discussion Questions

1. Do your absorbance spectra for the three colored dyes match the predictions that you made before coming to the laboratory? If not, can you think of a reason why they are different?

2. Do your spectra match those of the rest of the class? If not, consider the reasons.

3. How do the spectra of the three colored dyes differ from one another? Why?

4. Do any of the spectra have more than one absorbance peak? If so, what does this tell you about the composition of the dye?

5. Explain the features of the absorbance spectrum for your sample that combined two dyes.

6. If you were able to repeat the exercise with both a scanning and nonscanning spectrophotometer, how did the spectra compare? In principle, they should be the same—is this the case? If not, why not?

7. In your own words, what is the function of the blank? Is a blank always purified water?

8. What is the relationship between absorbance and transmittance? Refer to your data for red food coloring.

Laboratory Exercise 11: Concentration, Absorbance, and Transmittance

OVERVIEW

This exercise reviews basic principles of quantitative spectrophotometry. You will be provided with prepared standards containing known concentrations of analyte (red food coloring). You will measure the absorbance and transmittance of all the standards at a single wavelength. You will then plot and analyze the results in different ways in order to understand how a spectrophotometer is used to obtain information about the concentration of analyte in a sample. In this spectrophotometry exercise, you will thus perform the following tasks:

- Explore the relationship between the concentration of analyte in a sample, and the absorbance and transmittance of the sample.
- Examine how spectrophotometry can be used to obtain quantitative data.

LABORATORY PROCEDURE

1. Use whatever model of spectrophotometer is available.
2. Adjust the spectrophotometer to 517 nm.
3. Blank the instrument with purified water.
4. Measure the absorbance and transmittance of all of the red food coloring standards provided by your instructor, following the directions for the model of spectrophotometer you are using.
5. **Data analysis**
    5.1. Determine whether the relationship between transmittance and concentration is linear.
        5.1.1. Graph your red food coloring data with transmittance on the Y-axis and concentration on the X-axis. Remember to label the graph with a title and make sure the axes are labeled.
        5.1.2. Intuitively, you would expect the transmittance to decrease as the concentration increases. Examine your graph. Does your graph confirm this prediction?
            Yes ☐   No ☐
        5.1.3. The relationship between transmittance and concentration should slope downward, as in Figure 4.9.

    Transmittance versus concentration is not a linear relationship; it curves downward. Your graph of transmittance versus concentration should curve downward. Does it? Yes ☐   No ☐
    If you answered "No" to either question 5.1.2 or 5.1.3, consult with your instructor before continuing.

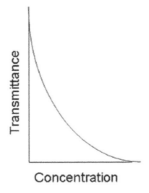

FIGURE 4.9   The relationship between transmittance and concentration.

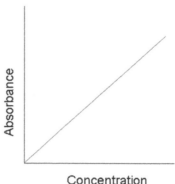

Concentration

FIGURE 4.10 **The relationship between absorbance and concentration.**

5.2.   Determine whether the relationship between absorbance and concentration is linear.

5.2.1.   Graph your data with absorbance on the Y-axis and concentration on the X-axis for red food coloring. Remember to label the graph with a title and make sure the axes are labeled.

5.2.2.   Intuitively, you would expect the absorbance to increase as the concentration increases. Examine your graph. Does your graph confirm this prediction?
Yes ☐   No ☐

5.2.3.   The relationship between absorbance and concentration should have the shape of Figure 4.10. This graph is called a **standard curve.** Observe that this is a linear relationship; it does not actually curve. Your graph of absorbance versus concentration should have this shape. Does it?
Yes ☐   No ☐

If you answered "No" to either question 5.2.2 or 5.2.3, consult with your instructor before continuing.

5.2.4.   It is easier to work with relationships that are linear than with ones that curve. Therefore, it is easier to work with absorbance than transmittance.

5.3.   Investigate the relationship between absorbance and transmittance.

5.3.1.   Recall Formula 4.2:

$A = -\log_{10}$ (transmittance)

Let's reproduce this relationship between absorbance and transmittance using your data.

5.3.2.   Using your calculator, take the log of each of the transmittance values (t, not %T) for red food coloring and record the values in this table:

Concentration (X-axis)	Transmittance	Log Transmittance (Y-axis)

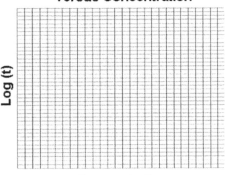

Concentration

5.3.3. Plot the log of transmittance (Y-axis) versus concentration (X-axis). Draw and label the general shape of your graph here:

Is your graph linear?

Yes ☐   No   ☐

Does log (transmittance) decrease as concentration increases?

Yes ☐   No   ☐

If you answered "No" to either question, consult with your instructor before continuing.

5.3.4. Using your calculator, take the negative log of the transmittance values. Plot the negative log of transmittance versus concentration.

Concentration (X-axis)	Transmittance	Log (transmittance)	−Log (transmittance) (Y-axis)

- Log of Transmittance versus Concentration

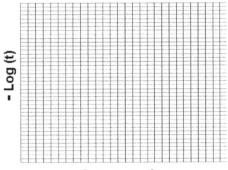

Concentration

5.3.5.   Graph –log (t) on the Y-axis versus concentration on the X-axis. Draw and label the general shape of your graph here. Is your graph linear?
Yes ☐   No   ☐

Does –log (transmittance) increase as concentration increases?
Yes ☐   No   ☐

Compare this graph to your graph of absorbance versus concentration. The two should be identical because A = –log (t). Are they identical?
Yes ☐   No   ☐

If you answered "No" to any of these questions, consult with your instructor before continuing.

Stop and Review

You should be able to answer these questions from memory:
Draw the shape of the graph of absorbance versus concentration.
Draw the shape of the graph of transmittance versus concentration.
What is the equation that relates absorbance and transmittance?
_____

LAB MEETING/DISCUSSION QUESTIONS

1.   Based on your graph of absorbance versus concentration, what do you predict would be the absorbance of a solution with 23 ppm red food coloring at 517 nm?
2.   Based on your graph, if a solution of red food coloring had an absorbance of 0.75 at 517 nm, what is its concentration?
3.   Based on your graph, if a solution of red food coloring had an absorbance of 0.25 at 517 nm, what is its concentration?
4.   Based on your graph, would 20 ppm red food coloring have twice the absorbance as 10 ppm red food coloring?
5.   Based on your graph, would 10 ppm red food coloring have half, or double, the transmittance as 20 ppm red food coloring?
6.   Compare your results to the rest of the class.

Laboratory Exercise 12: Preparing a Standard Curve with Food Coloring and Using it for Quantitation

OVERVIEW

This exercise is related to Laboratory Exercise 11 in which you investigated the relationship between absorbance, transmittance, and concentration of analyte. Now you will use spectrophotometry to determine the concentration of an analyte in samples. This time, you will prepare the standards yourself rather than using standards provided by your instructor. Thus, in this exercise you will perform the following tasks:

- Dilute a stock solution to prepare standards.
- Prepare a standard curve using your standards.
- Use your standard curve to determine the concentration of red food coloring in unknown samples provided by your instructor.
- Test your pipetting and calculation skills.

BACKGROUND

A. Standard Curves and Quantitation

The purpose of quantitative applications of spectrophotometry is to determine the concentration (or amount) of an analyte in a sample. Quantitative analysis with a spectrophotometer typically involves a **standard curve**, which, *for spectrophotometric analysis, is a graph of analyte concentration (X-axis) versus absorbance (Y-axis)*. To construct a standard curve, standards are prepared with known concentrations of analyte. The absorbances of the standards are determined at a specified wavelength, and the results are graphed (see Figure 4.11). Given a standard curve, it is possible to determine the concentration of an analyte in a sample based on the sample's absorbance.

B. Standard Curves: Linear Range

In Figure 4.11, you can see that at lower standard concentrations, a plot of concentration versus absorbance forms a straight line. But as the concentration of the standards increases, the plot begins to curve and eventually plateaus. This is because at a certain point, the samples absorb so much light that the capacity of the spectrophotometer is exceeded. Analysts usually work with standards and samples whose absorbances are in the linear range. Standards or samples with absorbances above the linear range are diluted before measuring their absorbance.

C. Diluting a Stock Solution

A **dilution** *is when one substance (often but not always water) is added to another to reduce the concentration of the first substance. The original substance being diluted is the* **stock**

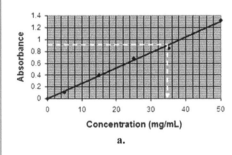

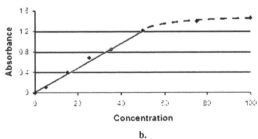

a.   b.

FIGURE 4.11   Standard curve. (a) Six standards of known concentration were prepared. Their absorbances were measured and the results were graphed. The line that best fits the data points was drawn. In this example, an unknown sample had an absorbance of 0.90. This represents a concentration of about 34 mg/mL. (b) A graph of concentration versus absorbance is linear in a certain range and then plateaus.

solution. For example, to make orange juice from concentrate, you pour the concentrated orange juice into a pitcher and add three cans of cold water to it. The concentrated stock solution is the orange juice concentrate, and the pitcher of orange juice is the diluted material. Dilutions are often required in the laboratory. Buffers, for example, can be prepared and stored as concentrated stock solutions and diluted when needed.

In a typical situation, the laboratory scientist knows the concentration of the stock solution and the desired concentration of the final solution. The question then is how much of the stock solution should the scientist use to get the right concentration of diluted solution. Suppose, for example, that you have 1 L of a 2 M stock solution of Tris buffer and you want to dilute some of it so that you have 100 mL of 1 M Tris. The $C_1V_1 = C_2V_2$ equation, as shown in Formula 4.3 and the example following, is a handy tool to calculate how much stock solution is required.

Formula 4.3 To Make a Less Concentrated Solution from a More Concentrated Solution

$$C_1V_1 = C_2V_2$$

where:

C_1 = concentration of the stock solution
V_1 = volume of stock solution required (usually the unknown)
C_2 = final concentration needed
V_2 = final volume needed
Notes:
   Any units can be used for volume, as long as the same units are used for V_1 and V_2.
      Any units can be used for concentration (e.g., molarity, percent), as long as the same units are used for C_1 and C_2.
   The $C_1V_1 = C_2V_2$ equation is used only when calculating how to prepare a less concentrated solution from a more concentrated solution. Do not use this equation in other situations.

Example

How would you prepare 100 mL of a 1 M solution of Tris buffer from a 2 M stock of Tris buffer?

Answer

Consider first, is this a situation where a less concentrated solution is being made from a more concentrated solution? Yes. Therefore, it is appropriate to apply the $C_1V_1 = C_2V_2$ equation:

C_1 = Concentrated solution = 2 M V_1 = Volume of concentrated stock necessary = ?

C_2 = Desired concentration = 1 M V_2 = Volume you want to prepare = 100 mL

Substituting into the equation $C_1 V_1 = C_2 V_2$.

$$2M(?) = 1M(100mL)$$

Solving for ?, the unknown:

$$? = \frac{100 \ \cancel{M} \ (mL)}{2 \ \cancel{M}} ? \quad = 50\,mL$$

This means you need to take 50 mL of the concentrated stock solution and bring it to a final volume of 100 mL.

> ## Planning Your Work: Standard Curves and Quantitation
>
> 1. The purpose of preparing a standard curve is to determine the concentration of analyte in your samples. In this exercise, what are the samples?
>
> 2. What will your standard curve look like if your pipetting technique is sloppy?
>
> 3. Fill in Table 4.3. This table will guide you in performing dilutions of the stock solution when you go to the laboratory. You cannot begin the exercise until the table is completed.
>
>   • Assume that the stock solution of red food coloring that comes from the original bottle has a concentration of 25,000 ppm.
>
>   • Calculate how much stock solution and how much water is required to make each dilution in the table. Use Formula 4.3, the $C_1V_1 = C_2V_2$ formula. The first two rows of Table 4.3 are filled in as examples.
>
>   • Your instructor may have you cut out the completed table and attach it to your lab notebook.

LABORATORY PROCEDURE

Part A: Dilute a Stock Solution to Make Standards

- If sufficient materials are available, everyone should prepare all of his or her own dilutions.
- You will be given a stock solution of McCormick red food coloring at a concentration of 25,000 ppm. This dye is concentrated and must be substantially diluted before its absorbance and transmittance are measurable by a spectrophotometer.
- Red food coloring in the concentration range of at least 1–20 ppm should yield absorbance readings that are in the linear range of the spectrophotometer. Newer spectrophotometers should provide linear readings at concentrations up to 20 ppm or even higher.
- There are many dilution strategies. The best strategy depends on the volumes you need, the glassware you have available, and the pipettes you are using. One strategy is outlined in this exercise, but you are welcome to devise your own strategy to best meet the circumstances in your laboratory.

    A.1.   Dilute the stock to make standards. Follow Table 4.3 and remember to mix well at each step.

        A.1.1.   Observe from the second column in Table 4.3 that you will need to obtain two beakers or flasks to hold 40 mL each and 10 test tubes (or small beakers or flasks) to hold 10 mL each. If you use test tubes, place them in a test-tube rack. Obtain an additional container for the blank (in this case, water).

        A.1.2.   Label glassware/tubes with labeling tape and a laboratory marker.

        A.1.3.   Prepare 10 ml of 1000 ppm stock from the original stock solution.

            A.1.3.1.   Measure 9.6 mL of purified water into its labeled container.

            A.1.3.2.   Pipette 0.4 mL of red food coloring stock solution using a measuring pipette or a micropipette. Rinse the food coloring out of the pipette or the tip with the dispensed water. The original food coloring stock is very viscous, so you will need to rinse several times.

            A.1.3.3.   You should now have a total volume of 10 mL in your container: 0.4 mL of red food coloring stock and 9.6 mL of water.

            A.1.3.4.   Mix thoroughly either with a vortex mixer, as directed by your instructor, or by tightly covering the container with Parafilm and inverting it five to ten times.

TABLE 4.3   Preparing Standards from a 25,000 ppm Stock

Final Concentration (C_2)	Final Volume (V_2)	Concentration of Stock Solution (C_1)	Volume of Stock Solution Needed (V_1)	Volume Water Required
1000 ppm stock	10 mL	25,000 ppm stock	$C_1V_1 = C_2V_2$ 25,000 ppm (?) = 1000 ppm (10 mL) ? = 0.4 mL	10 mL – 0.4 mL = 9.6 mL = volume of water needed (Thoroughly mix 0.4 mL of 25,000 ppm stock with 9.6 mL of purified water.)
100 ppm stock	40 mL	1000 ppm	$C_1V_1 = C_2V_2$ 1000 ppm (?) = 100 ppm (40 mL) ? = 4 mL	36 mL
10 ppm stock	40 mL	100 ppm		
100 ppm standard	10 mL	100 ppm		
75 ppm standard	10 mL	100 ppm		
50 ppm standard	10 mL	100 ppm		
25 ppm standard	10 mL	100 ppm		
20 ppm standard	10 mL	100 ppm		
10 ppm standard	10 mL	10 ppm		
8 ppm standard	10 mL	10 ppm		
6 ppm standard	10 mL	10 ppm		
4 ppm standard	10 mL	10 ppm		
2 ppm standard	10 mL	10 ppm		

A.1.4.   Using your 1000 ppm stock, prepare 40 mL of 100 ppm stock. As shown in Table 4.3, combine 4 mL of the 1000 ppm stock solution that you just made with 36 mL of water. Mix thoroughly.

A.1.5.   Using your 100 ppm stock, prepare 40 mL of 10 ppm stock.

A.1.6.   Use your 10 ppm, 100 ppm, and 1000 ppm stocks to prepare 10 mL of standards at concentrations of 2 ppm, 4 ppm, 6 ppm, 8 ppm, 10 ppm, 20 ppm, 25 ppm, 50 ppm, 75 ppm, and 100 ppm.

A.2.   Check your work: All the test tubes should have the same volume. You should be able to see the color in the standard tubes get progressively lighter as the concentration decreases. If not, something is wrong.

Part B: Prepare a Standard Curve

B.1.   Measure the absorbance for each standard (2 ppm–100 ppm) at 517 nm.

B.2.   **Data analysis**

B.2.1.   Graph the data with absorbance on the Y-axis and concentration on the X-axis. This is your standard curve.

B.2.1.1.   If your spectrophotometer has the capability, have it prepare the standard curve and print it out for you.

B.2.1.2. If your spectrophotometer cannot do this automatically, pre-pare your graph by hand.

B.2.2. Notice if there is a plateau in your data. If so, what standards fall in the linear range?

B.2.2.1. Remove any standards whose absorbance is so high as to be outside the linear range of the spectrophotometer because they cannot be used when making a standard curve.

B.2.2.2. Prepare another standard curve without the out-of-range standards.

Part C: Use the Standard Curve to Determine Analyte Concentration in Samples

C.1. Measure the absorbances of the samples provided by your instructor. These samples are your unknowns. The samples contain red food coloring at a con-centration that is unknown to you—although your instructor will know their concentrations.

C.2. ⊞**Data analysis**

C.2.1. Determine the concentration of red food coloring in the unknowns based on your standard curve. (See Figure 4.11a for an example.)

C.2.2. If the concentration of food coloring in an unknown is too high and does not fall in the linear range of the standard curve, dilute it and determine the absorbance. (For example, to dilute the sample so that it is 1/10 the original concentration, take 1 mL of sample and add 9 mL of water, for a total volume of 10 mL.)

C.2.3. If a sample was diluted, multiply by the reciprocal of the dilution to determine its original concentration. For example, if you diluted the sample 1/10 (e.g., 1 mL sample + 9 mL water), then multiply by 10 to determine the concentration in the undiluted sample.

C.2.4. Check the answer key provided by your instructor to find out the con-centration of red food coloring in your unknown samples. How close were your values to the expected values? Calculate percent errors for each value.

LAB MEETING/DISCUSSION QUESTIONS

1. What happens if the concentration of a standard is so high that its absorbance cannot be read accurately? What should you do if this is the case? Explain the term "linear range" in your own words.

2. Compare the class results for the unknowns.

3. Discuss your results for your unknowns. Remember, the point of preparing a standard curve is to find the concentrations of analyte in the unknowns. Compare your percent accuracy to the class average for percent accuracy. Which is closer to the correct value, your value or the class average? Explain.

Classroom Activity 10: Beer's Law and Calculating an Absorptivity Constant

OVERVIEW

This classroom activity deals with Beer's law and the calculation of absorptivity constants. You will use the same graph as you prepared in Laboratory Exercise 11 (or Laboratory Exercise 12), but you will now analyze it in more depth. This activity will also prepare you to determine the absorptivity constant for *o*-nitrophenol in Laboratory Exercise 13.

CLASSROOM ACTIVITIES

Beer's Law

1. Figure 4.12 shows standard curves for two dyes: allura red and sunset yellow. Allura red is a compound that is a major component of red food coloring. Sunset yellow is a compound that is a major component of yellow food coloring. Both standard curves were prepared at 517 nm, the same wavelength you used to prepare your standard curves in Laboratory Exercises 11 and 12. Observe that the line for allura red rises more steeply than the line for sunset yellow; in other words, the line for allura red has a steeper slope. The reason the slopes of the lines differ has to do with the chemistry of the compounds that are used in the food colorings. Allura red is a compound that has a strong intrinsic tendency to absorb light at 517 nm, while sunset yellow is a compound that does not capture as much light at this wavelength. *This intrinsic ability of a compound to absorb light at a particular wavelength* is called its **absorptivity constant**. Allura red has a higher absorptivity constant at 517 nm than does sunset yellow.

2. What would happen to the absorbance of a solution if you used a wider cuvette than normal? In this case, the light would have to travel through more solution. Intuitively, which of the following is correct?
    a. The absorbance would be the same in a wide cuvette as it is in a normal cuvette.
    b. The absorbance would be less in a wide cuvette rather than a normal cuvette.
    c. The absorbance would be greater in a wide cuvette rather than a normal cuvette.

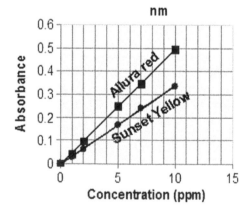

FIGURE 4.12   Comparing the intrinsic ability of allura red and sunset yellow dyes to absorb light at a wavelength of 517 nm. Allura red is used to make red food coloring. Sunset yellow is used to make yellow food coloring. Observe that at 517 nm, allura red, because of its chemistry, has more intrinsic tendency to absorb light than does sunset yellow. Therefore, the slope of the line for allura red is steeper.

3. Your answer to question 2 should be c. A wider cuvette has a longer **path length**, or *distance for the light to travel through the sample.* The longer the path length, the more of the analyte the light encounters and the greater the absorbance.

4. Two factors thus affect the steepness of the slope in your graph of absorbance versus concentration:

    a.   The absorptivity constant for the analyte, which we will abbreviate as α (the Greek letter "alpha").

    b.   The path length through the cuvette, which we will abbreviate as *b*.

5. The higher the absorptivity constant and the longer the path length, the steeper the slope of the line. The slope of the line on a standard curve can be abbreviated as *m*.

    The slope of the line is $m = \alpha b$.

6. Your graphs of absorbance versus concentration for red food coloring from Laboratory Exercises 11 and 12 should be linear. A line on a graph has an equation. Recall from algebra that the general equation for a line can be written as:

$$Y = mX + a$$

where:
    m is the slope.
    a is the Y-intercept.

7. On your graph of concentration versus absorbance, the Y-axis is **absorbance**. Therefore, substitute absorbance (A) into the equation:

$$A = mX + a$$

8. The X-axis is **concentration** of red food coloring. Therefore, substitute concentration (C) into the equation:

$$A = mC + a$$

9. Students who have questions at this point should check with their instructor before proceeding.

10. The Y-intercept is the value for Y, or absorbance, when the value for X, concentration, is zero. If the concentration of red food coloring in a sample is zero, what should the absorbance be? _____

    Your answer should be **zero** because if there is no analyte, there should be zero absorbance (relative to the blank).

    For a graph of absorbance versus concentration of any substance, the Y-intercept theoretically should be zero since, when the concentration of the analyte is zero, there is no absorbance by the analyte.

    What is the Y-intercept on your graph? Is it close to zero? (In practice, the Y-intercept often deviates slightly from zero.)

    Yes ☐   No   ☐

    If you answered "No," consult with your instructor before continuing.

11. Putting together steps 6–10, the equation for the line on your graph should be as follows:

$$A = mC + 0 \ \ \text{or} \ \ \text{simply} \ A = mC$$

In words:

$$\text{Absorbance} = \left(\text{Slope of line}\right)\left(\text{Concentration of analyte}\right)$$

12.   In steps 1–5, we talked about the slope of this line, m. We saw that two factors affect the steepness of the line: the absorptivity constant for the analyte and the path length. We rewrote the slope, *m*, as follows:

$$m = (\alpha b)$$

In words:

$$m = (\text{Absorptivity constant})(\text{Path length})$$

So, we can rewrite the equation for the line on your graph as follows:

$$A = (\alpha b)C$$

This equation, which shows the relationship between absorbance, concentration, absorptivity constant, and path length forms the basis for quantitative analysis by absorption spectrophotometry. It is generally referred to simply as Beer's law (see Formula 4.4).

Formula 4.4
Beer's Law

$$A = (\alpha b)C$$

where:
   A = absorbance
   α = absorptivity constant for the analyte at a specific wavelength
   b = path length through the cuvette
   C = concentration of analyte
   (αb) = slope of the line, m

Beer's law states that *the amount of light transmitted through a sample is reduced by three things*:
   • The concentration of absorbing substance in the sample (*C* in the equation).
   • The distance the light travels through the sample (*path length* or *b*).
   • The probability that a photon of a particular wavelength will be absorbed by the specific material in the sample (the *absorptivity* or α).

13.   You will now determine the absorptivity constant for red food coloring based on your graph of absorbance versus concentration from Laboratory Exercise 11 or 12:

   13.1.   First, determine the slope of your line on your graph of concentration versus absorbance.

The equation for the slope of a line is:

$$\frac{Y_2 - Y_1}{X_2 - X_1}$$

where:
   X_1 and X_2 = X-coordinates for any two points on the line
   Y_1 and Y_2 = corresponding Y-coordinates for the same two points
Be sure to include the units.

In this situation with red food coloring, absorbance has no units, so your units for the slope will be 1/ppm.

Figure 4.13 illustrates finding the slope of a line. (It is not data for red food coloring.)

14.   Next, determine the absorptivity constant for red food coloring at the wavelength you used based on the slope of your standard curve:

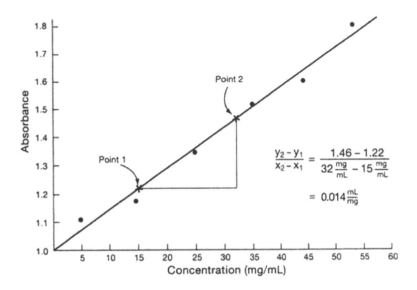

FIGURE 4.13   **An example of how to find the slope of a line on a standard curve.**

    a.   Assume the path length through your cuvette was 1 cm. (This is standard for normal cuvettes.) From Beer's law: $m = (\alpha b)$. Rearranging this equation:

$$\alpha = m/b$$

       Now you can substitute into this equation your slope and 1 cm for the path length. Solve for the absorptivity constant. _____

15.   Compare and discuss the class values for this constant. In principle, everyone in your class should have the same absorptivity constant for red food coloring because you all used the same standards to prepare your graph.

16.   For comparison with the results of your class, we have provided class data for one class:

  •   Values ranged from 0.0136 to 0.0910/ppm-cm for the absorptivity constant at 517 nm.
  •   The class average was 0.067/ppm-cm.
  •   A laboratory intern, using volumetric flasks and careful technique, obtained a value of 0.062/ppm-cm (using McCormick brand red food coloring and based on the assumption that the initial concentration of food coloring was 25,000 ppm).

Discuss your class data relative to these data.

17.   There is some indication that the dye can fade with exposure to light. How might you test this?

18.   The absorptivity constant for allura red at 517 nm in our laboratory is 0.0495/ppm-cm. Why do you think it might be lower than that of red food coloring, as shown in question 16?

Stop and Review

Make sure you understand the significance of an absorptivity constant.

Laboratory Exercise 13: Determination of the Absorptivity Constant for ONP

OVERVIEW

The purpose of this exercise is to determine the absorptivity constant for the compound o-nitrophenol (ONP) at two different pHs. This exercise is associated with Laboratory Exercise 26 in which you will use an assay to test for the activity of the enzyme ß-galactosidase. Completing this laboratory exercise will help you better understand the enzyme assay.

BACKGROUND

A compound will have a different absorptivity constant for every wavelength of light. For example, the absorptivity constant of allura red (refer to Figure 4.12) is higher at 517 nm than it is at 600 nm. This is because of the way the chemical structure of the compound interacts with light. For this reason, when an absorptivity constant is reported, the author states the wavelength at which the constant was measured.

It is less well known that pH can affect the absorptivity constant of a compound. You will explore the effect of pH on the absorptivity constant of a particular compound in this laboratory exercise in order to see this pH effect. The fact that absorptivity constants can vary with pH is of importance when you use spectrophotometry in assays (as you will in Unit VI).

In this exercise, you will prepare a series of standards at pH 7.0, and you will measure their absorbances. This is comparable to what you did in Laboratory Exercises 11 and 12, but this time, you will not be practicing with food coloring. After measuring the absorbances of the standards, you will add sodium carbonate to each one, causing their pHs to rise. You should see the color of the standards intensify at the higher pH. You will then measure the standards' absorbances again. You will graph both sets of data to determine the absorptivity constant for ONP at both pHs.

Concentration is expressed in this exercise in terms of molarity (M). *When concentration is expressed in terms of molarity, the absorptivity constant is called a* **molar absorptivity constant**. The concept of molarity is explored more fully in Unit V. For now, if molarity is unfamiliar to you, consult your instructor, but do not worry too much if the term is unclear.

Planning Your Work: Determination of the Absorptivity Constant for ONP

1. Review Classroom Activity 10 to be sure you understand the idea of an absorptivity constant.
2. Fill in Table 4.4. This table will guide you in performing dilutions of the stock solution when you go to the laboratory. You cannot begin the exercise until the table is completed.
    *   Your instructor will provide you with a stock solution of ONP at a concentration of 10 millimolar (mM).
    *   Calculate how much stock solution and how much Z buffer is required to make each dilution in the table. Use the $C_1V_1 = C_2V_2$ formula. The first two rows are filled in as examples.
    *   Your instructor may have you cut out the completed table and attach it to your lab notebook.
3. Optional: Your instructor may have you check the calculations in Table 4.5.

LABORATORY PROCEDURE

Part A: Diluting a Stock Solution to Make Standards

A.1.   Have one person check the pH of the stock solution of ONP (provided by the instructor), which is at a concentration of 10 mM in Z buffer. Everyone should record the pH, which should be close to 7 because the Z buffer is made to pH 7.0. If the pH is not close to 7, consult your instructor before proceeding.

A2.   Prepare 5 mL of the standards shown in Table 4.4, using Z buffer as the diluent.

TABLE 4.4   Calculations for Preparing Standards from a 10 mM Stock

Concentration Needed (C_2)	Volume to Make (V_2)	Concentration of Stock Solution to Use (C_1)	mL of Stock Solution Needed (V_1)	mL of Z Buffer Needed
2.0 mM	5 mL	10 mM stock	Example calculation: $$C_1V_1 = C_2V_2$$ 10 mM (?) = 2.0 mM (5 mL) ? = 1.0 mL of stock (With a micropipette, use 1000 µL.)	Combine 4.0 mL of Z buffer with 1.0 mL of the 10 mM stock of ONP (4 mL = 4000 µL)
1.0 mM	5 mL	10 mM	0.5 mL (0.5 mL = 500 µL)	4.5 mL (4.5 mL = 4500 µL)
0.80 mM	5 mL	10 mM		
0.50 mM	5 mL	10 mM		
0.20 mM	5 mL	10 mM		
0.10 mM	5 mL	10 mM		
0.05 mM	5 mL	10 mM		
0.01 mM	5 mL	10 mM		

Part B: Measure the Absorbances of All the Standards

B.1.   Measure the absorbances of the standards at 420 nm. Remember to use Z buffer as your blank. *Save each standard and the blank.*

Part C: Change the pH of the Standards by Adding Sodium Carbonate

C.1.   Add 0.5 mL of 1 M sodium carbonate to 2 mL of each standard and the blank. Mix thoroughly.

      C.1.1.   Observe the color as you add the sodium carbonate.

      C.1.2.   Determine if the color changes after you add the sodium carbonate.

C.2.   Measure the absorbances of the standards at 420 nm after adding sodium carbonate.

C.3.   Use pH paper to estimate the pH of one or more standards after adding the sodium carbonate.

D.   **Data analysis**

    D.1.   Observe that after you add the sodium carbonate to 2 mL of each standard, you have changed the concentration of the standard. Table 4.5 shows the new concentrations.

TABLE 4.5   Concentrations of Standards Before and After Adding Sodium Carbonate

Concentration Before Adding Sodium Carbonate	Concentration After Adding Sodium Carbonate
1.0 mM	0.80 mM
0.80 mM	0.64 mM
0.50 mM	0.40 mM
0.20 mM	0.16 mM
0.10 mM	0.08 mM
0.05 mM	0.04 mM
0.01 mM	0.008 mM

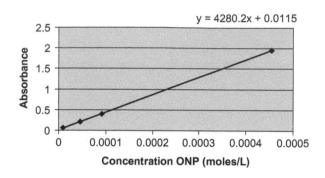

$y = 4280.2x + 0.0115$

FIGURE 4.14 Student intern data for absorptivity constant graphed with Excel. These data were obtained after the sodium carbonate was added. The pH of the standards was 9.5, as estimated with pH paper. The equation for the line is shown at the top. A 1 cm path length was used. How does this student's value for the absorptivity constant compare to your values? How does it compare to the published value?

D.2. Graph your data for concentration versus absorbance (concentration on the X-axis and absorbance on the Y-axis) before adding sodium carbonate.

D.3. Graph your data for concentration versus absorbance (concentration on the X-axis and absorbance on the Y-axis) after adding sodium carbonate. (Use the concentrations in Table 4.5.)

D.4. Calculate the absorptivity constant from both graphs to give you the absorptivity constant at both pHs. Keep track of the units as you perform your calculations.

LAB MEETING/DISCUSSION QUESTIONS

1. Put your absorptivity constants on the board. Compare the results for the entire class. What is the range of values for the absorptivity constants at both pHs?

2. What can you conclude about the effect of pH on the intrinsic ability of ONP to absorb light at 420 nm?

3. The published absorptivity constant for ONP at pH 9.0 is 4500 L/mole-1 cm at 420 nm. How do your values compare to the published value? How does the average class value compare? Observe the student data in Figure 4.14. How does the value demonstrated in this figure compare to the published value?

UNIT DISCUSSION: SPECTROPHOTOMETRY AND THE MEASUREMENT OF LIGHT

1. Electronic measuring instruments generate an electrical signal that is related to the magnitude of the property being measured. How does this idea relate to spectrophotometers? What is the property that is measured?
2. How will you ensure that all your spectrophotometric measurements in the laboratory are as accurate as possible?
3. How will you ensure that all your spectrophotometric measurements in the laboratory are as precise as possible?

Consider the overarching themes of good laboratory practices in the following table. Jot down notes on the diagram or on another piece of paper that describe how the ideas and skills that you practiced in this unit relate to these overarching themes.

Good laboratory practices
DOCUMENT
PREPARE
REDUCE VARIABILITY AND INCONSISTENCY
ANTICIPATE, PREVENT, CORRECT PROBLEMS
ANALYZE, INTERPRET RESULTS
VERIFY PROCESSES, METHODS, AND RESULTS
WORK SAFELY

Biological Solutions

DOI: 10.1201/9781003360742-5

UNIT INTRODUCTION

[Chemistry] laboratory work was my first challenge. … I still carry the scars of my first discovery—that test-tubes are fragile.

—Edward Teller, renowned physicist, with Judith L. Shoolery, *Memoirs:*
A Twentieth-Century Journey in Science and Politics **(2001), 42**

A. BIOLOGICAL SOLUTIONS ARE ESSENTIAL

Life evolved in water. Animal cells are bathed by blood and fluids, the compositions of which are rigidly controlled by the body. The organelles inside of cells similarly reside in a watery cytoplasm. Biological macromolecules (e.g., proteins, DNA, and RNA) are adapted to function in this aqueous cellular environment. It, therefore, makes sense that laboratory scientists must provide cells and biological macromolecules with a suitable aqueous environment. The preparation of aqueous solutions is the first step in nearly every biotechnology procedure. It is also one of the most critical steps; if these aqueous environments are prepared improperly, then biological systems may not function properly. Also, if these aqueous environments are prepared inconsistently, then experimental and assay data may be meaningless. Similarly, biotechnology products may be worthless if they were manufactured in an improperly prepared aqueous environment.

Chemists define a **solution** as a *homogeneous mixture in which one or more substances are dissolved in another*. **Solutes** *are the substances that are dissolved*. The **solvent** *is the substance in which the solutes are dissolved*. The solvent is almost always water in biological applications. We use the term **biological solution** *to refer broadly to the aqueous environments in which biological materials (e.g., DNA molecules, protein molecules, and intact cells) are contained in the laboratory or production facility*. The solutes will vary depending on the situation. For example, the composition of a biological solution to support the activity of an enzyme is different than the composition of a solution that permits intact, living cells to grow and reproduce. A solution in which an enzyme functions to catalyze a reaction is different than a solution used to store that enzyme.

There are, fortunately, many sources of biological solution "recipes" (procedures) for a vast number of purposes so that, in normal practice, we do not need to invent these recipes ourselves. We do, however, need to be able to find, interpret, and follow these recipes.

B. "AMOUNT" AND "CONCENTRATION" ARE NOT SYNONYMS

Amount refers to *how much of a substance is present*. **Concentration** *is a ratio where the numerator is the amount of the solute and the denominator is usually the volume (or sometimes mass) of the entire mixture* (see Figure 5.1). "Amount" and "concentration" are not synonyms in the context of solutions.

C. STRATEGY TO MAKE A SOLUTION AT A PARTICULAR PH (WITHOUT USING STOCK SOLUTIONS)

Box 5.1 describes a general strategy to make a biological solution. There are variations on this strategy, but it provides a basis for introducing solution-making procedures. Note that this procedure assumes the solution is to be buffered to a particular pH, as is commonly required for biological systems. In some cases, however, you will prepare solutions that are not buffered, and step 5 in Box 5.1 is omitted.

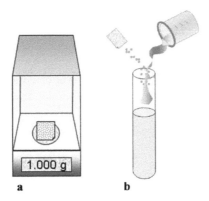

FIGURE 5.1   Amount versus concentration. (a) 1 g of solute is weighed out on a balance. 1 g is an amount. (b) The 1 g of solute is combined with water so that there is 10 mL of final volume. The concentration is now 1 g/10 mL. It would be common practice to report this concentration as 0.1 g/mL because 1 g in 10 mL of final volume is equivalent in concentration to 0.1 g in 1 mL of final volume.

TABLE 5.1   A Comparison of Two Recipes

Laboratory Recipe I	*Washing Buffer
Na_2HPO_4 6 g	50% v/v** ethanol, 125 mM NaCl, 10 mM Tris, 1 mM EDTA, pH 8.0
KH_2PO_4 3 g	
NaCl0. 5 g	
NH_4 Cl1 g	
Dissolve in water.	
Bring to a volume of 1 liter.	

Source: Rohland, Nadin, and Hofreiter, Michael "Ancient DNA Extraction from Bones and Teeth," *Nature Protocols* 2 (July 12): 1756–1762, 2007. http://www.nature.com/nprot/journal/v2/n7/abs/nprot.2007.247.html.
**See Table 5.4 for an explanation of "v/v."

STEP 1 IN BOX 5.1

Step 1 in Box 5.1 is about obtaining the required materials. Classroom Activities 11 and 12 explore this step in some detail.

STEP 2 IN BOX 5.1

Step 2 in Box 5.1 is about the solute(s). Consider Laboratory Recipe 1 that is shown in Table 5.1. This recipe is easy to interpret because it lists the amount of each solute required. Whenever you have a recipe like this one that lists the amounts of all the solutes, you do not need to perform calculations relating to the solutes (unless you decide to make a different volume than the one described in the recipe).

The recipes for biological solutions in the professional scientific literature are usually written more succinctly than Laboratory Recipe 1 in Table 5.1. Recipes are typically written in a "shorthand" style that assumes knowledge on the part of the person following the recipe. The recipe for washing buffer in Table 5.1 is an example from the literature. Moreover, the recipes in professional literature seldom show the *amount* of each solute required; rather, they specify the *concentration* of each solute. This is because the authors do not know how much volume you plan to make, so they do not know how much of each solute you will need. You must, therefore, learn to do the following:

- Interpret succinctly written solution recipes.
- Perform calculations to determine the amount of each solute required.
- Accurately perform the recipe's instructions.

Classroom Activity 11 introduces some of the considerations in interpreting such shorthand recipes, and subsequent laboratory exercises in this unit continue to work on this skill.

STEP 3 IN BOX 5.1

Step 3 is about weighing out the solute. Laboratory biotechnologists must often weigh small amounts of compounds because they often make small volumes of solutions with low solute concentrations. When weighing out small amounts, say less than 10 g, use an analytical balance. (If you do not have access to an analytical balance, consider making extra volume of solution.) When weighing out larger quantities of solute, use a balance designed to handle a larger capacity.

Box 5.1   General Procedure to Make a Solution at a Particular pH

Step 1. Read and understand the procedure; identify and collect materials and equipment needed.

Step 2. Determine the volume of solution required and, based on that volume, calculate the amount of each solute that is required (if the amounts are not specified in the procedure).

Step 3. Weigh out the amounts of solutes required. Record information about chemicals used. (Minimally, record exact chemical name, manufacturer, manufacturer's catalog number, and lot number. You may also record date container was opened, expiration date, and formula weight.)

Step 4. Dissolve the weighed-out solutes in less than the desired final volume of purified water.

Step 5. Bring the solution to the desired pH, if pH is specified.

Step 6. Bring the solution to the desired volume (BTV) with purified water.

Step 7. Perform quality control checks.

Step 8. Put in a labeled, covered storage container.

STEP 4 IN BOX 5.1

Step 4 is about mixing the solute in less than the desired final volume of water. You may be accustomed to stirring solutions with a stir rod. Magnetic stir plates are usually preferred in the workplace. A small metal cylinder (stir bar) encased in plastic is put into the solution, which is placed on a stir plate. When turned on, the stir plate causes the stir bar to rotate, thus mixing the sample. This device has the obvious advantage that the solution can stir while you do something else. Be sure, however, that the stir bars are clean before putting them into your solution. Do not handle stir bars with your fingers as that will contaminate them (use gloves or a laboratory tissue). It is good practice to keep solutions covered during the mixing step to avoid contamination, evaporation, and spillage. This is particularly important for hazardous substances.

STEP 5 IN BOX 5.1

Biological solutions generally must be at a particular pH. Bringing the solution to this pH is an important step that is considered in Laboratory Exercise 16.

STEP 6 IN BOX 5.1

The *solution is brought up to its final volume*; we call this step **bring to volume (BTV)**. Biologists routinely use a graduated cylinder for this purpose. If, however, you are preparing standards or other solutions where the utmost accuracy is required, use volumetric flasks.

It is essential that biological solutions are prepared properly. Improperly prepared biological solutions can lead to failed experiments, incorrect test results, and improperly produced products. Methods to help ensure quality are considered throughout this unit.

Labeling requirements vary in different workplaces. In a classroom setting, be sure to minimally label your solutions with the following:

- Your name (initials may be suitable; check with your instructor).
- The name of the solution (include concentration).
- The date of preparation.
- Storage temperature (if you know).
- As discussed in Unit I, it is important to label hazards (e.g., acidic or flammable).

Unless told otherwise, use labeling tape and laboratory markers (e.g., Sharpies) to label all containers. Avoid ballpoint pens because their ink can run when exposed to moisture. Do not write directly on glassware unless it is disposable. Keep all your solutions covered.

D. WATER: THE SOLVENT

Water, the solvent, is arguably the most important ingredient in biological solutions. It is certainly the ingredient required in the largest amount. Obtaining water of an acceptable purity is, therefore, an essential task in any biotechnology facility. There actually is no such thing as "pure water" because water is an excellent solvent and readily dissolves contaminants from a wide variety of sources. Effective methods are used to remove these contaminants, but regardless of how stringently water is purified, it is never totally free of contaminants.

Different applications require different levels of water purity. Recombinant DNA techniques, for example, may often be performed in the distilled water available at the grocery store. RNA, in contrast, requires water that is much purer. Cells in culture will die if extremely small amounts of contaminants are present. Pharmaceutical manufacturing requires water specially purified for use in drugs. In most teaching laboratories, you will not have the most stringently purified water, but you are likely to have a supply that is purified to some degree. The types of water you are likely to have are the following:

- **Tap water** is *what comes out of the sink's faucet.* It is purified to a level so that it is safe to drink, but it may have many contaminants that make it unsuitable for use in the laboratory (e.g., calcium and magnesium ions). Never use tap water to make biological solutions unless a procedure explicitly calls for it.
- **Deionized water** is *water from which ionic contaminants have been removed.* Deionized water may still contain other types of contaminants and so is not highly purified. Your laboratory may have a source of "house" deionized water that is sometimes acceptable for certain procedures, but it is generally not used in the professional workplace for preparing biological solutions.
- **Distilled water** is *prepared by heating water until it vaporizes.* The water vapor rises until it reaches a condenser where cooling water lowers the temperature of the vapor and it condenses back to the liquid form. The condensed liquid flows into a collection container. Most contaminants remain behind in the original vessel when the water is vaporized. A few contaminants cannot be removed from water by distillation (e.g., dissolved carbon dioxide, chlorine, and ammonia). Distilled water is appropriate for most biological solutions, but it is becoming less common as more cost-effective purification systems have been introduced.

- **Ultrapurified water** usually refers to *water that has been purified using two or more complementary methods, each of which removes certain types of contaminants.* In most professional laboratories, ultrapurified water is used for demanding applications. You may or may not have this type in your teaching laboratory; if you do, use it to make your solutions.

Purified water must be monitored to be certain that it meets quality requirements. Resistivity is routinely monitored to test the purity of water. Ionic contaminants are easy to detect because they make water more conductive, that is, more able to allow current to flow. **Resistivity** is, therefore, a measure of water purity; the more *water resists electrical current flow*, the fewer ionic contaminants it contains. Water purification systems typically have an attached resistivity meter. A user who wants to draw water flushes the system for a minute or so, discarding the rinse water. Resistivity increases during flushing until it reaches an acceptable level. The theoretical maximum ionic purity for water is 18.3 megohm-centimeters (megohm-cm) resistivity. Seventeen megohm-cm is considered an acceptable value for highly purified water. Nonionic contaminants do not affect resistivity, so even if the resistivity of water is better than 17.0 megohm-cm, it may contain contaminants.

Note about terminology: Because distillation used to be the norm, it is common for people to speak of purified water as being "distilled" regardless of how it was actually purified. This practice is not recommended. The term "DI" water, to refer to purified water, is common but confusing because it might stand for deionized or distilled water—or neither. Deionized water is generally less pure than distilled and should not be called "distilled." In this manual, the term "purified water" is used to mean water that has been distilled or that has been run through an ultrapurification system. Because the terminology of water purification is used inconsistently, solution-preparation procedures should explicitly state the source of water required.

E. PUTTING IT ALL TOGETHER IN THE LABORATORY

This unit introduces biological solution preparation, beginning with simple solutions and moving into more complex ones by the end of the unit.

- Classroom Activities 11 and 12 and Laboratory Exercise 14 introduce the preparation of four relatively simple solutions; each has only one solute and the pH of each is not adjusted. In Classroom Activity 11, you will calculate how much of each chemical you will need to prepare the four solutions. In Classroom Activity 12, you will learn how to use a chemical catalog to order the chemicals required. You will prepare the four solutions in Laboratory Exercise 14.
- Laboratory Exercise 15 is a challenge to test your solution-making skills. You will prepare a series of solutions with different concentrations of NaCl (table salt), and you will use a conductivity meter to test your success in preparing these solutions.
- Once you have mastered preparing solutions with one solute, we will add another feature; many biological solutions are buffered to maintain a particular pH. Laboratory Exercise 16 explores the preparation of two common buffers: Tris and phosphate.
- Laboratory Exercise 17 introduces the preparation of solutions with more than one solute, using a strategy that does not involve stock solutions.
- Laboratory Exercise 18 introduces a second strategy for the preparation of solutions with more than one solute, that is, the use of stock solutions.
- Laboratory Exercise 19 provides you with more practice making solutions, performing calculations, and documenting your work.
- Laboratory Exercise 20 is a simulation that explores some of the issues involved in solution preparation in a company setting.

Classroom Activity 11: Getting Ready to Prepare Solutions with One Solute: Calculations

OVERVIEW

This classroom activity is the first of two that are preparation for Laboratory Exercise 14, *Preparing Solutions with One Solute*. In this activity, you will perform the calculations necessary to determine how much of each compound is required to make the solutions in Laboratory Exercise 14.

BACKGROUND

A. The Solutions to be Prepared

Table 5.2 shows the four solutions that you will prepare in Laboratory Exercise 14. Each of these four solutions contains only one solute. The concentration and the volume desired are provided in the table. The table also provides an example of how this solution might be used in a biotechnology laboratory.

Observe that the recipes in column 2 of Table 5.2 do not provide the amounts of each compound required; rather, the concentrations and volumes are given. You must calculate the amounts of each solute required. Use Table 5.3 to show these calculations and your results. If you are not certain how to perform the required calculations, this background section provides a brief review.

The concentration of solute is expressed in four ways in Table 5.2:

- The concentration of sodium chloride (NaCl) is expressed as a percent.
- The concentration of calcium chloride ($CaCl_2$) is expressed in terms of molarity.
- The concentration of magnesium chloride ($MgCl_2$) is expressed in terms of millimolarity.
- The concentration of bovine serum albumin (BSA) is expressed as a simple ratio.

B. Calculations When Concentration Is Expressed as a Weight/Volume Ratio

A straightforward way to express concentration is as a weight/volume ratio, like this:

$$\frac{2 \text{ mg NaCl}}{1 \text{ mL}}$$

This means that a milliliter of solution contains 2 mg of NaCl. The following example illustrates how to calculate the amount of solute required to make a solution of a particular concentration (strength) and volume, when concentration is expressed as a simple weight/volume ratio.

TABLE 5.2   Solutions to Prepare

Name of Solution	Recipe as Given in Technical Manual	Volume Needed	Example of a Purpose for This Solution
Saline solution	0.91% NaCl (w/v)*	50 mL	Used to maintain intact cells so they do not shrink or swell.
Calcium chloride solution	1 M	50 mL	Used in the molecular biology laboratory to help prepare bacterial cells to take up DNA from an outside source.
Magnesium (as MgCl₂) solution	15 mM	20 mL	Cofactor for an enzyme that is used to amplify DNA in the polymerase chain reaction.
BSA standard solution	1 mg/mL	15 mL	Used as a standard to establish the response of a protein assay.

*See Table 5.4 for an explanation of "w/v."

TABLE 5.3   Calculations of Amount of Solute Required

Name of Solution	Recipe as Given in Technical Manual	Volume Needed	Amount Solute Needed (Show Calculations)	Amount Needed for Whole Class
Saline solution	0.91% NaCl (w/v)	50 mL		
Calcium chloride	1 M	50 mL		
Magnesium chloride	15 mM	20 mL		
BSA standard	1 mg/mL	15 mL (Do the calculation assuming that you will make 15 mL. However, if you weigh out a bit too much or too little BSA, you can adjust the volume.)		

Example

a. How much of the enzyme proteinase K (the solute) is required to make 250 mL of a solution with a concentration of 0.1 mg/mL?

b. How would you prepare this solution?

Answer

a. Calculating the amount of solute can be accomplished using a proportion equation:

$$\frac{0.1 \text{mg}}{1 \text{mL}} = \frac{?}{250 \text{mL}} \quad \textbf{? = 25mg}$$

Thus, 25 mg of proteinase K in 250 mL is the same concentration as 0.1 mg of proteinase K in 1 mL.

This problem can also be solved using a unit-canceling approach (which is likely the method you learned in chemistry class):

$$250 \,\text{mL} \left(\frac{0.1 \,\text{mg}}{1 \,\text{mL}} \right) = \textbf{25 mg}$$

The answer, 25 mg, is an amount. The concentration of solute is 0.1 mg/1 mL.

b. Weigh out 25 mg of solute; dissolve it in less than 250 mL of purified water; BTV 250 mL using a graduated cylinder or volumetric flask. Check, cover, and label the solution.

C. Calculations When Concentration Is Expressed as a Percent

Concentration is often expressed in terms of percent where:

$$\frac{\text{The numerator is the amount of solute}}{\text{The denominator is 100 units of total solution}}$$

Technically, the units in a percent expression should be the same in the numerator and denominator (e.g., if the amount of solute is expressed in grams, then the total amount of solution should also be expressed in grams). Consider, for example, the following expression:

$$\frac{2 \text{g NaCl}}{100 \text{g total solution}} = 2\% \text{ NaCl solution} \left(\text{w/w} \right)$$

The grams in the numerator and denominator cancel. The symbol "w/w" means that there are units of weight (g) in the numerator and denominator.

Biologists, in practice, often express percent solutions with different units in the denominator and numerator. This leads to three types of percent expressions (see Table 5.4). Note that "weight per volume" is the most common way to express a percent concentration in biology manuals. If a recipe uses the term "percent" and does not specify type, assume it is a weight-per-volume percent, unless both the solvent and solute are liquids at room temperature.

Example

How would you prepare 500 mL of a 5% (w/v) solution of NaCl?

Answer

1. Calculate the amount of solute required:

   The percent strength needed is 5%, and the volume needed is 500 mL.

   Expressed as a fraction, $5\% = \dfrac{5\ g}{100\ mL}$

   Determine the amount of solute required:

   Proportion method:

   $$\frac{5\ g}{100\ mL} = \frac{?}{500\ mL} \qquad ? = \mathbf{25\ g} = \text{amount of solute (NaCl) needed}$$

   Unit-canceling method:

   $$500\ mL\left(\frac{5\ g}{100\ mL}\right) = \mathbf{25\ g} = \text{amount of NaCl needed}$$

2. Weigh out 25 g of NaCl. Dissolve it in less than 500 mL of water.

3. In a graduated cylinder or volumetric flask, BTV 500 mL.

D. Calculations When Concentration Is Expressed in Terms of Molarity

Recall from chemistry class that a **mole** *of any element contains* 6.022×10^{23} *(Avogadro's number) atoms.* Some atoms are heavier than others, so a mole of one element weighs a different amount than a mole of another element. *The weight of a mole of a given element is defined to be equal to its atomic weight in grams,* or its **gram atomic weight.***

Compounds are *composed of atoms of two or more elements bonded together.* A mole of a compound contains 6.022×10^{23} molecules of that compound. The **gram formula weight (FW)** or **gram molecular weight (MW)** of a compound is *the weight in grams of 1 mole of the compound.* The FW is calculated by adding the atomic weights of the atoms that make up the compound. (If you are not sure how this is done, consult a beginning chemistry textbook.) Observe that a mole is an amount.

Molarity is a *concentration expression that is equal to the number of moles of a solute that are dissolved per liter of solution.* Thus, by definition, a 1 molar (1 M) solution of a compound has 1 mole of that compound dissolved in 1 L of solution. For example, a 1 M solution of sodium sulfate (anhydrous) contains 142.04 g of sodium sulfate in 1 L of total solution. Molarity is used to express concentration when the number of molecules in a solution is important.

A formula can be used to calculate the amount of solute required to make a solution of a particular volume and a particular molarity. The formula is shown in Formula 5.1 followed by an example illustrating its use.

---

* Weight and mass are not synonyms, and it is correct to speak of gram molecular mass in the context of molarity. It is, however, common practice to speak of "atomic weight," "molecular weight," and "formula weight" as we do in this manual.

TABLE 5.4 **Three Types of Percent Expressions**

Weight per Volume Percent (w/v)	Volume per Volume Percent (v/v)	Weight per Weight Percent (w/w)
Grams of solute / 100 mL solution	Milliliters of solute / 100 mL solution	Grams of solute / 100 g solution
Example: Make a 20% w/v solution of NaCl. Weigh 20 g of NaCl. BTV 100 mL to get a 20% w/v solution.	Example: Make a 70% v/v solution of glycerol in water. Measure 70 mL of glycerol. BTV 100 mL to get a 70% v/v solution.	Example: Make a 20% w/w solution of NaCl. 20 g of NaCl plus 80 g of water is a 20% by mass, w/w solution. The weight of the NaCl is 20 g. The total weight of the solution = 20 g + 80 g = 100 g. 20 g/100 g = (0.2) (100%) = 20%

Formula 5.1

Calculation of How Much Solute Is Required for a Solution of a Particular Molarity and Volume

$$\text{Grams of solute required} = (\text{Grams/1 mole})(\text{Molarity})(\text{Volume})$$

Alternatively, this formula can be written:

$$\text{Grams of solute required} = (FW)(M)(V)$$

Where grams/1 mole is the FW of the solute.

Note: To use this formula, volume must be expressed in units of liters. To convert a volume expressed in milliliters to liters, simply divide the volume in milliliters by 1000.

Example

Calculate the amount of solute required to prepare 150 mL of a 0.010 M solution of $Na_2SO_4 \bullet 10H_2O$.

Answer

1. Find the solute's FW.

   [It is possible to calculate the FW of a compound by adding the atomic weights of its constituents.

   It is often more accurate (and easier) to find the FW on the chemical container's label or in the manufacturer's catalogue. The FW of $Na_2SO_4 \bullet 10H_2O$ is 322.04 g. This is the grams/mole required.]

2. The molarity required is 0.010 M.

3. Express the volume required in units of liters. In this example, 150 mL = 0.150 L.

4. Use Formula 5.1 to determine how much solute is necessary:

$$\text{Grams of solute required} = (\text{Grams/mole})(\text{Molarity})(\text{Volume})$$

$$= \left(\frac{322.04\,g}{1\,mole}\right)\left(\frac{0.010\,mole}{1L}\right)(0.150\,L) \approx \mathbf{0.4831g}$$

E. Variations on a Theme: Millimolar and Micromolar Solutions

Molecular biologists often work with solutes that are present in very small concentrations.

- A **millimole (mmol)** is *1/1000 of a mole*.
- A **micromole (μmol)** is *1/1,000,000 of a mole*.
- A **1 millimolar solution** (**1 mM**) is *1 millimole of solute dissolved in 1 L of total solution*.
- A **1 micromolar solution** (**1 μM**) is *1 μmol of solute dissolved in 1 L of total solution*.

Consider, as an example, the amount of solute required to make 1 L of NaCl solution at various molar concentrations:

- **1 M NaCl** is *1 mole/L or 58.44 g of NaCl in 1 liter of total solution.*
- **1 mM NaCl** is *1 mmole/L or 0.05844 g of NaCl in 1 liter of total solution.*
- **1 µM NaCl** is *1 µmole/L or 0.00005844 g of NaCl in 1 liter of total solution.*

F. Variations on a Theme: Hydrates

Hydrates are *compounds that contain chemically bound water*. (The bound water does not make the compounds liquid; they remain powders or granules.) The weight of the bound water is included in the hydrate's FW. Calcium chloride, for example, can be purchased either "anhydrous" (no bound water) or as a dihydrate. Anhydrous calcium chloride, $CaCl_2$, has a FW of 111.0 g. The dihydrate form, $CaCl_2 \bullet 2H_2O$, has a FW of 147.0 (111.0 g + weight of two waters, 18.0 g each). When hydrated compounds are dissolved, the water is released and becomes indistinguishable from the water that is added as solvent.

Example

A solution recipe calls for 11.1 g of anhydrous $CaCl_2$ to be dissolved in water to a final volume of 100 mL. You look in your chemical stock room and find $CaCl_2 \bullet 2H_2O$, that is, calcium chloride dihydrate (FW = 147.0). How can you prepare a solution that will have the $CaCl_2$ concentration specified in your recipe?

Answer

Step 1. This problem can be solved with proportions if we know the FW of both forms of calcium chloride. The FW of anhydrous calcium chloride is 147.0 g minus the weight of two waters, or (18.0) (2) = 36 g.

$$147.0g - 36.0g = \textbf{111.0g}$$

Step 2. Use a proportion to calculate how much dihydrate is required. You know you need 11.1 g of the form whose FW is 111.0, so how much is required of the form with the FW of 147.0?

$$\frac{11.1g}{111.0} = \frac{?}{147.0} \quad ? = \textbf{14.7g}$$

Use 14.7 g of the calcium chloride in your stockroom and BTV 100 mL.

CLASSROOM ACTIVITIES

1. Observe that Table 5.3 shows the solutions you will be preparing in Laboratory Exercise 14.
2. Fill in the blank columns with the amount of solute required for each solution and then the amount required for the whole class.

    2.1. If you need to know the FW of a compound to do this exercise, obtain it from the label on the compound's container. (Your instructor will make the compounds available to you for this purpose.)

    2.2. Your instructor may direct you to copy Table 5.3 into your laboratory notebook and complete the calculations there. If so, be sure to fill out the top of the page (date, name of exercise, purpose, etc.) *before* putting in the table of calculations. Alternatively, your instructor may direct you to write your calculations on Table 5.3, cut it out, and attach it in your laboratory notebook.

    2.3. Remember that this table is what you *plan* to prepare in the laboratory exercise that follows this activity. When you come to the laboratory, be sure to record what you actually do.

**Classroom Activity 12: Getting Ready to Prepare Solutions
with One Solute: Ordering Chemicals**

OVERVIEW

This classroom activity is the second of two that are preparation for Laboratory Exercise 14. In the previous exercise, you calculated the quantities of each chemical needed. You will now learn how to order these chemicals using a chemical catalog—although your instructor will have already ordered the compounds you require, so you do not have to wait for them to arrive. In this exercise, you will also plan your work for the upcoming laboratory exercise.

BACKGROUND

Chemical manufacturers sell the compounds that we use to make biological solutions. Chemical catalogs can seem confusing at first because you usually have to make choices when ordering chemicals. The following are some of the decisions you will need to make:

1. **Form.** Some chemicals are available both in purified solid form and as a predissolved solution at a particular concentration. The latter may be convenient for specific applications, but for general use, select the solid, purified form.

2. **Anhydrous versus hydrated.** When a chemical comes in different hydrated forms, you can usually use whichever is cheaper or is already in your stockroom. Note that some anhydrous chemicals, notably $CaCl_2$ and $MgCl_2$, are very **hygroscopic**; *they absorb moisture from the environment*. This is problematic because a hygroscopic chemical changes over time once its container is opened. Hydrated forms of these chemicals may be preferable.

3. **Cost.** Chemical costs vary because of these factors:

   • Some chemicals are more difficult to manufacture and, therefore, cost more than others. When following a recipe, use the specified chemical, not a less expensive substitution. You may, however, be able to use a less expensive grade of the same chemical.

   • Some chemicals come in different **grades** that *vary in purity*. The more contaminants the manufacturer removes, the more costly the grade of chemical. The grade you need depends on the application.

   • Some grades of chemicals are not only more highly purified, they are also more extensively tested for contaminant levels. Generally, the more testing that is performed, the more expensive the grade.

   • Although it is usually less expensive on a per-gram basis to order larger quantities at once, this may not be the best strategy. Many chemicals have a limited shelf life and should be purchased only when needed. Some chemicals are hazardous to store or dispose of, and their quantities should be strictly limited.

4. **Molecular biology grade.** Molecular biologists can often purchase compounds that have been tested for contaminants that are known to interfere with molecular biology procedures. Molecular biology grade would be relevant for biotechnology applications that involve manipulating DNA, RNA, or proteins.

5. **Other grades. Reagent grade** is *a common term for chemicals that are suitable for most laboratory work*. **USP (United States Pharmacopeia)** is *a grade acceptable for pharmaceutical use*. **ACS (American Chemical Society) grade** is *likely to be acceptable for routine biological use*. Special grades for chromatography and spectrophotometry work are available for some compounds and may be preferable for those applications.

CLASSROOM ACTIVITIES

A. Using a Chemical Supplier's Catalog

A.1. Fill in the blanks in Table 5.5 using a chemical manufacturer's catalog:

A.1.1. Your instructor may direct you to a particular manufacturer that is used by your institution. Otherwise, the Sigma-Aldrich Company catalog is comprehensive and readily available online. Type "Sigma" into a search engine to find the website. Type the name of the chemical of interest into the company's search engine.

A.1.2. Column 1 in Table 5.5: Observe that this column is filled out already.

A.1.3. Column 2: Classify each solute as one of the following types of compounds:

 • Salt   • Alcohol   • Enzyme   • Protein (other than an enzyme)

A.1.4. Column 3: Is this compound a liquid or a solid at room temperature? (For example, ethyl alcohol is a liquid and table salt is a solid.)

A.1.4.1. Note that some compounds that are normally a solid at room temperature can be purchased already dissolved in water. If a compound is normally a solid at room temperature, then select a solid form of the compound.

A.1.5. Column 4: Some compounds may be purchased in an anhydrous form or in a hydrated form, as explained in the previous text. Record if this chemical is available in more than one hydrous form.

A.1.6. Columns 5 and 6: These columns are self-explanatory.

A.1.7. Column 7: Determine the cost of a sufficient amount of the chemical for the entire class.

A.1.8. Column 8: Describe the chemical (e.g., anhydrous solid, crystalline solid, highly purified).

A.1.9. Column 9: Enter anything else about this chemical that you think should be noted.

A.2. Check whether your selections of chemicals match those that your Instructors previously purchased. If not, why not? Would your selections also work?

B. Further Preparations

B.1. Your instructor may have you prepare for Laboratory Exercise 14 by putting tables into your laboratory notebook in which you will record the following information regarding each chemical compound that you will use:

 • Name of the chemical manufacturer   • Name of the chemical
 • Catalog number   • Lot number
 • FW given on the container (If no FW is provided, check in the manufacturer's catalog.)

B.2. You may want to prepare a table to record the *actual amount of solute that you weigh out*. For example, suppose you calculate that you need to weigh out 0.5000 g of a particular compound. When you come to lab, you actually weigh out 0.5003 g and consider that to be "close enough." You can use this table to record the actual amount that you weighed out. You may want to discuss what is "close enough" with your instructor or classmates.

TABLE 5.5   Chemical Catalog Information

Column 1	Column 2	Column 3	Column 4	Column 5	Column 6	Column 7	Column 8	Column 9
Compound	Type	Liquid or solid?	Anhydrous or hydrated?	Amount needed for whole class	Manufacturer's catalog number	Cost	Grade or description	Considerations
Sodium chloride								
Calcium chloride								
Magnesium as magnesium chloride								
BSA	Protein (not enzyme)	Solid, but can be purchased in solution	NA	Assuming 20 students or pairs, less than 1 g	Sigma A2153-10 g or A7906-10 g A4503-10 g	$89.70 $80.60 $103.00 (Prices current at the time of writing.)	Lyophilized powder	BSA is isolated from a natural source, bovine serum. It, therefore, must be purified before use. All three of the grades shown on this table appear to be suitable for use as a protein standard.

Laboratory Exercise 14: Preparing Solutions with One Solute

OVERVIEW

This laboratory exercise introduces good practices for making biological solutions with one solute. In this exercise, you will test the water sources in your laboratory, make four commonly used biological solutions, and test the solutions' quality. It is assumed that you have prepared for this exercise by performing Classroom Activities 11 and 12.

BACKGROUND

Controlling the Quality of Biological Solutions

It is important to follow quality practices to ensure that biological solutions are prepared properly. Consider the following:

- **Documentation.** All steps of solution preparation require documentation that includes the procedure(s) followed; the details of preparation (weights and volumes); problems or deviations from the procedure; components of the solution, including the manufacturer's catalog and lot numbers; the names of those who prepared the solution; date of preparation; handling after preparation; equipment used; equipment maintenance and calibration records; and safety considerations. As students, you will put this documentation in your laboratory notebook. Forms are often used for this purpose in workplaces.
- **Traceability.** Traceability ensures that every component of every solution can be identified. Keep records of all compounds used to make the solution. Label the solution's storage container.
- **Standard operating procedures (SOPs).** SOPs can be used to tell everyone how to prepare a particular solution. A solution SOP will usually show any calculations required or example calculations; what type of purified water is to be used; whether the solution should be brought to a particular pH and, if so, how; whether the pH should be checked; whether the solution should be sterilized and, if so, how; and other details.

There is often some ambiguity in published solution recipes. For example, it may be unclear as to whether a specified pH refers to a particular dissolved solute or whether it is the pH of the final solution. SOPs in a particular institution, therefore, are essential to ensure consistency and clarity in solution preparation. Even in laboratories that do not adhere to a formalized quality system, it is good practice to develop standard procedures for solution preparation. For solutions that are not made routinely, record calculations and details of preparation in your laboratory notebook.

- **Instrumentation.** Balances, pH meters, and other instruments used in the preparation of a solution must be properly maintained and calibrated; maintenance and calibration records must be kept for each instrument. Record which instruments you use.
- **Stability and expiration date.** You may not know the expiration dates for your solutions in a class setting, but this information is used in many workplaces to ensure that solutions are not stored too long.
- **Testing.** How can you tell whether a solution contains the proper components at the proper concentrations? It is possible to check the pH, clarity, and volume of a solution, but these properties are not sufficient to determine if a solution has the correct constituents.

Sometimes it is necessary to check a new solution by running a procedure or a test with the old and new solutions side by side. This method of testing the quality of a new solution is ideal—although time consuming. It relies on having a batch of old solution available that is known to be of acceptable quality.

Testing the conductivity of the new solution is another method of testing its quality. Conductivity is a measure of a solution's ability to conduct an electrical charge. Salts, like those used in biological solutions, are able to carry charges. Acids and bases also ionize in water and increase the conductivity of a solution. Conductivity is, therefore, a measure that can be used as one indication of whether the concentration of solutes is correct. If a solution is made consistently, its conductivity should be consistent. An unexpected conductivity reading is a warning that there is a problem. We will use conductivity for the purpose of checking solution consistency throughout the exercises in this unit.

LABORATORY PROCEDURE
Part A: Testing Water

A.1.   Obtain samples of all types of water that are available in your laboratory.

A.2.   Check the conductivity of each type of water available, using specific directions for your meter provided by your instructor (see Figure 5.2).

    A.2.1.   It is best not to hold the probe in your hand while making a measurement; mount the probe using a ring stand and clamp.

    A.2.2.   Avoid allowing the probe to touch the bottom of the vessel, but make sure the probe is sufficiently immersed. Depending on the probe, the recommended immersion depth might be 3 or 4 cm.

    A.2.3.   Record your values in a table in your laboratory notebook.

A.3.   Use your conductivity results to decide which source of water you should use to make solutions. Why? Confirm with your instructor that you are using the proper source of water.

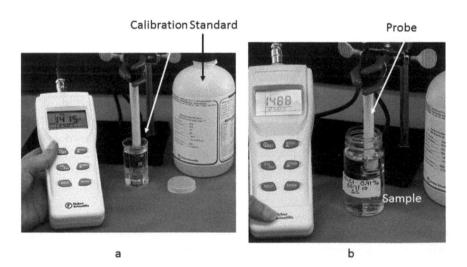

FIGURE 5.2   **Using a conductivity meter.** (a) This is a handheld meter. The probe is clamped to a ring stand. It is calibrated by placing the probe in a solution of known conductivity and then adjusting the meter to read the value of the standard. This standard is 1415 μS/cm, so the meter is adjusted to that value. (b) The probe has been rinsed and placed in the sample. The conductivity of the sample is read; it is 14.88 mS/cm. The units are not legible in this image, but were displayed by the meter. Note that the meter provided units of μS/cm for the standard and switched to units of mS/cm for the sample. Be careful to note the units provided by your meter! Note: Most meters display the units of conductivity as microsiemens/cm (μS/cm) or millisiemens/cm (mS/cm). The derivation of these units is beyond the scope of this manual, but it is not essential that you understand the origin of the units when using the meters for the purpose of quality control. It is essential, however, to record and keep track of the units.

Part B. Preparing Salt Solutions (NaCl, CaCl$_2$, and MgCl$_2$)

B.1. Find the chemical bottles for NaCl, CaCl$_2$, and MgCl$_2$. Prepare a table in your lab notebook that includes the information in Table 5.3, and record each of the following items for each solute:
   • Name of the chemical manufacturer • Name of the chemical   • FW given on the container
   • Catalog number • Lot number

B.2. Using labeling tape and a laboratory marker, label three beakers that are of a sufficient size to contain your three solutions with a little extra room. Include the following:
   • Solution name • Your name or initials • Safety information provided on the original bottle

B.3. Label storage bottles for each solution. Your label should include at least the following information:
   • Solution name • Date of preparation • Name or initials • Safety information

B.4. With an analytical balance (if available) weigh out the three salts—NaCl, CaCl$_2$, and MgCl$_2$—one at a time, and place each into its labeled beaker (see Figure 5.3a–d).

   B.4.1. For solutes where less than 1 g is to be weighed, use weighing paper or a small weigh boat. Do not use a beaker or other heavy vessel.

   B.4.2. If you use weighing paper, fold the sheet into halves or quarters so it is creased. This will help you "pour" the solute from the paper into its beaker, using the paper like a funnel.

   B.4.3. Be sure to record exactly how much solute you weighed out.

   B.4.4. Cover the beaker with Parafilm or aluminum foil.

B.5. The next step is mixing; refer to Figure 5.3e–f for an example.

   B.5.1. Beginning with the NaCl solution, add 50%–75% of the purified water required.

   B.5.2. Obtain a clean magnetic stir bar. Do not handle it with bare fingers. You may want to wash the stir bar with dishwashing detergent, followed by a complete rinse in purified water to ensure that the stir bar is clean. Place the stir bar into the beaker that contains the NaCl.

   B.5.3. Cover the beaker with Parafilm and place it on a stir plate.

   B.5.4. Turn on the magnetic stirrer. If the stir plate also has a heater, make sure the heater is off.

B.6. BTV (see Figure 5.3g–k).

   B.6.1. When the NaCl is completely dissolved, pour the solution into a 50 mL volumetric flask or a 50 mL graduated cylinder. A volumetric flask will be more accurate, but these are expensive and often are not available in the required sizes.

   B.6.2. If using a graduated cylinder, use the smallest size of graduated cylinder that will accommodate the entire solution. A 50 mL graduated cylinder is preferable for preparing 50 mL of solution, but a 100 mL cylinder can be used if necessary.

   B.6.3. Rinse the beaker and stir bar with a small volume of purified water to remove all the salt and use this rinse water to help bring the solution to its final volume.

   B.6.4. Bring the solution to its final volume with purified water, being careful not to overshoot the volume. If you do overshoot the volume, check with your instructor to determine how to proceed.

B.7. Move the solution to a labeled storage bottle. Alternatively, cover it securely with Parafilm.

B.8. Mix the CaCl$_2$ and BTV 50 mL following the procedure you used for the NaCl solution.

B.9. Mix the MgCl$_2$ and BTV 20 mL.

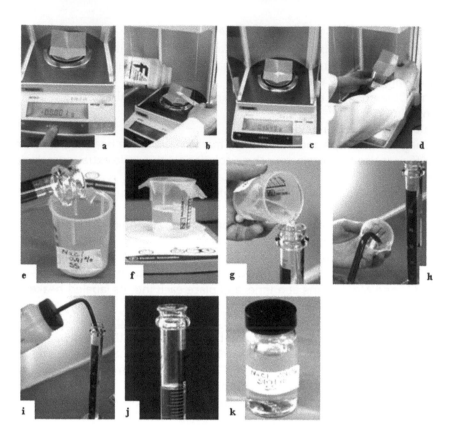

FIGURE 5.3 **Preparing 50 mL of 0.91% NaCl in water.** (a) The analyst previously turned on and warmed up the balance, made sure it was clean, and set it to 0.0000. She then placed a creased filter paper onto the balance pan. She is now taring the balance; the balance display will read 0.0000 when ready. (b) The analyst is slowly adding chemical to the paper until the display reads the desired value. If she accidentally adds a little too much NaCl, she will remove and discard it, rather than returning it to the stock bottle. (c) The final reading is made with the balance draft shields closed. The analyst is weighing out 0.4550 g of NaCl. The display reads 0.4549 g, which is acceptable because the last decimal place is assumed to be estimated. (d) The weighed-out NaCl is poured into its labeled beaker. The analyst carefully avoids spilling the solute. (e) She adds about 50% of the final volume of water required to the NaCl. (f) The analyst has placed a clean magnetic stir bar in the beaker (visible in the back of the beaker). She has placed the beaker on a stir plate. When she turns on the stir plate, the stir bar will rotate to mix the solution. (g) The NaCl has gone into solution and is ready to be brought to its final volume. Observe that the analyst has placed a second stir bar in the palm of her hand as she pours. The palmed stir bar sticks to the stir bar inside the beaker to prevent it from falling into the graduated cylinder. (h) The analyst uses a small volume of water to rinse any remaining NaCl out of the beaker. She pours this rinse water into the graduated cylinder. (i and j) The solution is brought to its final volume of 50 mL. k. The solution is stored in a labeled, covered vessel.

Part C. Preparing the BSA Solution

BSA is powdery and difficult to handle. Also, only 15 mL is required, a relatively small volume. We will demonstrate a slightly different strategy to prepare the solution of BSA than was used for the other three solutions (see Figure 5.4).

C.1. Label a tube or other small vessel with all the information required for your BSA solution.

C.2. If available, use a small, metal (to help avoid static charge) weigh boat and weigh out between 15 and 20 mg of BSA. Record the weight of the BSA.

a   b   c

FIGURE 5.4   Preparing the BSA solution: 1 mg/mL in water. (a) The analyst wants to prepare a solution of BSA that is at a concentration of 1 mg/mL, and she wants 15 mL. To make 15 mL, she needs 15 mg of BSA. However, BSA is difficult to handle, and 15 mg (0.0150 g) is hard to weigh out exactly. Therefore, she weighs out an amount that is close to 15 mg, in this case, 0.0158 g. She calculates that she needs 15.8 mL of water to get the proper concentration of BSA. (b) She dispenses 15.8 mL of water using a 25 mL serological pipette, allowing it to mix with the BSA in the weigh boat. (c) She pours the BSA solution directly into a labeled tube for storage. In this particular situation, the BSA readily dissolved and did not require mixing on a stir plate. Compare and contrast this strategy to the one in Figure 5.3.

C.3.   Calculate how much water is required to obtain a concentration of 1 mg/mL of BSA based on the weight of the powder you weighed out; consult your instructor or a colleague if you are not sure how to do this calculation.

C.4.   Add the required amount of purified water to the weigh boat. A 25 mL serological pipette, if available, can be used for this purpose. (If the weigh boat is very small, add only a portion of the water required to the weigh boat.)

C.5.   Allow the BSA to dissolve.

C.6.   Carefully pour the solution into the BSA container. Cover and mix the solution.
        Note: You are not bringing the BSA solution to volume; rather, you are adding the total volume of water to the powder. When you use this strategy, you are assuming that the powder will have little, if any, effect on the final volume. This strategy is acceptable when the concentration of the solute is very low. It is also the most practical method of preparing solutions with a small volume.

D. Quality Control Check

D.1.   Check the conductivity of each of your solutions.

D.2.   Compare your values to the expected values. There are several ways to arrive at expected values:

    D.2.1.   Your instructor may provide properly made solutions as standards. In this case, measure the conductivity of these standard solutions and call them the "expected values."

    D.2.2.   Another possibility is to use the values shown in Table 5.6. These values were obtained using a calibrated conductivity meter.

    D.2.3.   A third possibility is to use your class average as the "expected value." This has the drawback that the class may or may not be correct.

    D.2.4.   Record the expected values next to the values you obtained.

D.3.   **Data analysis**

    D.3.1.   Calculate the percent error for each of your solutions' conductivities, assuming that the "expected" value is the "true" value. (Is this a good assumption?)

LAB MEETING/DISCUSSION QUESTIONS

1. Suppose you are bringing your solution to volume and you accidentally add too much water.
   a. What should you do?
   b. Would you expect the conductivity of the solution to go up or to go down?
2. How should you decide which type of balance to use to weigh out the solutes for each of these solutions?
3. How did the conductivity of your solutions compare to the class's results? What was your percent error for each one? Is this acceptable?
4. **Work flow** is *the organization and order of one's work in the laboratory or other workplace*. Paying attention to work flow will help you to avoid making mistakes and will allow you to work more efficiently. As you become more experienced, you will organize your own work in ways that you personally find to be effective. How could you organize your work to be most efficient and accurate when making a series of solutions?
5. Put your class data for the four solutions on the blackboard. Compare your results. You can also compare your results to the class results shown in Figure 5.5.
   a. What was the range and standard deviation of values for each of the four solutions for your class? What was the range and standard deviation of values for each of the solutions for the example data in Figure 5.5?
   b. Was the conductivity of one of the solutions more variable than the others? If so, speculate as to why and what this variability might mean in terms of the quality of the solution.
   c. Our lab manager's values are shown in Table 5.6. How do her values compare with yours?
   d. What could you do in the future to increase the consistency of your solutions?

FIGURE 5.5   **Example of class results.**

TABLE 5.6   Lab Manager's Conductivity Values (Average of Three Runs)

Name of Solution	Concentration	Conductivity	Water Source	Conductivity
BSA standard	1 mg/mL	29 μS/cm	Tap water	733 μS/cm
Saline solution	0.91% NaCl (w/v)	14,600 μS/cm	Deionized water	17 μS/cm
Calcium chloride	1 M	107,300 μS/cm	Water purified with Milli-Q multistep treatment system	1.5 μS/cm
Magnesium chloride	15 mM $MgCl_2$	2,677 μS/cm		

6.   The *U.S. Code of Federal Regulations* (CFR) outlines the following requirement for companies that perform laboratory testing of potential drug compounds:

All reagents and solutions in the laboratory areas shall be labeled to indicate identity, titer or concentration, storage requirements, and expiration date. Deteriorated or outdated reagents and solutions shall not be used.

(21 CFR 58.83)

This would be difficult to achieve in a teaching laboratory (why?), but how might you meet these requirements in a workplace?

Laboratory Exercise 15: Preparing Solutions to the Correct Concentration

OVERVIEW

In this laboratory exercise, you will challenge your ability to prepare NaCl solutions to the proper concentrations. You will use conductivity as a method of evaluating the accuracy of concentration of NaCl in your preparations. Solutions with a higher concentration of salt should have a higher conductivity. Also, the relationship between concentration and conductivity should be linear (within a certain range, as determined by your conductivity meter). You will, therefore, perform the following tasks in this laboratory exercise:

- Explore the concepts of concentration and amount.
- Practice preparing simple salt solutions to a particular volume and concentration.
- Continue to explore conductivity as an indication of solution quality.

Planning Your Lab Work: Preparing Solutions to the Correct Concentration

1. What is the purpose of this exercise?
2. What information will you record in your laboratory notebook?
3. Complete the calculations in Table 5.7. Copy the information into your laboratory notebook, or attach the completed table in your notebook, as directed by your instructor. (The FW of NaCl is 58.44.)
4. Answer the following questions after completing the calculations:
   a. Which of the solutions have the same concentration?
   b. Which of the solutions have the same volume?
   c. Which of the solutions have the same amount of solute?
   d. Which solutions do you expect to have the same conductivity?
5. If your instructor tells you to do so, write a brief procedure to prepare each solution in your laboratory notebook. Remember, this is what you *plan* to do. Record what you actually do when you come to lab.
6. What numerical data will you collect?
7. How will you analyze your numerical data?
8. What will you have learned after you analyze your numerical data?

   Note: Your instructor may direct you to prepare only a subset of the solutions in Table 5.7 in order to save time and equipment.

LABORATORY PROCEDURE

It is preferable to perform this procedure individually to test your own skills, but it is likely you will need to work in pairs or groups to share balances, meters, and glassware efficiently.

1.   Prepare all the solutions in Table 5.7.
2.   Measure the conductivity of each of your solutions.
3.   **Data analysis**
    3.1.   Prepare a table in your lab notebook in which each solution is listed according to its concentration, from the lowest concentration at the top to the highest concentration at the bottom. If two solutions have the same concentration, they should be next to each other in the table. In this table, also record the conductivity of each solution.
    3.2.   Prepare three graphs of your data using Excel (or a similar program). Put NaCl concentration on the X-axis and conductivity on the Y-axis for each graph. One graph is for the solutions of lower concentration (e.g., ≤ 0.01 g/mL), and the second graph is for solutions of higher concentrations. Prepare a third graph with the values for all of the solutions. If your solutions were prepared properly, your points will fit on a straight line on the first two graphs. Depending on your conductivity meter, the third graph may not be completely linear. Evaluate your success.

TABLE 5.7   Solutions to Prepare

Solution	Concentration	Final Volume Required	Solute Amount Required
A	1.000 M	100 mL	
B	1.0 g/L	100 mL	
C	2.0 g/100 mL	100 mL	
D	3.0 g/100 mL	100 mL	
E	5.0 g/100 mL	100 mL	
F	0.10 g/mL	100 mL	
G	0.070 g/mL	100 mL	
H	0.010 g/mL	100 mL	
I	0.0010 g/mL	100 mL	
J	0.00010 g/mL	100 mL	
K	0.000010 g/mL	100 mL	
L	1.0 mg/mL	100 mL	
M	1.0%	100 mL	
N	0.10%	100 mL	
O	1.000 M	200 mL	
P	1.0 g/100 mL	200 mL	
Q	1.0%	200 mL	

3.3. Use Excel to calculate an R^2 value for each of the lines on your graphs. The R^2 value tells you how well your points form a straight line. The closer the R^2 value is to 1.0000, the better your data fits a straight line. The better your points fit a straight line, the more consistent you were in your solution preparation. Figure 5.6 illustrates an Excel graph for a student's data.

3.4. Determine the slopes of your lines. Is there a different slope for the lower concentrations and the higher ones? In principle, all three graphs should have the same slope. However, some conductivity meter probes have a narrower range than others, resulting in slightly different slopes for the three graphs.

LAB MEETING/DISCUSSION QUESTIONS

1. Put the slopes for the entire class on the board. Compare the class data.
    a. What is the range of values?
    b. Do the slopes vary between the lower concentration, higher concentration, and all concentration graphs?
    c. How much variability do you see?
    d. You were all working with the same solutions, so in principle, the results of each group should be the same.
       Note, however, that if you used more than one conductivity meter, and if the meters were not calibrated, then the slopes may be different, depending on the meter. In a workplace, the conductivity meters would be calibrated to accepted standards, much as balances are. In a classroom, you may use inexpensive conductivity meters that are not readily calibrated to accepted standards.
2. Compare your values to the class data. Did you prepare each solution properly? How can you tell?
3. Based on class data, calculate the standard deviations (SD) for the solutions that ought to have the same conductivity. (Refer to Appendix 5 for more information about standard deviation and its calculation.) If there are any outliers,

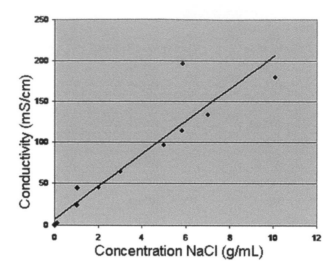

FIGURE 5.6   Student results plotted using Excel. Observe that there is one point clearly off the line at about 200 mS/cm. The student who obtained this result later reviewed her notes in her laboratory notebook and realized that she had made a simple arithmetic mistake in her calculations for that concentration. She was, therefore, able to recognize an improperly prepared solution based on its conductivity and was able to use her documentation to find the problem. Observe also that all the concentrations were plotted on one graph, so the values for the lower points are difficult to see.

FIGURE 5.7   Student results summarized as the slopes of their lines. Student groups are indicated by their initials (far left). Groups provided slopes for two graphs: One graph used data for NaCl concentrations less than or equal to 0.01 g/mL; the other had data for concentrations higher than 0.01 g/mL. The columns labeled "R^2" indicate how closely the points formed a straight line. Observe in the second column that one group had a slope of 491.698 mS/g/mL and an R^2 value of 0.76507. This group therefore recognized that there was a problem in their solution preparation and repeated the exercise, with much better results.

discuss with your teacher what to do with them. If the value for the SD is low, what does this tell you about the class's results? Conversely, what does a high value for SD tell you about the class's results?

4.   Figure 5.6 shows graphed results for one pair of students. Observe that one value is clearly unexpected; it lies significantly far from the line. What does this tell the pair about their solutions? How might such a result be handled in a workplace?

5.   Figure 5.7 summarizes a class's data by showing the slopes of the lines for each group. Observe that there was some variation in the slopes obtained by this class, particularly for lower NaCl concentrations. The slope value 491.698 mS/g/mL is particularly different than the other values.* Observe that the R^2 values are also shown on the blackboard. The R^2 value is an indication of how close all the points are to being right on the line. A perfect line, where every point is exactly on the line, has an R^2 value of 1.000. Excel will automatically calculate the R^2 value for you when you add a trend line to your data. All the R^2 values for the class data are greater than 0.9—except for the data with the aberrant value of 491.698 mS/g/mL where the R^2 was 0.76587. The group that obtained this result, therefore, repeated the activity. When they plotted their new data, they obtained a slope value of 1255 mS/g/mL and an R^2 value better than 0.99. Discuss this situation in terms of quality control. Would they have known their solutions were made improperly if they had not checked the conductivities of each salt solution?

6.   Compare your results to the sample class data in Figure 5.7.

7.   How can conductivity be used for quality control purposes in a professional laboratory setting?

---

* The units of "mS" that the students used here are equivalent to mS/cm. The students used "mS" in this case because of the way the units are expressed on the particular conductivity meter they used.

Laboratory Exercise 16: Working with Buffers

OVERVIEW

This laboratory exercise introduces the preparation of biological buffers that are used to maintain the proper pH in a biological solution. In this exercise, you will perform the following tasks:

- Prepare HCl and NaOH to use when bringing buffers to the proper pH.
- Practice proper glove use when handling hazardous chemicals.
- Practice using the $C_1V_1 = C_2V_2$ equation for dilution calculations.
- Prepare Tris buffer, which is commonly used in molecular biology.
- Prepare phosphate buffer, which is commonly used in biology.

BACKGROUND

A. Working with Buffers

The purpose of buffers is to maintain a solution at the proper pH. You will work with two commonly used buffers: Tris and phosphate. Tris [tris(hydroxymethyl)aminomethane] is commonly used because it buffers over the normal biological range (pH 7 to pH 9), is nontoxic to cells, and is relatively inexpensive. Tris is supplied as a crystalline solid. Tris that is not conjugated to any other chemical is called "Tris base." Tris base has a FW of 121.1 and a pKa of 8.0, which means that it has the greatest capacity to buffer a solution at a pH of around 8.

A 1 M solution of Tris can be prepared by dissolving 121.1 g of Tris base in 1 liter of water. However, the pH of such a solution will be more than 10. This pH is far from the pKa for Tris, and the solution will have little buffering capacity. Moreover, few biological systems function at such a high pH. Therefore, before the Tris solution is useful as a biological buffer, its pH must be lowered. There are several strategies to obtain a buffer that has both the correct pH and the correct concentration. The two strategies that we will use in this exercise are as follows:

Strategy 1. The buffer is titrated with a strong acid or base.

The buffer is dissolved in water, but it is not brought to its final volume. If the pH of the solution is lower than the desired pH, then a strong base (often NaOH) is added to raise the pH. If the pH of the solution is above the desired pH, then a strong acid (often HCl) is added to lower the pH. There are situations where the sodium from NaOH or the chloride from HCl interferes with the system of interest. In such cases, other acids or bases are substituted. Once the desired pH is reached, the solution is brought to its final volume.

This strategy is summarized in Box 5.3 and you will follow this procedure in the laboratory to prepare Tris buffer.

Strategy 2. Two stock solutions are prepared, one of which is more acidic and the other more basic.

The two stock solutions are combined in the proper proportion to get the desired pH and concentration. You will prepare phosphate buffer using this method.

Box 5.3   General Procedure to Bring a Solution to the Correct pH Using a Strong Acid or a Strong Base

Step 1. Determine the amount of solute(s) required to get the correct concentration.

Step 2. Mix the solute(s) with most, but not all, the solvent required. Do not bring the solution to volume. (To ensure consistency, an SOP will usually specify the volume of water in which to initially dissolve the solute(s). When in doubt, start with about 50% to 75% of the final volume desired.)

Step 3. Place the solution on a magnetic stir plate, add a clean stir bar, and stir (see Figure 5.8). Place pH electrodes into the solution. (Be careful not to crash the stir bar into the electrode.)

Step 4. Check the pH.

Step 5. Add a small amount of acid or base, whichever is needed to bring the solution toward the desired pH. If the recipe does not specify which acid or base to use, it is usually safe to add HCl or NaOH.

Step 6. Stir again and then check the pH.

Step 7. Repeat steps 5 and 6 until the pH is correct.

Step 8. Bring the solution to the proper volume; then recheck it and record the pH.

NOTES ABOUT ADJUSTING THE PH OF A SOLUTION

- **Temperature.** Adjust the pH of the solution when it is at the temperature at which you plan to use it. Note that some solutions change temperature during mixing.
- **Factors that can affect the temperature of a solution.** Magnetic stirring devices can generate heat. Some chemical reactions are endothermic or exothermic (use up or generate heat) and so, as chemicals are combined, their temperature may change. Restore the buffer to the correct temperature before making the final adjustment of its pH.
- **Overshooting the pH.** If you accidentally overshoot the pH, remake the solution. It is not correct to compensate by adding some extra acid or base because the additional acid or base adds extra ions to the solution; these ions will change the composition of your solution and may adversely affect it. Also, if you add more acid or base sometimes (because of a mistake), but not other times, there will be inconsistency that can cause problems that are difficult to diagnose.
- **Overshooting the volume.** If you accidentally overshoot the volume, begin again to avoid changing the concentration of solute(s) and to avoid introducing inconsistency.

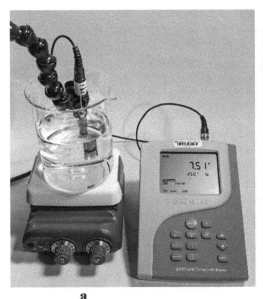
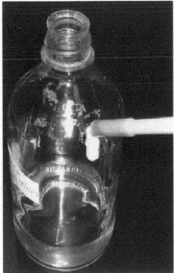

a   b

FIGURE 5.8 Checking the pH of a solution. (a) The solution is on a stir plate and has a stir bar inside it so that it can mix. After adding acid or base, allow time for the solution to thoroughly mix before checking the pH. (b) Removing a stir bar from a solution by using a magnetic stir-bar-removing rod.

Safety Briefing

Working with Acids and Bases

- Remember that HCl and NaOH can burn your skin and attack laboratory materials.

- Always handle concentrated HCl in a fume hood.

- Do not peer under a fume hood sash (window) in order to see your solution better. The sash is a shield that is there to protect you. If you lower your face to see better, then you defeat the safety purpose of the hood.

- You must wear personal protective equipment (eye protection, lab coat, and gloves) when handling acids and bases. For concentrated acids and bases, wear goggles instead of safety glasses.

- Use caution when handling diluted HCl (e.g., ≤3 M), even though it may be handled outside of the fume hood.

- Change gloves frequently—it is not enough to simply put on gloves to protect your skin from hazardous materials. Change gloves whenever they might have come in contact with a hazardous substance, such as after pouring acids or bases from one vessel to another.

- When working with concentrated acids, put the acid into water and not vice versa to avoid hazardous splashing and overheating.

- Be aware that when you dissolve NaOH, there can be significant heat generated. Check to make sure your calculations are correct, so you do not accidentally have an excessive concentration— this is a situation where it is hazardous to be 10 times off in your calculations.

- If you get acids or bases on your skin, rinse it thoroughly and immediately. For larger spills, use a safety shower and get medical assistance.

- If you get acids or bases in your eyes, use an eyewash station immediately and get medical assistance.

Planning Your Lab Work: Working with Buffers

1. Read the exercise and find places where calculations are required. Perform these calculations and record them in your laboratory notebook (unless told otherwise by your instructor).

2. Write a procedure to make each solution required (3 M HCl; 2 N NaOH; 1 M Tris, pH 8.0; and 0.1 M sodium phosphate, pH 7.0). Your instructor will tell you where to record your procedure. It is important to have your procedures written out before you attempt to make these solutions. Remember also that this is what you *plan* to do. Record what you actually do when you come to lab.

3. Plan your work with concentrated acids and bases *before* coming to the laboratory. Rehearse the steps in your mind.

Working with Tris Buffers

- Tris buffers change pH significantly with changes in temperature. Recall from Laboratory Exercise 9 that the pH of a Tris buffer will decrease approximately 0.028 pH units with every 1°C increase in temperature. Make final pH adjustments to Tris buffers when the solution is at the temperature at which it will be used.

- Recall from Laboratory Exercise 9 that the pH of buffers changes with dilution. The pH of Tris buffer decreases about 0.1 pH units for every tenfold dilution. Generally the change in pH of Tris when it is diluted is not a problem, so Tris buffers are often mixed as concentrated stock solutions. If this dilution effect is a problem, then the use of stock solutions should be avoided.

- Tris buffers require a compatible electrode for accurate pH determination.

- We will use HCl to bring Tris to the right pH, but other acids can also be used for specific applications. For example, boric acid, citric acid, and acetic acid can be used to prepare Tris-borate, Tris-citrate, and Tris-acetate buffers, respectively.

LABORATORY PROCEDURE

A. Prepare 3 M HCl from 6 M Stock

Once you begin this procedure, complete it without interruption.

A.1.   Calculate how much 6 M HCl is needed to make 100 mL of 3 M HCl.

    A.1.1.   Because 6 M HCl is a concentrated stock solution, use the $C_1V_1 = C_2V_2$ equation to determine how much of it is required. Subtract to determine how much water is needed.

A.2.   Label a storage bottle for the 3 M HCl you will prepare.

    A.2.1.   Include a hazard warning on the label.

    A.2.2.   Be sure your bottle has a cap to avoid spillage. (Do not store acids and bases using Parafilm.)

    A.2.3.   Place the bottle in the chemical fume hood.

    A.2.4.   Place a stir plate in the chemical fume hood.

A.3.   Using a graduated cylinder, measure out the required amount of water, as calculated in step A.1. Pour the water into a 250 or 400 mL beaker and place the beaker with water in the fume hood. Label the beaker to indicate that it contains water.

A.4.   Put on gloves and safety goggles. (You should already be wearing eye protection and a lab coat.)

A.5.   Label a 250 or 400 mL beaker with an acid safety warning.

A.6.   In the fume hood, pour about 50 or 60 mL of the concentrated stock solution of 6 M HCl into the labeled beaker. *Observe all the safety rules described in the briefing.* (Pour the acid into a beaker first because it is difficult to pour acid from its original stock bottle directly into a graduated cylinder or volumetric flask.)

    A.6.1.   Change gloves.

    A.6.2.   Securely close the stock bottle and put it away as directed by your instructor.

    A.6.3.   Change gloves.

A.7.   Slowly measure out the required amount of 6 M HCl by pouring it from the beaker into a graduated cylinder.

    A.7.1.   Change gloves.

A.8.   Place the beaker with water on a stir plate and turn the stirrer on to its lowest setting.

A.9.   Mix gently while slowly adding the acid to the water.

    A.9.1.   Change gloves when the acid and water have been mixed.

A.10.   Allow the acid solution to come to room temperature.

A.11.   Pour the acid solution into its labeled storage bottle.

    A.11.1.   Change gloves.

A.12.   With a paper towel, carefully remove any drips from the outsides of the storage and stock bottles and from the surface of the stir plate. If there is only a drop or two, you can discard the paper towel in ordinary trash. If there is significant moisture on the paper towel, let it dry in the fume hood.

    A.12.1.   Change gloves.

A.13.   Remove the storage bottle from the fume hood.

A.14.   Leave the beaker and graduated cylinder that had contained the acid in the hood to allow the remaining acid to evaporate. Later, rinse them with ample tap water to remove any traces of acid.

A.15.   Discard your gloves and wash your hands before documenting your work in your lab notebook.

Observe that this procedure is written to minimize the amount of handling of acid that you will perform. The acid solution is not brought to volume, but rather relies on calculated amounts of acid and water. Also, some of the acid remains behind in the beaker and graduated cylinder. This means that the final concentration of acid is not exactly 3 M. In this situation, safety considerations are more important than achieving a concentration of exactly 3 M.

B. Prepare 2 N NaOH from Pellets

Use all safety measures given in the safety briefing, but you do not need to use a fume hood to prepare this solution because it is not volatile.

B.1.   Calculate how much NaOH is required to make 100 mL of 2 N NaOH.

    B.1.1.   NaOH is a base. It comes from the manufacturer as solid pellets. It has an FW of 40.0 g. Note that for this compound, 2 N* = 2 M.

    B.1.2.   Use Formula 5.1 for the calculation.

B.2.   Label a storage bottle for the 2 N NaOH you will prepare.

    B.2.1.   Include a hazard warning on the label.

    B.2.2.   Be sure your bottle has a cap to avoid spillage.

B.3.   Put on gloves.

B.4.   Weigh out the required amount of NaOH.

    B.4.1.   Change gloves.

B.5.   Label a 250 or 500 mL beaker. Place a clean stir bar in the beaker. Add one-half to three-fourths of the required water (50–75 mL). Place the beaker on a stir plate.

B.6.   Slowly add the NaOH pellets to the beaker while gently mixing. Stir until completely dissolved.

    B.6.1.   While the solution is mixing, be sure it is covered and labeled. You do not want someone to be injured by this caustic solution.

    B.6.2.   Allow the solution to come to room temperature after it has dissolved.

    B.6.3.   Change gloves.

B.7.   Pour the solution into a graduated cylinder or volumetric flask.

    B.7.1.   Change gloves.

B.8.   Rinse the beaker and stir bar with a small volume of water and use this rinse water to help bring the solution to final volume.

    B.8.1.   Change gloves.

B.9.   Pour the NaOH into a labeled storage bottle with a cap.

    B.9.1.   Change gloves.

B.10.   Rinse the used glassware thoroughly with tap water.

B.11.   Discard your gloves and wash your hands before documenting your work in your laboratory notebook.

C. PREPARE 100 ML OF A STOCK SOLUTION OF 1 M TRIS, PH 8.0

C.1.   Calculate how much Tris base is required to make 100 mL of a 1 M solution.

    C.1.1.   The FW of Tris base is 121.14.

    C.1.2.   Use Formula 5.1 for the calculation.

    C.1.3.   Follow the directions in Box 5.3 to make the completed solution.

C.2.   Check the temperature, pH, and conductivity of your completed solution.

C.3.   Save this solution in a labeled, capped storage bottle; you will need it in Laboratory Exercise 18.

D. Prepare 200 mL of 0.1 M Sodium Phosphate, pH 7.0

D.1.   Follow the directions in Box 5.4 to prepare 200 mL of 0.1 M sodium phosphate buffer, pH 7.0.

    D.1.1.   Observe that this standard recipe has you first make 1 L each of stock solutions A and B. This is much more than you will need. Your instructor may direct you to prepare solutions A and B in a group or to prepare less of each. If you prepare less than 1 L of each solution, calculate how much solute is required.

D.2.   Check the temperature, pH, and conductivity of your solution.

D.3.   Save or dispose of this solution as directed by your instructor.

---

* "N" stands for "normal" or "normality," which is an expression of concentration. For NaOH, a 1 normal solution has the same amount of solute as a 1 molar solution. Thus, to calculate the amount of solute required to make 2 N NaOH, use the same calculation method as you would if the recipe called for 2 M NaOH. Consult a chemistry textbook if you would like to better understand this concentration expression.

Box 5.4   PREPARING 0.1 M Sodium Phosphate Buffer

Step 1. Prepare Solution A, 0.2 M $NaH_2PO_4 \bullet H_2O$ (monobasic sodium phosphate monohydrate, FW = 138.0):
Dissolve 27.6 g $NaH_2PO_4 \bullet H_2O$ in purified water and bring to volume of 1000 mL.
Step 2. Prepare Solution B, 0.2 M Na_2HPO_4 (dibasic sodium phosphate):
Use either 28.4 g Na_2HPO_4 (anhydrous dibasic sodium phosphate, FW = 142.0) or
53.6 g $Na_2HPO_4 \bullet 7H_2O$ (dibasic sodium phosphate heptahydrate, FW = 268.1)
Dissolve in purified water and bring to volume of 1000 mL.
Step 3. Combine Solution A and Solution B in the amounts listed in the table to achieve the desired pH.
Step 4. Bring the mixture of Solution A and Solution B to a final volume of 200 mL.

VOLUME SOLUTION A (mL)	VOLUME SOLUTION B (mL)	pH
93.5	6.5	5.7
92.0	8.0	5.8
90.0	10.0	5.9
87.7	12.3	6.0
85.0	15.0	6.1
81.5	18.5	6.2
77.5	22.5	6.3
73.5	26.5	6.4
68.5	31.5	6.5
62.5	37.5	6.6
56.5	43.5	6.7
51.0	49.0	6.8
45.0	55.0	6.9
39.0	61.0	7.0
33.0	67.0	7.1
28.0	72.0	7.2
23.0	77.0	7.3
19.0	81.0	7.4
16.0	84.0	7.5
13.0	87.0	7.6
10.5	89.5	7.7
8.5	91.5	7.8
7.0	93.0	7.9
5.3	94.7	8.0

(Sources: Chambers, J.A.A., and Rickwood, D. eds, Biochemistry LabFax, Bios Scientific Publishers, Oxford, UK, 1993; and Calbiochem Corporation, Buffers: A Guide for the Preparation and Use of Buffers in Biological Systems, Calbiochem, San Diego, CA, no date given.)

LAB MEETING/DISCUSSION QUESTIONS

1. Compare the conductivity of your buffers with the rest of the class. How do your values compare? What is the range of values for the entire class? What can you conclude about the consistency of your class results? Would you use these solutions in a workplace?
2. Compare your class results to those of the class in Figure 5.9. How close is your average conductivity for the Tris solution to that class's average? How does the variability in your class compare to that shown in Figure 5.9?

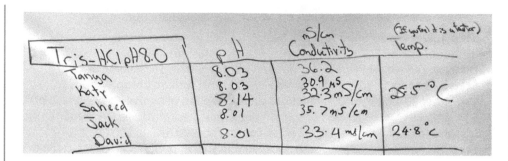

FIGURE 5.9 **Class results for Tris buffer (prepared to be 1 M, pH 8.0).**

3. How close was your phosphate buffer to pH 7.0? If your value was not exactly 7.0, do you think you should bring it to 7.0 using an acid or base?

4. You will need 1 M Tris buffer stock at pH 8.0 for a later laboratory exercise. Is your stock solution prepared properly, or do you need to make it again?

Laboratory Exercise 17: Preparing Breaking Buffer

OVERVIEW

This laboratory exercise involves preparing a solution with more than one solute. You will prepare breaking buffer, which is used in the purification of proteins from cells. You will learn to perform the calculations required to make this solution using a strategy that does not involve preparing stock solutions. In Laboratory Exercise 18, you will prepare TE buffer using another strategy that does involve stock solutions.

BACKGROUND

Most of the solutions you have prepared so far in this manual contain only one solute. In reality, most biological solutions contain more than one solute. Consider the following recipe for breaking buffer, which, as the name suggests, is used to break open cells in order to extract a protein.

Breaking Buffer

0.2 M Tris, pH 7.6 (room temperature)

0.2 M NaCl

0.01 M magnesium acetate (MgAc)

5% glycerol

This is the recipe written as it might appear in a manual. This buffer contains four solutes. The recipe lists the final concentration of each solute; Tris is present in the final solution at a concentration of 0.2 M, NaCl at a final concentration of 0.2 M, and so on. The final volume of the solution is not given because the author does not know how much you need. Water is not listed as a component, but it is assumed that the solution will be prepared in purified water.

The first instinct people sometimes have when looking at a recipe like this is to prepare separate solutions of 0.2 M Tris, 0.2 M NaCl, and so on, and then to mix them together. This instinct, however, is wrong. If you combine any volume of 0.2 M Tris with any volume of 0.2 M NaCl or any of the other solutes, they will dilute one another, giving the wrong final concentrations.

There are two correct strategies to use when preparing this solution. In this laboratory exercise, you will practice one of these strategies, as shown in Box 5.5. In Laboratory Exercise 18, you will practice the other strategy to make another biological solution.

Planning Your Lab Work: Preparing Breaking Buffer

1. What is the purpose of this exercise?

2. Prepare a table in which to record the chemicals used, along with their manufacturers, catalog numbers, and lot numbers.

3. Complete the calculations in Table 5.8. You may either copy the information into your laboratory notebook or paste the completed table into your notebook. Be sure to show your calculations.

4. Write a procedure in your laboratory notebook describing how to prepare the solution. Remember, this is what you *plan* to do. Record what you actually do when you come to lab.

5. How will you check the quality of your solution?

TABLE 5.8   Calculations of Amounts of Each Solute Required to Make 100 mL of Breaking Buffer*

Solute	Calculations for Making 100 mL
0.2 M Tris (FW Tris Base = 121.1)	
0.2 M NaCl (FW NaCl = 58.44)	
0.01 M MgAc (FW MgAc = 214.4)	
5% glycerol	

*This buffer normally contains 5 mM dithiothreitol (DTT) to help stabilize protein structure. DTT is somewhat expensive and toxic and, therefore, is not included in this laboratory exercise.

Box 5.5   Strategy to Prepare Breaking Buffer without the Use of Stock Solutions

Step 1. Decide what volume to make. (To simplify this, everyone will prepare 100 mL.)
Step 2. Calculate how much of each solute is needed to make 100 mL of breaking buffer at the specified concentration.
Step 3. Prepare 0.2 M Tris but do not bring it to the final volume.
Step 4. Adjust the pH of the Tris solution to 7.6, since that is specified in the recipe.**
Step 5. Weigh out the MgAc and NaCl and dissolve them directly in the Tris buffer.
Step 6. Glycerol comes as a liquid. Measure out the amount needed.
Step 7. BTV and record the final pH.

LABORATORY PROCEDURE

1. Weigh out the calculated amounts of Tris, NaCl, and MgAc to make 100 mL of breaking buffer.
2. Measure out the calculated amount of glycerol to make the buffer.
3. Dissolve the Tris in about 75% of the final volume of the solution, around 75 mL.
4. Adjust the pH of the Tris solution to 7.6, following the instructions in Box 5.3, but do not bring to volume yet.
5. Slowly add each of the other solutes to the Tris solution, one at a time, allowing each to completely mix.
6. When the solution is completely mixed, bring it to volume.
7. Check and document the pH and the conductivity of the solution.

LAB MEETING/DISCUSSION QUESTIONS

1. Put the pH and conductivity data for the entire class on the board. Compare the class data. How much variability do you see? You were all working with the same solution, so in principle, your values should be the same. (Note, however, that if you used more than one conductivity meter, and if the meters were not calibrated, then the conductivity values may be slightly different from meter to meter.)
2. Do you think you prepared the solution properly? How can you tell?
3. One class had the following average values for breaking buffer: pH 7.65 and conductivity, 27 mS/cm. How do your values compare? If there is variation within your class's values, or between your values and the values in this other class, how should this variation be handled?

**There is ambiguity in the recipe; some people would adjust the pH after adding all the solutes rather than before. We follow a common convention in this manual and adjust the pH of the Tris before combining all the solutes.

Laboratory Exercise 18: Preparing TE Buffer

OVERVIEW

This laboratory exercise involves preparing another solution that contains more than one solute. Tris-EDTA (TE) is a commonly used buffer consisting of Tris combined with ethylenediaminetetraacetic acid (EDTA). Tris is used to maintain a solution of DNA at the proper pH. EDTA binds divalent cations (e.g., Ca^{++} and Mg^{++}) that are cofactors for DNase. **DNase** is *an enzyme that breaks down and destroys DNA*. If the cations are bound to EDTA, then DNA in a TE solution is protected from degradation by DNase.

In this exercise, you will prepare TE buffer using a different strategy than you used to make breaking buffer in Laboratory Exercise 17. You will first prepare stock solutions, and then you will combine the stock solutions in the proper proportions to obtain a solution with each solute at the right concentration. You will therefore perform the following tasks:

- Prepare a stock solution of EDTA.
- Use the EDTA stock solution and the stock solution of Tris prepared in Laboratory Exercise 16 to make TE buffer.
- Write an SOP to make TE buffer to ensure that this solution is always made consistently.

BACKGROUND

To make TE, the Tris and EDTA are prepared individually as concentrated stock solutions that are combined together in the proper amounts. As the concentrated stocks are combined, they dilute one another to the proper final concentration. An example of the type of calculations required is shown in the example, "Preparing a Multi-Solute Buffer Using Stock Solutions" Below.

There are various reasons to use stock solutions. Stock solutions can be convenient. Powdered EDTA, for example, takes a long time to dissolve in solution. It is, therefore, advantageous to have a concentrated, dissolved stock solution of EDTA that can be used over and over again. Tris buffer is frequently used in many molecular biology solutions, and having a stock of it on hand can save time. It is not a good idea to make stocks of solutions that will be used infrequently or that are unstable, as they may degrade over time. The powdered form of a chemical is likely to be more stable than the dissolved form.

Planning Your Work: Preparing TE Buffer

1. Be sure you understand the strategy of preparing a multisolute solution by combining stock solutions. Read the following example carefully. Check with your instructor if you have questions.

2. Complete the calculations to make this solution. Prepare your own table in which you record these calculations. You may either copy the information into your laboratory notebook or attach the completed table in your notebook. Be sure to show your calculations.

3. Write a procedure to prepare the solution. Remember, this is what you *plan* to do. Record what you actually do when you come to lab. Note that EDTA takes a long time to dissolve, so plan to do something else while it is stirring.

4. How will you check the quality of your solution?

Preparing a Multisolute Buffer Using Stock Solutions

Example (You will not actually make this buffer)

How would you prepare Buffer Solution A, using stock solutions?

Buffer Solution A	
Tris buffer, pH 7.5	1 M
SDS (a detergent)	1%

ANSWER

This solution has two components: (1) Tris buffer at a pH of 7.5, and a final concentration of 1 M and (2) SDS (a detergent) at a final concentration of 1%. There are two solutes, so prepare two stock solutions, both at a higher concentration than is needed at the end.

There is no rule as to the concentration of the stock solutions. The concentration needs to be higher than the final concentration required but low enough so that all the solute will dissolve.

In this example, let's make stock solutions of 2.0 M Tris and 5% SDS.

There is also no rule as to how much volume of each stock to make.

In this example, let's make 1 L of the Tris stock and 100 mL of the SDS stock.

First, make the stock solutions.

To make 1 L of the stock of 2.0 M Tris buffer, pH 7.5, FW Tris base = 121.1, perform these steps:

Step 1. Determine how much solute is needed using Formula 5.1:

$$\text{Solute required} = (\text{Grams/mole})\ (\text{Molarity})\ (\text{Volume})$$

$$(121.1\,\text{g/mole})\ (2\,\text{mole/L})\ (1\,\text{L}) = \textbf{242.2 g}$$

Step 2. Weigh out 242.2 g of Tris base.

Step 3. Dissolve in about 750 mL of purified water.

Step 4. Bring to pH 7.5.

Step 5. BTV 1 L.

To make 100 mL of the stock solution of 5% SDS in water (w/v), perform these steps:

Step 1. Weigh out 5 g of SDS.

Step 2. BTV 100 mL.

Next, use the stock solutions to prepare buffer Solution A.

The recipe does not specify the volume, so you can make any volume you want. Let's make 100 mL. Use the $C_1V_1 = C_2V_2$ equation twice to calculate the amount of each stock solution needed; this equation was introduced in Laboratory Exercise 12.

Tris	**SDS**
$C_1V_1 = C_2V_2$	$C_1V_1 = C_2V_2$
2 M (?) = 1 M (100 mL)	5% (?) = 1% (100 mL)
? = 50 mL	? = 20 mL

Step 1. Combine 50 mL of the Tris stock and 20 mL of the SDS stock.

Step 2. BTV 100 mL.

The two stock solutions are thus combined with one another and with water in such a way that they are diluted to the desired concentration.

SCENARIO
WORKING AS A MEDIA SPECIALIST

Here's what I'd like you to do today

You are a media preparation specialist in the research and development section of Cleanest Genes Corp. You have been assigned to, first, practice making TE buffer using the recipe that is found in a professional biotechnology manual. Your supervisor then wants you to write an SOP to prepare this buffer, which will be used by others in your company. It is essential that all biological solutions are made correctly and consistently, and your SOP will help ensure that this is the case in your company.

LABORATORY PROCEDURE
Part A: Prepare 50 mL of TE Buffer

TE Buffer
10 mM Tris, pH 8.0
1 mM EDTA

A.1.   Prepare 100 mL of a stock solution of 0.5 M EDTA.
  A.1.1.   EDTA only dissolves at a pH greater than 7. Slowly adjust the pH of the EDTA stock solution to pH 8, stirring until the EDTA dissolves.
  A.1.2.   To adjust the pH, either use 2 N NaOH or add NaOH pellets directly to the solution initially, and then switch to 1 N NaOH when the pH is close to 8.0. *Be patient; this takes a while.*
Safety reminder: Do not allow the NaOH pellets to contact your skin.
A.2.   While the EDTA is going into solution, prepare 100 mL of a stock solution of 1 M Tris buffer that is pH 8.0. You may also use your Tris stock solution from Laboratory Exercise 16.
A.3.   Make 50 mL of TE buffer that has the components shown in the recipe using your stock solutions of Tris and EDTA. Keep in mind that this is a situation where you are beginning with concentrated stock solutions and diluting them to a final concentration.
  A.3.1.   Use the $C_1V_1 = C_2V_2$ equation to determine how much Tris stock is required.
  A.3.2.   Use the $C_1V_1 = C_2V_2$ equation to determine how much EDTA stock is required.
A.4.   Perform quality control tasks.
  A.4.1.   Check the pH and conductivity of the TE buffer after mixing.
  A.4.2.   Check the conductivity of your EDTA and Tris buffer stock solutions.
A.5.   **Data analysis**
  A.5.1.   Compare your pH and conductivity results to those of the class and to values for the standard provided by the instructor.
  A.5.2.   Based on your analysis, determine a range for pH and conductivity that is acceptable.

Part B: Write an SOP
B.1.   Write an SOP to make TE buffer.
B.2.   If time permits, prepare TE buffer following the SOP and evaluate the consistency of class results.

LAB MEETING/DISCUSSION QUESTIONS
1.   What are the key issues in preparing this buffer consistently and correctly?
2.   Examine your class data. Is the variability in your class results acceptable? If you wrote and followed an SOP, did the use of an SOP improve your consistency?
3.   TE is used routinely in procedures that involve dissolving and manipulating DNA. What might be some consequences of using improperly prepared solutions of TE during experiments that involve DNA?
4.   Table 5.9 includes data both from our student intern and from commercially prepared solutions.
  a.   Compare your data to that in the table. Do your results for each solution (Tris base stock, EDTA stock, and TE buffer) fit within the range of values shown in the table?

TABLE 5.9   TE Buffer Results

	Student Intern's Values		Purchased Solutions		
1 M Tris base	pH 8.0	27.8 mS/cm	**1 M Tris base** Amresco E199	pH 8.0	38.3 mS/cm
0.5 M EDTA	pH 8.0	57.6 mS/cm	**0.5 M EDTA** Amresco E177	pH 8.0	50.7 mS/cm
TE buffer bottle 1	pH 8.0	863 µS/cm tested again next day: 784 µS/cm	**1X TE buffer** Promega V6231	pH 8.0	805 µS/cm
TE buffer bottle 2	pH 8.0	768 µS/cm tested again next day: 763 µS/cm			

b.   Calculate the average class value for each solution. Do the class averages fit in the range of values shown in the table?

c.   If your values or your class's average values do not fit within this range, do you think there is a problem in the preparation of your solutions? How could you find out if your solutions are made properly or not?

5.   Think about the procedure to make this solution. Where does variability likely arise? How might this variability be reduced?

Laboratory Exercise 19: More Practice Making a Buffer

OVERVIEW

The purpose of this laboratory exercise is to give you more practice making a buffer solution. You will create a buffer recipe of your own. You will make the buffer and will document how you prepared it.

LABORATORY PROCEDURE

Work individually and create a buffer solution with concentrations of solutes that you select within the following parameters:

- Your buffer must be between pH 6.8 and 8.0, and the pH must be specified in your procedure and your recipe.
- Your buffer must contain Tris base at a concentration between 0.1 and 2.0 M. Be sure to specify the concentration in your procedure and your recipe.
- Your buffer must contain one or more salts at a concentration between 0.01 and 0.1 M. You will have $CaCl_2$, $MgCl_2$, KCl, and NaCl available. Be sure to specify the concentration(s) in your procedure and your recipe.
- Your final volume must be between 100 and 500 mL.
- Use purified water, of course.
- Determine the final pH and conductivity of your solution after it is prepared.
- Document your work completely in your laboratory notebook, including the concentration of the solution components, any calculations that you performed, and full information about each chemical.

Laboratory Exercise 20: Making a Quality Product in a Simulated Company

OVERVIEW

The purpose of this laboratory exercise is to simulate working in a biotechnology company. Companies make products, so in this laboratory exercise, you will make a product. For simplicity, your product will be phosphate buffer (which you also prepared in Laboratory Exercise 16). Thus, in this exercise you will perform the following tasks:

- Explore the roles of research and development scientists in a company.
- Explore the roles of production technicians in a company.
- Explore the roles of quality control (QC) technicians in a company.
- Consider the many factors that go into making a quality product.

BACKGROUND

This exercise is intended to introduce you to the concerns and the mindset of making a quality product in a company. As you do the exercise, you will see how complex it is to design, develop, and produce a high-quality product. In the first part of the task, you will pretend to be a research and development (R&D) team in charge of developing a standard procedure to prepare a certain material. You will then pass along the information regarding the material from R&D to production. "Changing hats," you will next be the production crew and make the material. In the third part of the exercise, you will act as the QC team and will do the quality control on the product.

Scenario

WORKING IN A START-UP BIOTECHNOLOGY COMPANY

You and a small group of colleagues have begun a new biotechnology company. For seven years, you have struggled to develop a good product that will help people and will support your company financially. You have sacrificed weekends, family time, and friendships to develop this product. Now you are working on putting together a kit that will be used in medical labs to diagnose Alzheimer's disease. This disease is common, devastating, and difficult to diagnose before the symptoms are severe. You have a good product, and you are optimistic that it will be successful.

LABORATORY PROCEDURE

Work in teams of two or three students.

Your task is to develop the procedures to make and do quality control on a 0.1 M phosphate buffer solution, pH 7.20. Imagine that this is the reaction buffer in the diagnostic kit you will sell. If this buffer is not made correctly, then the reaction that is used to diagnose Alzheimer's disease will not occur, and your kit will give a negative result whether or not the patient has the disease. Obviously, it would be disastrous if the kit failed in this way. So you must ensure that the buffer is made correctly every time. We will break this task into phases. Note that, while this is an imaginary product and imaginary company, it is true that consistently and properly made buffers are often critical components of assays (tests) used throughout the biotechnology industry.

Part A: Research and Development

Answer these questions to the best of your ability:

1. How is the buffer to be made? (The recipe is in Box 5.4.)
2. What raw materials are required? How can you be certain that the chemicals are pure and correct? How can these raw materials be tested to ensure they are right? Who should test them? When? What documentation must be associated with the raw materials?

3. What equipment will need to be specified? How should this equipment be tested to make certain it is calibrated and accurate? Who should do this testing? When should it be done? How should it be documented? What types of standards should be used for calibration?

4. Once the product is made, how can the QC team be certain it is correctly made? What are the specifications for the product? How can these specifications be tested? Who should do the testing? When? Consider the options open to you (pH, conductivity, volume, spectrophotometry, others).

5. How should the product be packaged and labeled? How can the documentation associated with the product be traced to the product?

6. What about stability? Will your product require an expiration date? How can you test its stability?

7. Is your documentation system effective enough that you could trace every piece of equipment used and every chemical used to make the product?

8. What is passed along to the production team? Consider procedures, forms, specifications, and information regarding documentation. How should the production team be trained to ensure that they make the product correctly?

After you have answered all of these questions, prepare and test the product *more than once*; develop the required forms, procedures, and documentation. Pass along all the required pieces to the production crew.

Part B: Production

You will now change roles and become the production team. If the R&D team did a good job, your tasks as the production crew should be laid out. You must follow the directions given to you to calibrate and check equipment, clean glassware, check the raw materials, and make the product. Along the way, you must complete all required documentation.

Part C: Quality Control

This phase should be laid out for you by the R&D team. Test your product to see whether it meets its specifications. Document the results.

Part D: Evaluation

You now must summarize and evaluate the project you performed. As a group, write an assessment of your success in making your product. Include at least the following items:

- A written summary of what you experienced. How did you approach this task? What strategies did you use? Divide your report into three parts: R&D, production, and QC.
- Attach your documentation to your report.
- Explain the traceability for the product, that is, how you know which raw materials and equipment were involved in production of each product.
- Include the specifications for your product.
- Include a sample of your final product along with a QC report.
- Evaluate the effectiveness of your team.

As an individual, write a separate statement about how you *felt* working on this product. For example, did you like doing the R&D or production or QC tasks? Do any of these feel like a good fit for you in your future career? How did you like working as part of a team? Try to ignore personalities here; it is always true that some people get along better than others. Try to focus on yourself and how you feel working as a team member. The purpose of this part to help you evaluate what you are looking for in your future jobs and what is comfortable to you.

UNIT DISCUSSION: BIOLOGICAL SOLUTIONS

1. Consider this situation: An R&D team in a biotechnology company is working on a new product. The product is an enzyme that will be useful for breaking down cellulose for making biofuels. The team purifies their enzyme through a series of steps. At each step they measure the activity of the enzyme using an enzyme assay. The enzyme has a pH optimum of 6.5, although it functions in a pH range from 6.0 to 7.5. They routinely perform the enzyme assay in a buffer at pH 6.5. Unbeknownst to the team, however, a technologist in the buffer preparation department has accidentally made one batch of the buffer that is pH 7.2 instead of 6.5. The R&D team uses the buffer, thinking it is pH 6.5. A few weeks later, the technologist in the buffer preparation department realizes the mistake and makes a new batch of buffer that is pH 6.5. The technologist gives the new batch to the R&D team but does not tell them that the old batch was pH 7.2.

   a. Speculate as to the consequences of this error.

   b. What should the technologist have done?

   c. Can you think of any systems the company might have put in place to prevent this sort of mistake from ever occurring?

2. Consider the overarching themes of good laboratory practices in the diagram that follows. Jot down notes on the diagram or on another piece of paper that describe how the ideas and skills that you practiced in this unit relate to these overarching themes.

Good laboratory practices
DOCUMENT
PREPARE
REDUCE VARIABILITY AND INCONSISTENCY
ANTICIPATE, PREVENT, CORRECT PROBLEMS
ANALYZE, INTERPRET RESULTS
VERIFY PROCESSES, METHODS, AND RESULTS
WORK SAFELY

Assays

UNIT

VI

DOI: 10.1201/9781003360742-6

UNIT INTRODUCTION

If we had called this "idiosyncratic Southern blot profiling," nobody would have taken a blind bit of notice. Call it "DNA fingerprinting," and the penny dropped.

—Sir Alec Jeffreys, who conceived of the assay method that uses DNA for identification of individuals, 1996, http://anil2970.tripod.com/d001.html

A. ASSAYS PROVIDE INFORMATION ABOUT SAMPLES

The products of a laboratory are knowledge, data, and information. Assays are one of the primary means by which laboratory analysts obtain data or information. We broadly define an **assay** as *any test used to analyze a characteristic of a sample, such as its composition, purity, or activity.* As is true for measurements, a "good" assay might be defined as one that provides data that can be trusted when making decisions or reaching conclusions. This unit provides an introduction to the nature of assays and how we ensure that assay results are "good."

A vast number of assays and tests are performed by biologists. Research scientists, for example, use assays to look for effects of treatments on their experimental subjects. Forensic scientists use DNA "fingerprinting" to obtain information about the identity of persons. Quality control analysts use assays to evaluate raw materials and products.

Assays can be classified as being either qualitative or quantitative (or both). **Quantitative assays** *answer this question: How much or what concentration of a substance is present in a sample?* **Qualitative assays** *answer questions about the nature or characteristics of substances in a sample.* This unit explores both types of assays.

Assays can also be classified according to what instruments are involved in their performance. For example, a number of assays take advantage of spectrophotometers to obtain qualitative and quantitative information about samples. Most of the assays in this unit are spectrophotometric.

Figure 6.1 shows the basic features of an assay. Assays are performed on samples where a **sample** is *a part of the whole that represents the whole.* When a clinical laboratory technician analyzes the level of glucose in a blood sample, for example, the sample represents all the blood in the patient. The samples in an assay undergo a procedure (method) in which a property of interest becomes detectable. Often a laboratory instrument is involved in testing the samples. Information from the samples may be compared to information from a reference standard or reference material that is well characterized with respect to the property of interest. The data collected in the assay are analyzed, displayed, and/or stored.

B. METHOD DEVELOPMENT

Analysts often use assay methods that other people have invented, developed, and optimized. If you use a previously developed method, you may not realize how much effort is required to create a "good" assay method that produces trustworthy results. First, scientists research the basic concept for the assay. The concept must involve an interaction or effect that can be detected and that is relevant to the property being analyzed. Once the concept for an assay is created, scientists must experiment with a variety of factors to optimize the assay's performance and to eliminate interferences and ambiguous results. For example, suppose a scientist is trying to discover new protein drug products to treat cancer. The scientist is searching for agents that interact with a particular molecule on the surface of cancer cells. The scientist, therefore, invents an assay that tests in the laboratory how tightly various proteins bind to the target molecule. The scientist is assuming that proteins that bind more tightly are potential drugs. In this example, the scientist will need to optimize the conditions under which the proteins and targets are allowed to bind, will need to find a way to detect and quantify the tightness of the binding, and so on.

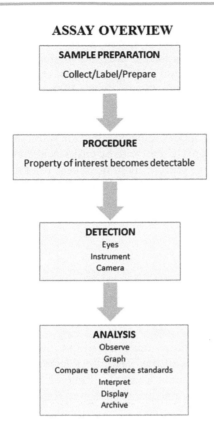

ASSAY OVERVIEW

SAMPLE PREPARATION
Collect/Label/Prepare

PROCEDURE
Property of interest becomes detectable

DETECTION
Eyes
Instrument
Camera

ANALYSIS
Observe
Graph
Compare to reference standards
Interpret
Display
Archive

FIGURE 6.1   Overview of an assay.

Even after a method has been successfully developed, each analyst who uses it still needs to ensure that the assay provides trustworthy results in his or her own laboratory with his or her specific samples and equipment. As you work through the exercises in this unit, you will not invent any new assays, but you will evaluate the accuracy and precision of existing assay methods. You will also explore assay optimization in your own laboratory.

C. "CONTROL" AND "CONTROLS"

At the heart of any quality system lies the requirement for *control*. This is true regardless of the type of the product. The product might be a tangible item, like a drug or a widget. It might also be an intangible item, such as an assay result or an experimental result that discloses a truth about the natural world. It is always necessary to control how the product is obtained in order to ensure its quality. We have already begun to explore the idea of control in this manual:

- You have seen that it is necessary to control the maintenance, calibration, and operation of measuring instruments to obtain reliable measurements with suitable accuracy and precision.
- You have seen that it is necessary to control the composition of biological solutions so they support the biological system of interest.
- Indeed, the entire basis for constructing a laboratory is to provide a space where the environment is controlled by the experimenter or analyst.

In the previous examples, the word "control" is used as a verb. "Control" can also be a noun and we will use it as such in this unit. Assays require the use of positive and negative controls; the **controls** are *known substances*. A **positive control** is *known to contain the* **analyte,** *the substance of interest.* A **negative control** is *known not to contain the analyte.* Controls are run alongside the samples of interest when an assay is performed. The responses of the samples of interest are compared to the responses of the controls. For example, in Laboratory Exercise 21, you will assay samples for the presence of starch. The assay involves adding a particular reagent to each sample and looking for a color change that occurs when starch is present and interacts with the reagent. Suppose you add the reagent to the samples and there is no color change. Does this mean the samples contain no starch? Perhaps. But perhaps the reagent was made improperly, and this is the reason no color change occurred. Thus, the samples could mistakenly appear to be negative for starch even if they do contain it. A mistake could also cause a sample to appear to be positive for starch even if the sample really contains none. Controls help to prevent these mistakes. You know there is a problem with the assay if the negative control (which contains no starch) turns color when exposed to the reagent. Similarly, you know there is a problem with the assay if the positive control (which is known to contain starch) fails to turn color when the reagent is added.

D. THE EXERCISES IN THIS UNIT

The exercises in this unit involve learning how to perform assays with an understanding of what makes them accurate, precise, and trustworthy. It is easy to perform assays in a "cook-book" way, but this may result in errors.

Laboratory Exercise 21 introduces relatively simple qualitative assays with particular attention to positive and negative controls. Laboratory Exercises 22–28 rely on a spectropho-tometer to provide information about a sample. Some of these spectrophotometric assays provide qualitative information, some quantitative, and some both.

A NOTE ABOUT TERMINOLOGY AND ASSAYS

- You will have positive and negative controls when you perform assays.
- You will sometimes test substances whose composition is known in order to practice and explore an assay. For example, in Laboratory Exercise 21, you will practice the starch assay. You will be given various substances that are known to contain or not to contain starch to learn how to perform the assay. We will simply refer to these as "known substances." These known substances are much like controls; however, you are using them for the purpose of learning how to perform the assay. Controls are used every time an assay is run, even when you are skilled.
- In some exercises your instructor will provide you with **unknowns**, where *the instructor knows what is in the substance, and your task is to find out what they contain.* These unknowns take the place of "real" samples that you would be testing in a workplace.
- We will generally follow the common practice of referring to all of these sub-stances collectively as your "samples."

Laboratory Exercise 21: Two Qualitative Assays

OVERVIEW

This laboratory exercise introduces two assays: the iodine test for starches and the biuret test for proteins. You will use these tests for *qualitative* purposes, to detect whether starch and/or proteins are present in various food-related samples.

Your instructor will provide you with positive and negative controls, a variety of food extracts of known composition, and one or two samples whose composition you will not be told; these are the "unknowns." You will use the food extracts of known composition to explore how the biuret and iodine assays work. You will then use the iodine and biuret assays to test whether or not the unknowns contain protein and/or starch. Your instructor may also have you bring in your own food samples to test. If so, you will need to consider how to best prepare your samples for testing.

In this laboratory exercise, you will thus perform the following tasks:

- Practice the general steps in performing the starch and biuret assays.
- Use positive and negative controls.
- Learn about qualitative assays.
- Explore issues relating to sample preparation (optional).

BACKGROUND

A. The Iodine and Biuret Tests

The work of cells is performed by biological macromolecules. Biological macromolecules are classified into categories: carbohydrates, proteins, nucleic acids (DNA and RNA), and lipids. Each category of macromolecule has distinguishing characteristics. Proteins, for example, are composed of amino acid subunits. Each subunit has an amino group ($-NH_2$) and a carboxyl group (-COOH). There are assays that take advantage of these distinctive features of each category of macromolecule. You will use two such assays: one to test samples for the presence of proteins and the other to test them for the presence of starch (a type of carbohydrate). You will not be analyzing the amount or concentration of either protein or carbohydrate; therefore, you are using these assays for qualitative purposes.

The biuret test detects the chemical bonds that hold together amino acids within a protein. The biuret test uses biuret reagent, which is copper sulfate ($CuSO_4$). The peptide bonds in proteins form a complex with copper in the biuret reagent to produce a violet color. This color change thus indicates the presence of proteins.

The test for starch is called the "iodine test" because it uses the reagent iodine–potassium iodide. Starches are plant-derived compounds that consist of coiled chains of glucose. The iodine in the reagent interacts with the coiled starch molecules and changes color from yellowish brown to bluish black. Glycogen, a common carbohydrate found in animals, differs slightly from starch. Glycogen does change colors with the iodine reagent but not to the same extent as starch. Individual glucose molecules do not cause a color change in the iodine reagent.

B. Sample Preparation

Sample preparation is an important step in performing assays. In this exercise, your instructor will provide you with samples that are ready for you to test without further preparation. You might, however, want to test a sample that does require additional preparation, for example, a chunk of hamburger or a piece of lettuce. If you want to test solid food items, think about how to prepare them. The iodine and biuret reagents must be able to easily interact with the molecules in the foods. This will likely require grinding or mashing the food and then causing it to be dissolved or suspended in liquid.

Planning Your Work: Simple Qualitative Assays

1. What is the purpose of this exercise?
2. What procedural information will you record in your laboratory notebook?
3. Copy or tape Table 6.1 into your laboratory notebook as directed by your instructor.
4. What data will you collect?
5. How will you analyze your data?
6. What will you have learned after you analyze your data?
7. Explain the positive and negative controls in this exercise. What are they? What is their purpose?
8. Obtain your own food samples to test if instructed to do so by your teacher.
9. Based on your knowledge of foods, predict which will contain starch and which will contain protein, then record your predictions in Table 6.1. Be careful, however, not to let your predictions affect your observations and analysis of the results.

TABLE 6.1   Biuret and Iodine Tests

Sample or Control	Predicted Result of Iodine Test	Actual Result of Iodine Test	Predicted Result of Biuret Test	Actual Result of Biuret Test
+ control for iodine test, starch extract, low concentration				
+ control for iodine test, starch extract, high concentration				
+ control for biuret test, protein extract, low concentration				
+ control for biuret test, protein extract, high concentration				
Onion juice				
Potato juice (shake well)				
Sucrose, dissolved				
Glucose, dissolved				
Water				
Carrot juice				
Apple juice				
Egg albumin, dissolved				
Honey				
Amino acids, dissolved				
Bovine serum albumin (BSA), dissolved				
Salad-oil solution				
Unknown from instructor				
Unknown from instructor				
Your food sample				
Your food sample				
Your food sample				

LABORATORY PROCEDURE

1. Obtain enough test tubes for all the controls, samples, and unknowns provided by your instructor. (If you brought in food samples of your own to test, leave them alone for now.)
2. Label each test tube using a system you understand; line up the test tubes in a rack.
3. Place 1 mL of each control, sample, or unknown into its labeled test tube.
    3.1.   Use a system to ensure that you add each substance into the correct test tube.
        3.1.1.   Move the tube back a row on the rack so you know you added sample to it.
        3.1.2.   You may also want to check off the substance when you add it to a tube.
    3.2.   Be sure to use a clean pipette or tip for every substance. Do not accidentally cross-contaminate your samples with residues of another substance.
4. Add five drops of iodine reagent to each test tube.
    4.1.   Cover the tube securely with Parafilm and mix by inverting it.
    4.2.   Alternatively, mix with a vortex mixer at a low setting.
5. Record any color changes in Table 6.1.
    5.1.   Be descriptive.
    5.2.   If possible, use a camera to better document color changes.
6. Repeat steps 1–5, but this time use biuret reagent instead of iodine reagent in step 4.
7. If you brought in your own food samples, prepare them for analysis.
    7.1.   Grind, mince, or mash the food. A mortar and pestle or blender might be used for this purpose.
    7.2.   Try to extract the food substances into water and/or a mixture of 50% water and 50% alcohol.
    7.3.   Your instructor may provide further guidance, or you may have other ideas depending on the nature of your samples.
8. Test your food samples first using the iodine test; then, in a separate tube, test them using the biuret test.
9. Test all positive and negative controls alongside your food samples. Do not omit the controls, even though you tested them earlier.
10. **Data analysis**
    10.1.   Make sure that all your controls responded as expected. If not, consult with your instructor *before* leaving the laboratory.
    10.2.   Assuming your control results are as expected, determine which unknown samples contain starch and which contain protein. Compare your answers to the answer key provided by your instructor.

LAB MEETING/DISCUSSION QUESTIONS

1. Did the controls all provide results as expected? If not, are any of your results valid?
2. Compare your sample results to the class data. Did your class obtain consistent results?
3. Did you identify the unknowns correctly?

4. Were any of your results ambiguous? If so, are your results trustworthy? Are there ways that you could modify these assays to make the results less ambiguous?
5. Speculate as to some applications of these assays.
6. Sample preparation can be one of the most important steps in an assay. If you provided your own food samples, discuss the issues associated with their preparation.
7. How accurate were your predictions of the results of these assays?
8. Do you think your predictions could have influenced your analysis of your results? It is, in fact, difficult to eliminate unconscious expectations from one's analysis of results, especially with results that can be subjective (such as color interpretation). Can you think of any strategies to eliminate the influence of unconscious expectations?

Laboratory Exercise 22: UV Spectrophotometric Assay of DNA: Quantitative Application

OVERVIEW

Biologists routinely work with DNA and have devised spectrophotometric assays for this molecule. You will explore two quantitative ultraviolet (UV) spectrophotometric assays for DNA in this laboratory exercise. In Laboratory Exercise 23, you will perform a qualitative UV spectrophotometric DNA assay.

The concentration of DNA in samples can be assayed using a standard curve, much as you did for the dye in red food coloring in Laboratory Exercise 12. In the case of DNA, we work at a wavelength of 260 nm, which is in the UV range, because DNA does not absorb visible light.

Although you will practice using a standard curve to assay DNA concentration, this method tends not to be used routinely. There is, instead, a widely used, rapid spectrophotometric method to determine the concentration of DNA in a sample. The rapid method depends on the published absorptivity constant of DNA at 260 nm. (Review Classroom Activity 10 and Laboratory Exercise 13 for a discussion of absorptivity constants.)

In this spectrophotometry exercise, you will thus perform the following tasks:

- Prepare a standard curve for DNA at 260 nm and use it to determine the concentration of DNA in "unknowns" provided by the instructor.
- Calculate the absorptivity constant for DNA at 260 nm based on your own standard curve data.
- Compare your absorptivity constant to the published value.
- Estimate the concentration of DNA in samples by using the rapid spectrophotometric method.
- Compare the accuracy and precision of the two methods of quantifying DNA concentration.

BACKGROUND

Most biological molecules do not intrinsically absorb light in the visible range, but they do absorb UV light. Biologists take advantage of UV absorbance to estimate the concentration and purity of DNA, RNA, and proteins in a sample. In this laboratory exercise, you will compare two UV methods of determining DNA concentration:

- **Method 1** *involves constructing a standard curve and using it to assay the concentration of DNA In one's samples.* Figure 6.2 shows an example of standard curve for DNA that can be used to determine the concentration of DNA in samples.
- **Method 2** is *a rapid spectrophotometric assay method based on absorptivity constants from the literature.* Based on these absorptivity constants, analysts have calculated that the following rules apply to DNA:
  - If a sample containing pure double-stranded DNA has an absorbance of 1 at 260 nm, then it contains approximately 50 μg/mL of double-stranded DNA.
  - If a sample containing pure single-stranded DNA has an absorbance of 1 at 260 nm, then it contains approximately 33 μg/mL of DNA.

Biological samples are often turbid, which can cause an unwanted increase in the absorbance of a sample. To compensate for slight turbidity, a background correction is sometimes applied. Proteins and nucleic acids do not absorb light at 320 nm. Therefore, if a sample absorbs light at 320 nm, the absorbance is assumed to be caused by turbidity. The absorbance at 320 nm can be subtracted from the readings at 260 nm. In this laboratory exercise, you will check to see if this turbidity correction improves the accuracy of your results.

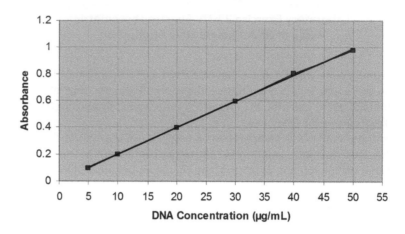

FIGURE 6.2   A standard curve for DNA prepared at 260 nm. This standard curve is consistent with the literature value, that is, an absorbance of 1 corresponds closely to 50 µg/mL of DNA. (The DNA is Sigma Catalog #D6898: deoxyribonucleic acid sodium salt from herring testes, Type XIV. Data were obtained using a calibrated Beckman DU 800 spectrophotometer.)

Planning Your Work: UV Spectrophotometry of DNA

1. Review Classroom Activity 10 (on absorptivity constants) and Laboratory Exercise 12 (using a standard curve to determine the concentration of analyte in unknowns).
2. Two assay methods will be compared in this exercise. What two methods are compared? What is their purpose?
3.
    a. Calculate the absorptivity constant for DNA based on the graph in Figure 6.2. After you complete the laboratory exercise, you will do the same calculation for your own data.
    b. Based on the data graphed in Figure 6.2, if the absorbance of a DNA sample is 0.5, what is the concentration of DNA in the sample?
4. You will be preparing standards in Part A of this laboratory exercise. Prepare a table in your laboratory notebook with your dilution calculations. You can use Table 4.3 for guidance in how to lay out such a table.
    a. Assume that you will begin with a stock solution of pure DNA at a concentration of 0.1 mg/mL.
    b. You will prepare five standards with concentrations from 5 µg/mL to 75 µg/mL.
5. If your instructor tells you to do so, write a procedure to perform this laboratory exercise in your laboratory notebook before coming to class. Remember, this is what you *plan* to do. Record what you actually do when you come to lab. Alternatively, your instructor may ask you to prepare a flowchart on a separate piece of paper that outlines the steps you will perform.
6. What numerical data will you collect?
7. How will you analyze your numerical data?
8. What will you have learned after you analyze your numerical data?

Mixing Is Critical

In Figure 6.3a, you can see my DNA standard curve when I forgot to mix my DNA as I made the standards. Oops. I repeated the standard curve, this time mixing as I prepared each standard (see Figure 6.3b). In graph

6.3a, the highest value on the Y-axis is less than 0.25 absorbance units, which was obtained with the 50 µg/mL DNA standard. In contrast, in graph 6.3b observe that the absorbance of the 50 µg/mL DNA standard was somewhat greater than 1. This means that the line on the second graph is steeper because more DNA is present. Mixing can have a dramatic effect on the results.

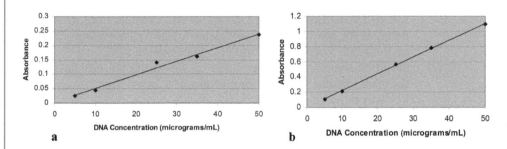

a   b

FIGURE 6.3   The effect of mixing. (a) Without mixing as standards are prepared. (b) With mixing.

LABORATORY PROCEDURE

Part A. Method 1: Using a Standard Curve to Quantify DNA

A.1.   Use the stock solution of DNA provided by your instructor to prepare five standards with concentrations from 5 to 75 µg/mL; refer to the table you prepared before coming to lab.

A.1.1.   Use water as your diluent and blank.

A.1.2.   DNA is a suspension that is unlikely to be homogeneously distributed in your stock bottle. Therefore, mix the stock solution by covering it securely and inverting it at least three or four times before withdrawing DNA. (We suggest mixing DNA suspensions by inversion because, at times, intact DNA is required. Intact DNA is damaged by rough handling as, for example, occurs during vortex mixing.)

A.1.3.   Even if you follow the calculations in your table exactly, record what you actually do in the lab.

A.2.   Mix each standard by inversion and measure its absorbance at 260 nm (A_{260}) with a UV-compatible cuvette. (We recommend quartz cuvettes and not plastic "UV compatible" ones.)

A.3.   Graph your standard curve.

A.4.   Use your standard curve to determine the concentration of DNA in one or more unknowns.

A.4.1.   Dilute any unknown whose absorbance is greater than 2 and recheck its absorbance.

A.4.2.   For diluted samples, remember to multiply the concentration times the reciprocal of the dilution. (For example, if the sample is diluted by removing 10 µL of sample and adding 90 µL of water, a 1/10 dilution, then multiply the concentration read from the standard curve by 10.)

A.5.   **Data analysis**

A.5.1.   Determine the absorptivity constant for DNA at 260 nm based on your data. Compare it to the value you calculated from Figure 6.2.

A.5.2.   Calculate the percent error for your unknowns based on the answer key provided by your instructor.

Part B: Method 2: A_{260} Quick Estimation of DNA Concentration

B.1.   Measure the A_{260} of the same unknown samples as you used in Part A.

B.1.1.   Dilute any unknown whose absorbance is greater than about 2.

B.2.   Measure the absorbance of the same unknown samples at 320 nm.

B.3.   **Data analysis**

B.3.1.   Determine the DNA concentration in your samples based on their absorbances at 260 nm. Use the published relationship that (for double-stranded DNA) an absorbance of 1 is roughly equivalent to 50 μg/mL of DNA.

B.3.2.   Subtract the absorbance at 320 nm for each sample from its absorbance at 260 nm. Determine each value for DNA concentration based on this result. Compare the results to the values in B.3.1.

B.3.3.   Based on the answer key provided by the instructor, calculate and compare the percent errors for the values obtained from:
- Your standard curve
- The A_{260} estimate
- The $A_{260} - A_{320}$ estimate

LAB MEETING/DISCUSSION QUESTIONS

1.   Compare your results with those for the rest of the class.
   a.   Compare your absorptivity constants.
   b.   Compare your values for the unknowns.
   c.   Compare your percent errors for the various methods.

2.   Discuss the accuracy and precision of the quick estimation method based on the class results.

3.   Are there positive and negative controls in these assay methods? If so, what are they?

4.   An absorptivity constant is always the same for a given substance at a particular wavelength. Thus, given a pure sample of a specific type of DNA, the constant should be the same in every laboratory. However, this assumes that every spectrophotometer is calibrated in the same way.
   a.   Are the spectrophotometers in your laboratory routinely calibrated?
   b.   Discuss the accuracy of quantitation that is based on published absorptivity constants.

Laboratory Exercise 23: UV Spectrophotometric Assay of DNA and Proteins: Qualitative Applications

OVERVIEW

In Laboratory Exercise 22, you explored the use of UV spectrophotometry to rapidly estimate the concentration of DNA in a sample. In this exercise, you will explore a rapid UV spectrophotometry method to estimate the purity of DNA in a sample; this is a qualitative assay.

Qualitative applications of spectrophotometry take advantage of spectral features of the analyte (features that you see when you plot absorbance versus wavelength). As you practice and analyze the results of this qualitative DNA assay, you will perform the following tasks:

* Prepare an absorbance spectrum for DNA and compare it to that of a protein.
* Estimate the purity of DNA samples based on a commonly used rapid assay method.
* Evaluate the usefulness of this rapid method of estimating purity.

BACKGROUND

A. Estimation of the Purity of a Nucleic Acid Sample

There are many situations where it is important to be able to estimate the purity of DNA in a preparation. Suppose, for example, a researcher wants to use enzymes to cut DNA isolated from a tissue sample. If the DNA is not sufficiently purified from the tissue, impurities may adversely affect the function of the enzymes.

Recall from Unit IV that an absorbance spectrum is a plot of an analyte's absorbance versus the wavelength of light shined on the sample. In this exercise, you will learn a widely used, rapid spectrophotometric method to estimate DNA purity. This method is based on spectral features of DNA. Figure 6.4 shows the absorbance spectrum for pure DNA. Observe that the spectrum peaks at close to 260 nm, but there is still absorbance at 280 nm. If an analyst measures the absorbance of pure, double-stranded DNA at both 260 nm and 280 nm and then takes the ratio of the two values, the result will be 1.8. Thus, for pure DNA, the ratio is expressed as

$$A_{260}/A_{280} = 1.8$$

The absorbance spectrum for proteins looks much like that of DNA, but it is different in that the absorbance peak is close to 280 nm. If an analyst takes the ratio of absorbances at 260 and 280 nm for a sample of pure protein, the result will be about 0.6. Thus, for pure protein, the ratio is expressed as

$$A_{260}/A_{280} = 0.6$$

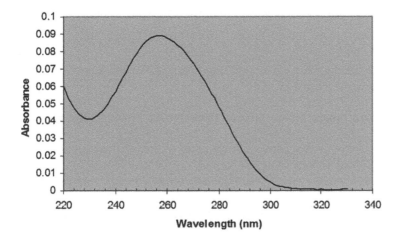

FIGURE 6.4   Spectral features of DNA.

Therefore, a ratio of 1.8 is desired when purifying DNA. A ratio less than 1.8 is often caused by protein contaminants, although other contaminants also absorb light at around 280 nm. (Phenol, which used to be routinely employed for DNA isolation, reduces the A_{260}/A_{280} ratio dramatically.) Thus, molecular biologists will often use the A_{260}/A_{280} ratio to quickly judge whether a DNA preparation is sufficiently pure for further use.

B. Potential Inaccuracies

Molecular biologists frequently use the rapid spectrophotometric methods to assess the concentration and/or purity of their DNA preparations. In fact, these calculations are performed so routinely that many spectrophotometers are designed to perform them automatically and display the resulting values. It is important, however, to be aware that values obtained by these rapid methods are only approximations and may not be particularly accurate or precise. The A_{260}/A_{280} ratio qualitative method is particularly inaccurate. The reasons for inaccuracy include the following:

- The A_{260}/A_{280} method is based on the spectral characteristics of "average" proteins and nucleic acids. In reality, proteins and nucleic acids vary from one another and so their spectra may vary from one another.
- These UV methods assume that the spectrophotometer is accurately calibrated. If the wavelength in the spectrophotometer is displaced by as little as 1 nm, its values may be significantly affected. (See, for example, Keith L. Manchester, "Value of A260/A280 Ratios for Measurement of Purity of Nucleic Acids," *BioTechniques* 19, no. 2 [1995]: 208–10.)
- These UV methods are affected by the pH and ionic strength (salt concentration) of the buffer, resulting in variability. (See, for example, Wilfinger, William W. "Effect of pH and Ionic Strength on the Spectrophotometric Assessment of Nucleic Acid Purity," *BioTechniques* 22, no. 3 [1997]: 474–80.)
- These methods assume that the spectrophotometer has a narrow spectral bandwidth, which is a characteristic typically found only in more expensive instruments (See, for example, Thermo Scientific, "Spectral Bandwidth as a Source of Error in DNA Measurements," http://www.thermo.com/com/cda/products/product_application_details/0,1063,10342,00.html.)

Planning Your Work: UV Spectrophotometry of DNA and Proteins

1. What is the purpose of this exercise?
2. What procedural information will you record in your laboratory notebook?
3. Copy or tape Table 6.2 into your laboratory notebook, as directed by your instructor. Complete the calculations in the fifth and sixth columns.
4. Consider how you will analyze your data. Use a planning process that you find useful, for example, a flowchart, diagram, or explanation.

LABORATORY PROCEDURE

Use quartz or UV-compatible cuvettes in the UV range. Handle quartz cuvettes with care; they are expensive and fragile.

1. Prepare an absorbance spectrum between 330 nm and 220 nm for DNA.
    1.1. Review Laboratory Exercise 10, if necessary.
    1.2. If your spectrum goes off scale, it means the sample is too concentrated and should be diluted.
2. Prepare an absorbance spectrum between 330 and 220 nm for a protein sample.
3. Evaluate the effect of protein impurities on the A_{260}/A_{280} ratio; refer to Table 6.2.
    3.1. You will be given solutions of DNA at a concentration of 100 µg/mL and protein at a concentration of 1 mg/mL.

TABLE 6.2   DNA Contaminated by Protein

Tube Number	μg of DNA	μL of DNA Stock (100 μg/mL)	μg of Protein	μL of Protein Stock (1 mg/mL)	μL of Water	% DNA	A_{260}	A_{280}	A_{260}/A_{280}
1	50	500	0	0	500	100			
2	50	500	5	5	495	≈91			
3	50	500	20	20	480	≈71			
4	50	500	50	50	450	50			
5	50	500	150	150	350	25			
6	50	500	250	250	250				
7	50	500	350	350	150				

3.2.   Prepare five to seven samples contaminated with increasing amounts of protein, as indicated in Table 6.2.

3.3.   Measure the A_{260} and the A_{280} for each sample.

3.4.   Calculate the A_{260}/A_{280} ratio for each sample. Record your values in Table 6.2.

4.   Prepare an absorbance spectrum between 330 nm and 220 nm for one of the DNA samples with a high percentage of protein contamination.

5.   **Data analysis**

5.1.   Graph the effect of protein contamination on the A_{260}/A_{280} ratios by plotting the percent DNA on the X-axis and the A_{260}/A_{280} ratio on the Y-axis.

5.2.   Compare the spectra of pure DNA, pure protein, and the mixture. Which features of the spectra vary? Where do they overlap?

LAB MEETING/DISCUSSION QUESTIONS

1.   Compare your results with those of the rest of the class for the three spectra, for the A_{260}/A_{280} ratio of pure DNA, and for the effect of protein contamination on the A_{260}/A_{280} ratio of DNA samples.

2.   Discuss the rapid spectrophotometric purity method based on the class results. The way people use this method is to measure the A_{260}/A_{280} ratio of a sample of their preparation. If the ratio is around 1.8 (e.g., 1.7–1.9), they assume their preparation is pure DNA. If the ratio is lower (e.g., ≤1.7), they suspect they have DNA contaminated with another substance, usually protein. Based on your results in this exercise, comment on the trustworthiness of this rapid method of assessing DNA purity.*

3.   Were there positive and negative controls when you performed this exercise? Are there positive and negative controls when this method is used routinely to estimate the purity of a DNA extract?

Advice from the senior technician

Observant students noticed that their DNA and RNA spectra were squiggly at the top (arrow in the Figure 6.5). Upon closer inspection, they noticed that the absorbance values on the Y-axis were above 4, which is clearly too high to be meaningful.

They diluted the DNA and RNA stocks 1/10, and the spectra look much better. Pay attention to odd results and look carefully at your data—see what it is telling you.

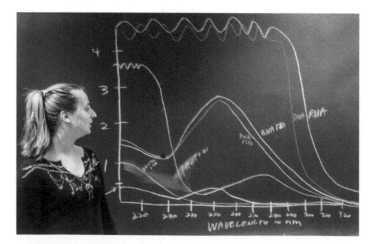

FIGURE 6.5   Sometimes dilution is required. In this case, the sample was not sufficiently diluted to be in the proper range of the spectrophotometer.

---

* Eppendorf is a company that sells a wide variety of molecular biology products. They tested the effect of protein contamination on A260/A280 ratios and published their results in *Application Note No. 279 Detection of contamination in DNA and protein samples by photometric measurements.* March 2013. https://www.eppendorf.com/product-media/doc/en/59828/Eppendorf_Detection_Application-Note_279_BioPhotometer-D30_Detection-contamination-DNA-protein-samples-photometric-measurements.pdf. Their results are consistent with the results obtained in our classes.

Laboratory Exercise 24: The Bradford Protein Assay: Learning the Assay

OVERVIEW

Scenario

USING AN ENZYME ASSAY TO TEST SAMPLES FOR QUANTITY AND PURITY

You work in the research and development department of a company that is planning to manufacture the enzyme beta-galactosidase (β-gal) for use in molecular biology procedures. Your task in the next four laboratory exercises is to learn assay methods to test both the purity of your enzyme preparations and the quantity of enzyme in them. You will use existing assay methods. Even though assays exist, however, you still need to learn the methods and verify that they produce trustworthy results in your laboratory.

The first assay you will learn as part of this scenario is the Bradford protein assay. The Bradford assay Is one of several assays commonly used to determine the concentration of protein in a sample. In Laboratory Exercise 24, you will learn how to perform the Bradford assay. In Laboratory Exercise 25, you will explore the Bradford method in more detail to get a sense of what is involved in ensuring that an assay provides trustworthy results.

Laboratory Exercise 26 introduces an enzyme assay that is specific for β-gal and does not detect other proteins. This assay allows you to quantify the amount of active enzyme in a sample. In Laboratory Exercise 27, you will use the Bradford assay together with the β-gal assay to compare the purity of two samples of your product and to analyze the amount of enzyme activity present in the samples. The flow chart in Figure 6.6 shows

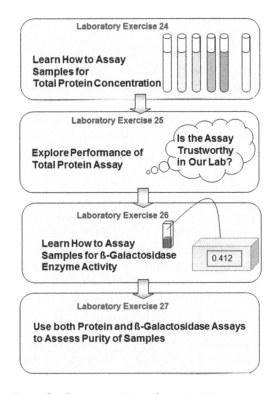

FIGURE 6.6   **Overview of Laboratory Exercises 24-27.**

how the next four laboratory exercises fit together in series as they might be used in our scenario.

As you explore the Bradford assay in this laboratory exercise, you will achieve the following objectives:

- Learn how to perform and analyze the results of a protein assay.
- Deepen your understanding of how to achieve accuracy and precision when performing assays.
- Begin to understand the overall process of purifying a product in a research and development setting.

BACKGROUND

A. Colorimetric Assays

Most biological materials, such as enzymes, DNA, and RNA, are naturally colorless (do not absorb visible light), but they do absorb UV light as you saw in Laboratory Exercises 22 and 23. UV spectrophotometric assays are useful, but they have some disadvantages. UV spectrophotometers are more expensive than those that only measure light in the visible range. Also, some UV assays are not sensitive to low concentrations of analyte. Scientists have, therefore, developed spectrophotometric assays that do not require that the analyte naturally absorbs either UV or visible light. These assays are called "colorimetric." A **colorimetric assay** is *one in which a colorless analyte is exposed to another compound(s) and/or to conditions that cause it to develop color. The amount of color that occurs is proportional to the concentration of the analyte.*

$$\text{Colorless substance} + \text{Reagent} \left(s \right) \rightarrow \text{Colored product}$$

There are hundreds of colorimetric methods that vastly expand the potential for using a spectrophotometer for qualitative and quantitative purposes. Colorimetric assays also extend the range of detection for many substances. It is often possible to detect an analyte that is present in very low concentration by using a colorimetric method.

Observe that you used colorimetric assays in Laboratory Exercise 21. You added iodine and biuret reagents to food extracts to see if the foods contained starch and/or protein. In that exercise, the assays were used qualitatively simply to detect whether or not the samples contained any starch or protein. You used your eyes as a detector. Using your eyes, you can see if color is present or not and determine, to some extent, if one sample is more intensely colored than another. But eyes alone are not suitable for a quantitative assay. This is where spectrophotometers greatly extend our sensory capabilities by measuring degrees of light intensity and color with accuracy and precision.

B. Protein Assays

Several colorimetric methods are commonly used to measure the total amount of protein in a sample. The biuret method, as you used in Laboratory Exercise 21, is one such method, but not the one you will use now. You will use another protein assay method that was developed by M.M. Bradford ("Rapid and Sensitive Method for the Quantitation of Microgram Quantities of Protein Utilizing the Principle of Protein-Dye Binding," *Analytical Biochemistry*, no. 72 [1976]: 248–54). The basis of the Bradford assay is the observation that certain amino acids, especially arginine, selectively bind to a dye, Coomassie Brilliant Blue G-250, also called "Bradford reagent." Bradford reagent changes color from reddish brown to blue, and the absorbance maximum shifts from 465 nm to 595 nm when the dye binds to these amino acids under acidic conditions. The more amino acids present, the more the color changes. This change can be quantified using a spectrophotometer set to a wavelength of 595 nm.

The Bradford assay has been packaged into easy-to-use kits by manufacturers. The company provides the dye solution at the correct pH and concentration so that it is simply pipetted into the samples. You will use such a kit purchased from Bio-Rad Laboratories.* We have slightly altered the Bio-Rad procedure for classroom use, and you will consider the consequences of these modifications in Laboratory Exercise 25.

C. Standards

When you run a protein assay, you prepare a standard curve, much like you did in Laboratory Exercise 12. Recall that spectrophotometry **standards** *contain a known amount or concentration of the analyte of interest*. The analyst determines the amount or concentration of analyte in samples by comparing their absorbance with that of the standards.

Every protein interacts in its own way with the Bradford reagent. For the utmost accuracy, therefore, you should prepare your standards from a stock solution containing your protein of interest at a known concentration. For example, in our scenario, you are interested in the enzyme beta-galactosidase. Ideally, you should have a stock solution of pure β-gal of known concentration from which to prepare standards. In practice, the protein of interest may not be available in sufficient quantity or inexpensively enough to use it over and over again to make standards. Analysts, therefore, often substitute a readily available protein, like bovine serum albumin (BSA), when making their standards. Purified β-gal is more expensive than BSA so you will routinely use BSA to make your standards. You will look at the impact of this decision in Laboratory Exercise 25.

You will be pipetting small volumes (5–20 µL) when preparing standards. You should strive to minimize pipetting errors. One method is to put the water or other diluent in the test tube first and then pipette the stock solution (in this case, BSA) carefully into the diluent, rinsing the tip by pipetting up and down several times. Discard the tip after rinsing it.

D. The Blank

A blank is required whenever using a spectrophotometer. In Unit IV, you used water as the blank because the food-coloring samples were dissolved in water. In the UV laboratory exercises earlier in this unit, you similarly used the solvent in which the analyte was dissolved as a blank. Preparing the blank for a colorimetric assay is a little more complex. Colorimetric assays involve adding various reagents to the samples and standards and perhaps heating, cooling, or otherwise manipulating them. All the reagents involved in the assay must also be added to the blank. The samples, any standards, and the blank must be handled identically when performing the assay. If, for example, the samples are subjected to heating or cooling, then the standards and the blank must also be treated in these ways, right alongside the samples. A properly prepared blank will provide a control for any spontaneous color that might arise that is unrelated to the analyte.

E. Dilution of Samples

The standards and samples in a quantitative assay must be diluted in such a way as to obtain values in the linear range of the standard curve. (There are a few spectrophotometric assays that are not linear, but we will not consider them in this manual.) The amount of dilution that the samples require can change depending on the circumstances. For example, suppose cells have been broken apart so that a protein can be purified from them. Breaking the cells is followed by a series of purification steps, resulting in protein that is progressively more pure and may be more or less concentrated. You will often not know if samples need to be diluted and, if so, how much. You may have to find the right dilution by trial and error. It sometimes happens that an analyst performs a colorimetric assay and discovers that none of the sample dilutions are in the linear range of the assay. When this occurs, the original sample needs to be diluted again to bring it into the right range. The entire assay must be repeated from the beginning, along with the preparation of new standards and a new blank.

---

* For helpful information about this assay, see the Bio-Rad instruction manual at https://www.bio-rad.com/webroot/web/pdf/lsr/literature/LIT33.pdf.

Planning Your Work: The Bradford Protein Assay

1.   What is the purpose of this exercise?
2.   Prepare a flowchart for this exercise that summarizes the steps you will perform in the laboratory. An example is shown in Figure 6.7, but we strongly encourage you to make your own version before looking at the one we have prepared.
3.   Copy or tape Table 6.3 into your laboratory notebook, as directed by your instructor, and complete all required calculations. As described in step 2 in the procedure, you will be given a BSA stock solution at a concentration of 1 mg/mL. You will dilute this initial stock solution to a concentration of 100 µg/mL, and you will use the 100 µg/mL stock to make five standards. The final volume for all standards should be 4 mL. (The first line is completed as an example.)
4.   What data will you collect?
5.   How will you analyze your data?
6.   What will you have learned after you analyze your data?

LABORATORY PROCEDURE

1.   Turn on the spectrophotometer and set the wavelength to 595 nm.
2.   Prepare the standards and blank according to Table 6.3.
     2.1.   Starting with a BSA stock solution of 1.0 mg/mL, make 5 mL of a stock with a concentration of 100 µg/mL. Use water as the diluent.
     2.2.   Using the stock of 100 µg/mL, make a series of five standards in water with final concentrations ranging from 1 to 25 µg/mL. Each standard should have a total volume of 4.0 mL.
     2.3.   Prepare a test tube with 4.0 mL of water for the blank.
3.   Prepare the unknowns. Your instructor will give you one or two samples of BSA of unknown concentration, and you will determine their concentration of protein. To perform the Bradford assay, your sample must have a concentration of protein that is in the range of the standards you have prepared, 1–25 µg/mL. You do not know how much protein there is in the unknowns and it is, therefore, prudent to prepare several dilutions. You might, for example, prepare a 1/10 and a 1/100 dilution of each unknown. An example of how these dilutions might be prepared is shown in Table 6.4.
     3.1.   If none of your dilutions lie within the linear range of the assay, you will need to try different dilutions and repeat the entire assay, standards and all.
4.   Conduct a check.
     4.1.   Make sure you have five standards, a water blank, and three different dilutions of your BSA unknown sample(s).
     4.2.   Make sure all the test tubes contain a volume of 4.0 mL by holding the rack of tubes up to your eyes and checking that the total volumes in every tube are the same. Do not proceed if the volumes differ because the assay will not be accurate, and you will be wasting reagents and time if you continue.

TABLE 6.3   Diluting BSA Stock to Make Standards

Standard Number	Concentration of BSA in Standard (µg/mL)	Volume of Stock (100 µg/mL)	Volume of Diluent (Water)	Total Volume
1	1.0	40 µL	3960 µL	4000 µL
2	5.0			4000 µL
3	10			4000 µL
4	20			4000 µL
5	25			4000 µL
Blank	0	0		4000 µL

TABLE 6.4   Dilutions

Sample	Volume Sample (mL)	Volume Water (mL)
Unknown 1. Undiluted	4	0
Unknown 1. Diluted 1/10	0.40	3.60
Unknown 1. Diluted 1/100	0.040	3.96
Unknown 2. Undiluted	4	0
Unknown 2. Diluted 1/10	0.40	3.60
Unknown 2. Diluted 1/100	0.040	3.96

5. Add 1.0 mL of Bio-Rad dye to each of the assay tubes, including the blank. Mix the tubes gently but thoroughly after the addition of the dye by using either a vortex mixer set to a low power setting or by gently flicking the tubes. Avoid foaming.

6. After 5–15 minutes, check to make sure that color is developing.
   6.1. Solutions without protein should remain reddish brown. Solutions with protein should turn blue; the more concentrated the protein, the bluer the solution should be.
   6.2. Expect to see a progression of blue color in the standard tubes; if not, something is wrong.

7. After a period of 10 minutes to 1 hour, read the A_{595} for all the standards and samples.
   7.1. Record the time elapsed.
   7.2. Blank the spectrophotometer.
   7.3. Start with the most dilute solution and progress to the most concentrated.
      7.3.1. For the most accuracy, use the same cuvette for the blank and every sample.
   7.4. Be careful to thoroughly remove each solution from the cuvette after reading its absorbance.
   7.5. Save each solution by putting it back into its tube in case you need to read it again.
   7.6. Record your absorbance values in Table 6.5.
      7.6.1. Securely attach Table 6.5 to your laboratory notebook or copy it into your notebook before obtaining data.
      7.6.2. Alternatively, tape the printout in your laboratory notebook if your spectrophotometer can record values automatically.

8. Graph the spectrophotometer readings of your BSA standards as a function of concentration of protein present. The X-axis is BSA (in µg/mL), and the Y-axis is A_{595}. Alternatively, your spectrophotometer may graph your data automatically.

9. **Data analysis**
   9.1. Determine the concentration of protein in your unknown(s) by using your standard curve.
      9.1.1. Add the results to Table 6.5.
      9.1.2. If the samples were diluted, multiply by the reciprocal of the dilution to determine their concentration.
   9.2. Compare your results to the answer key provided by your instructor and calculate your percent error(s).

LAB MEETING/DISCUSSION QUESTIONS

1. Evaluate the linearity of your standard curve. If your standard curve was not perfectly linear, can you still use it to determine the total protein concentration in your samples? (Note that, based on the manufacturer's literature, this protein assay may begin to become nonlinear above concentrations of 10 µg/mL of protein [when assaying BSA]. However, you may not see nonlinearity at this concentration.)

TABLE 6.5   Bradford Assay Data Table

Sample or Standard	A_{595}	Dilution (if any)	Concentration of Protein (μg/mL)
Standard 1, 1 μg/mL		Not applicable (NA)	NA
Standard 2, 5 μg/mL		NA	NA
Standard 3, 10 μg/mL		NA	NA
Standard 4, 20 μg/mL		NA	NA
Standard 5, 25 μg/mL		NA	NA
Unknown 1			
Unknown 2			
Unknown 3			
Unknown 4			
Unknown 5			
Unknown 6			

3. Did you successfully determine the concentration of protein in your unknown sample(s)? Calculate the percent error(s) for your unknowns. Are your results "close enough"? Compare your results to those of the class.
4. What problems do you think could make this assay method inaccurate? How can you avoid these problems?
5. What were the positive and negative controls in this assay?
6. Examine the flowchart for this procedure in Figure 6.7. How does this flowchart compare with the one you prepared before performing the assay? How would you modify this flowchart and your own, based on your experiences in the laboratory?
7. If you are going to perform this assay again, how will you organize your work for maximum efficiency?

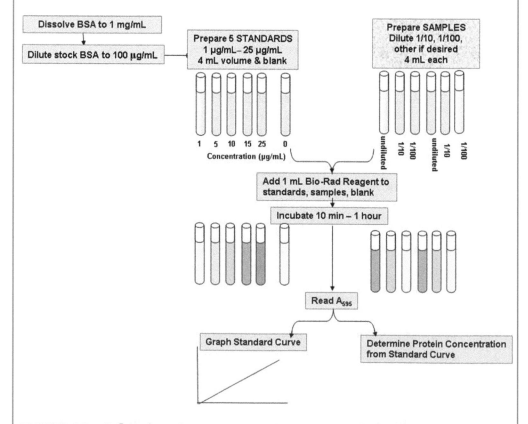

FIGURE 6.7   **A flowchart for the Bradford assay as performed in our imaginary company.**

Laboratory Exercise 25: The Bradford Protein Assay: Exploring Assay Verification

OVERVIEW

In the previous laboratory exercise, you performed a Bradford protein assay "cookbook style." In this exercise, you will begin to explore the assay in more depth. Virtually all assays, including the Bradford assay, have complexities; the better you understand the assay, the better you will be able to ensure that your results are trustworthy.

Time does not permit us to do a thorough study of the Bradford assay, but you will look at two aspects of the assay that affect its accuracy: the presence of interfering substances and the use of BSA as a standard instead of the protein of interest. You will also determine the linear range of the assay. To be most efficient, different groups will perform each of these tasks.

Although you had an existing procedure to follow when first learning how to perform the Bradford assay, you will now need to write your own procedures (also called "protocols") for your study. Each group will prepare its own protocol before coming to lab. The results of all the groups will be combined at the end.

As you explore the Bradford assay, you will achieve the following objectives:

- Practice devising and completing an independent investigation.
- Explore the complexities involved in developing and validating an analytical method.
- Practice working with colleagues to solve a problem.
- Develop a deeper understanding of protein assays.
- Develop a deeper understanding of the goals of assays.
- Practice interpreting data.
- Explore the complexities involved in developing a process to prepare a product.

BACKGROUND

A. Method Validation

Method validation is a formal process that establishes that an analytical method gives reliable results and specifies the conditions under which the assay works. The U.S. Food and Drug Administration (FDA) requires that pharmaceutical companies validate all the methods they use. Although not every company is regulated by the FDA, the principles of validation are applicable wherever analytical tests are performed (this would include nearly every laboratory). The *United States Pharmacopeia* (see the Bibliography, Appendix 3, for references relating to method validation) explains method validation as the process by which it is established, by laboratory studies, that the performance characteristics of a method meet the requirements for its intended applications.

Determining that an analytical method is suitable requires analyzing various parameters. The parameters that are typically measured during assay validation include the following:

- Accuracy
- Precision
- Specificity
- Limit of detection
- Linearity
- Range

These parameters are explained in Table 6.6. You will not evaluate all these parameters for the Bradford assay in detail, so we will call our study one of "verification" rather than "validation."

TABLE 6.6   Analytical Performance Parameters

Accuracy
Definition—The accuracy of an analytical method is the closeness of test results obtained by that method to the true value.
Determination—The accuracy of an analytical method may be determined by applying that method to samples to which known amounts of analyte have been added.
Precision
Definition—The precision of an analytical method is the degree of agreement among individual test results when the procedure is applied repeatedly to multiple samplings of a homogeneous sample. The precision of an analytical method is usually expressed as the standard deviation or relative standard deviation. Precision may be a measure of either the degree of reproducibility or repeatability of the analytical method under normal operating conditions. In this context, reproducibility refers to the use of the analytical procedure in different laboratories. Intermediate precision expresses within-laboratory variation, as on different days, or with different analysts or equipment within the same laboratory. Repeatability refers to the use of the analytical procedure within a laboratory over a short period of time using the same analyst with the same equipment.
Determination—The precision of an analytical method is determined by assaying a sufficient number of aliquots of a homogeneous sample to be able to calculate statistically valid estimates of standard deviation.
Specificity
Definition—The specificity of an analytical method is its ability to measure accurately and specifically the analyte in the presence of components that may be expected to be present in the sample. Specificity is a measure of the degree of interference (or absence thereof) in the analysis of complex sample mixtures.
Determination—The specificity of an analytical method is determined by comparing test results from the analysis of samples containing impurities and degradation products with those obtained from the analysis of samples without impurities or degradation products.
Limit of Detection
Definition—The limit of detection is the lowest level of analyte in a sample that can be detected, but not necessarily quantitated, under the stated environmental conditions.
Linearity and Range
Definition of linearity—The linearity of an analytical method is its ability to elicit test results that are directly, or by a well-defined mathematical transformation, proportional to the concentration of analyte in samples within a given range.
Definition of range—The range of an analytical method is the interval between the upper and lower levels of the amount of analyte that can be determined with precision, accuracy, and linearity using the method as written.
Determination of linearity and range—The range of the method is validated by verifying that the analytical method provides acceptable precision, accuracy, and linearity when applied to samples containing analyte at the extremes of the range.

B. Modification of the Assay

In this manual, we have modified the procedure from that recommended by Bio-Rad, a manufacturer of the Bradford reagent. Bio-Rad's procedure is shown along with our modified procedure in Figure 6.8. The reason we modified Bio-Rad's procedure is to increase the assay volume to 5 mL so that it can be used in laboratories whose spectrophotometers require a larger volume. Bio-Rad has performed extensive validation of this assay. However, because we modified the assay procedure, we cannot assume that their results will match ours. You will compare some of your results to theirs in this exercise.

C. Interfering Substances

Proteins that have been isolated from natural sources are frequently present in solution along with detergents, buffers, preservatives, and other such materials. These other materials can affect the accuracy of a protein assay. An **interfering substance** is *a component*

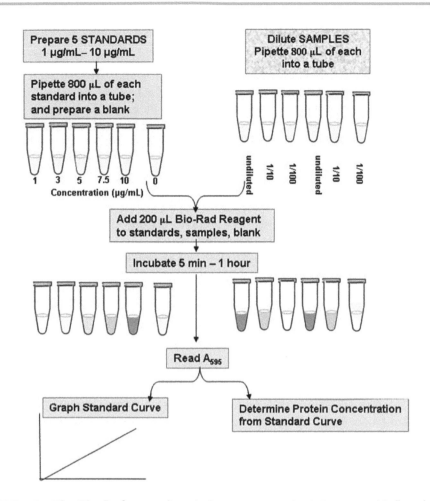

FIGURE 6.8 **The Bio-Rad procedure (microassay version).** Compare this flowchart to the modified procedure we are using, as illustrated in Figure 6.7. Observe that the volumes are different in the two procedures. The ratio of dye to protein, however, is the same in our modified procedure and in the Bio-Rad assay. For example, in the 10 μg/mL standard test tube in the Bio-Rad procedure, there are 8 μg of protein. This amount of protein is combined with 200 μL of dye. The ratio of protein to dye is 1 μg:25 μL. In our modified procedure, in the 10 μg/mL standard, there are 40 μg of protein, which are combined with 1000 μL of dye. The ratio is also 1 μg:25 μL. If the ratio of protein to dye were different, the linear range of the assay would shift. This ratio also affects how sensitive the assay is to interference.

in the solution, other than the material of interest, which either enhances or reduces the amount of color formed in the assay. An interfering substance can cause inaccuracy in an assay. Each type of protein assay is sensitive to interfering substances. Certain common detergents are interfering substances for the Bradford assay. The Bio-Rad kit directions contain extensive information about substances that interfere with the assay.

Interfering substances affect the specificity of an assay. Studying the effects of interfering substances is, therefore, part of method validation.

Planning Your Work: Verification of the Performance of the Bio-Rad Assay

1. What is the purpose of this exercise?

2. Write a protocol to complete one of the verification tasks, as assigned by your supervisor (instructor). Your supervisor may have you write the protocol in your laboratory notebook, or he or she may have you write it on a word processor. In either case, be sure to keep good notes in your laboratory notebook as to what you actually do, your results, and your conclusions.

LABORATORY VALIDATION PARAMETERS

Part A: Interfering Substances

Write and follow a protocol to investigate the effects of different concentrations of the detergent sodium dodecyl sulfate (SDS) on the Bio-Rad assay. Use the same concentrations of standards as in Laboratory Exercise 24, but this time, add SDS to each standard. Begin with detergent at a concentration of 0.0025% w/v as your lowest concentration. According to the literature from Bio-Rad, this concentration of SDS should not cause interference. Proceed up to 1% w/v, a concentration at which we expect to see interference in this assay.

Part B: Use of BSA as the Standard

Write and follow a protocol to determine the accuracy of using BSA as the standard when another protein is the analyte. In our scenario the enzyme, β-galactosidase, is our analyte of interest. Assay the amount of β-galactosidase in your samples at least twice. One time use BSA to make the standard curve, and another time use β-galactosidase to make the standard curve. Assess the accuracy of your results using the two methods. Dissolve the β-galactosidase enzyme stock solution, given to you by your instructor, in water to avoid any interference by buffers.

Part C: Linear Range

Write and follow a protocol to determine the linear range for the assay when the volume is 5 mL, as it is in our modified procedure. Prepare a series of standards of BSA, each in duplicate. Determine the lowest concentration that can be reliably quantified—the lower limit of quantitation. Find the maximum concentration where the assay provides linear results when graphed. Bio-Rad states that the assay, using the method shown in Figure 6.8, provides linear results between 1.2 and 10 µg/mL when used for BSA. Above 10 µg/mL, Bio-Rad says that the assay slowly becomes less linear. See if your results confirm that range.

LAB MEETING/DISCUSSION QUESTIONS

During your lab meeting, combine the results from all three groups in order to better understand the Bio-Rad assay in your laboratory. These questions may help guide your discussion:

- What is the precision of the method overall?
- How accurate is the method overall?
- Does the use of BSA to make the standards affect the accuracy of the method?
- Did the presence of SDS influence the results?
- Did you see any effect of changing the assay volume?

Your supervisor may ask for a written progress report, summarizing what you have learned so far about the assay parameters.

Laboratory Exercise 26: The Beta-Galactosidase Enzyme Assay

OVERVIEW

The Bradford assay detects nearly all proteins in a sample, regardless of the type of protein. Often, however, analysts want to look only at a specific protein of interest. There are many spectrophotometric assays that detect only a single protein. This laboratory exercise introduces an assay that is specific for the enzyme β-galactosidase. In this exercise, you will learn to perform the β-gal assay. In the next laboratory exercise, you will use the β-gal assay in conjunction with the Bio-Rad protein assay to compare the purity of two enzyme preparations. We use β-gal as a model protein for learning enzyme assays, partly because it is readily available at a moderate cost. β-galactosidase is also an interesting model protein because it is widely used in molecular biology. For example, β-galactosidase is used in detection systems that allow scientists to determine whether or not cells have taken up a gene of interest.

As you explore the β-gal assay, you will achieve the following objectives:

- Learn how to perform and analyze the results of an enzyme assay.
- Deepen your understanding of accuracy and precision in assays.
- Develop an understanding of how enzymes function to catalyze reactions.
- Begin to understand the overall process of purifying a product in a research and development setting.

BACKGROUND

A. β-Galactosidase

The bacterium *Escherichia coli* (*E. coli*) uses the enzyme β-galactosidase to break down lactose, the sugar found in milk. β-galactosidase catalyzes the cleavage of lactose into two monosaccharides that can be metabolized to produce energy. When β-galactosidase cleaves its natural substrate, lactose, the reaction looks like this:

β-galactosidase
Lactose → Glucose + Galactose

The natural products of this reaction are colorless. A colorimetric assay system for β-galactosidase has, therefore, been developed in which a synthetic sugar is cleaved to produce a colored product. The artificial sugar is *o*-nitrophenyl-β-galactoside (ONPG). The cleavage product of ONPG is *o*-nitrophenol (ONP). (Recall that you studied ONP in Laboratory Exercise 13.) The reaction looks like this:

β-galactosidase
ONPG → ONP + Galactose

The more enzyme that is present in a sample, the more ONP that forms. ONP is yellow and can be quantified using a spectrophotometer at a wavelength of 420 nm.

B. Calculation of the Activity of β-Galactosidase in a Sample

You prepared a new standard curve each time you performed the Bradford assay. One could also prepare a standard curve for the β-gal assay each time it is performed; this was the task in Laboratory Exercise 13. As you saw in that laboratory exercise, a β-gal assay standard curve is prepared with ONP because the colored product in the assay is ONP. It is common practice when performing this assay, however, to rely on a published absorptivity constant for ONP rather than to make a standard curve each time. It is obviously much faster to perform the assay without preparing a standard curve. This common shortcut will reduce the accuracy of the assay. In many cases, however, the absolute accuracy of the assay result is not important because the assay is used to compare the amount of β-gal in different samples. You will see how this assay is used to compare two samples in Laboratory Exercise 27.

The amount of β-gal present in a sample is expressed in units that are simply called "units" (U) of enzyme activity. In the case of β-gal,

$$1U = \text{Amount of enzyme that will convert } 1\mu\text{mol of}$$
$$\text{substrate to product in 1 minute at } 37°C$$

The formula for calculating the amount of β-gal in a sample is based on this definition of activity and the published absorptivity constant for ONP, which is 4500 L/mol-cm (refer to Laboratory Exercise 13). The result is shown as Formula 6.1. To use this formula, place the value you obtain for absorbance at 420 nm in the numerator. Multiply that number by 0.380 μmol (derived from the literature value for the absorptivity constant for ONP, as shown in Box 6.1) and divide by the number of minutes that you allowed the reaction to proceed.

Note that to be in the linear range of the assay, the substrate needs to be present in excess. As a guideline, your final absorbance at 420 nm should be between 0.3 and 0.9. If the absorbance is not in this range, then the results may be outside the linear range of the assay.

Formula 6.1

Calculating the Units of Beta-Galactosidase Activity Present

$$U \text{ of } \beta\text{-gal} = \frac{A_{420} \times 0.380\mu\text{mol}}{\text{Minutes at } 37°C}$$

Example 6.1 (Refer to Figure 6.9)

1. An analyst starts with an undiluted β-galactosidase solution and wants to know the enzyme activity in this solution expressed in terms of "enzyme units per milliliter."

   a. The analyst dilutes the enzyme solution in order to be in the linear range of the assay. The dilution involves mixing 10 μL of sample with 990 μL of dilution buffer for a 1/100 dilution (see Figure 6.9a).

   b. The analyst removes 10 μL of the diluted sample and adds it to 1 mL of Z buffer (see Figure 6.9b).

   c. The analyst pre-equilibrates the substrate (ONPG), the sample, and a blank to 37°C.

   d. The analyst adds 0.2 mL of substrate to the sample and blank and incubates them at 37°C (see Figure 6.9b). The enzyme in the sample reacts with the substrate to produce ONP. After 11 minutes have elapsed, the analyst stops the reaction by adding sodium carbonate (see Figure 6.9b).

   e. The analyst measures the absorbance of the ONP in the sample. In this example, the $A_{420} = 0.412$, see (Figure 6.9c).

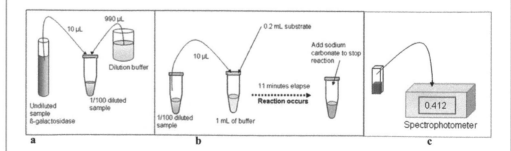

FIGURE 6.9   **Example of the β-galactosidase assay procedure.**

2. The analyst plugs the results into Formula 6.1:

$$\text{U of } \beta\text{-gal} = \frac{0.412 \times 0.380\,\mu\text{mol}}{11\,\text{min}} = 0.0142\,\mu\text{mol/min}$$

$$= \mathbf{0.0142\,U}$$

0.0142 U is the number of units of enzyme activity (an amount) that were present in 10 μL of diluted sample.

3. The analyst uses a proportion equation to calculate the number of units of β-gal on a per milliliter basis:

$$\frac{1.42 \times 10^{-2}\,\text{U}}{10\,\mu\text{L}} = \frac{?}{1000\,\mu\text{L}} = \mathbf{1.42\,U/mL} \;\left(\text{Observe that this is a concentration.}\right)$$

4. The diluted sample was not the original sample; the analyst needs to account for the 1/100 dilution:

$$1.42\,\text{U/mL} \times 100 = \mathbf{142\,U/mL}$$

Therefore, there were 142 U/mL of β-galactosidase in the original undiluted sample.

Example 6.2

Tracy, a student intern, performed the β-gal assay. The enzyme arrived from the manufacturer as a dried powder inside a small vial. The manufacturer's label stated that the enzyme had an activity of 622 U/mg solid. The milligrams of solid is primarily β-gal, but this preparation may also contain unknown impurities, such as other proteins, other organic materials, and perhaps even small quantities of inorganic contaminants.

*   Tracy dissolved the enzyme to a concentration of 1 mg/mL, so she expects each milliliter of her preparation to have roughly 622 U—assuming no activity has been lost during shipping, storage, or handling.
*   She prepared three dilutions of the enzyme: 1/10, 1/100, and 1/1000.
*   Tracy performed a β-galactosidase assay on her enzyme preparation; her absorbance results are shown in Table 6.7.

Let's consider these results. The 1/10 dilution gave an absorbance of 3.2422, which is definitely too high. Tracy, therefore, did not use this value for further calculations. Neither the 1/100 dilution nor the 1/1000 dilution are in the presumed linear range of the β-galactosidase assay, which is said to be at absorbance values between 0.3 and 0.9. However, she observed that the absorbance value for the 1/100 dilution is 10 times higher than the value for the 1/1000 dilution, which is consistent with a linear assay response. After consulting with an instructor, Tracy continued with the calculations for β-galactosidase activity.

For the 1/100 dilution:

*   Plugging into Formula 6.1 provided this result:

$$\frac{1.0234 \times 0.380\,\mu\text{mol}}{10\,\text{min}} = \mathbf{0.0388892\,U}$$

TABLE 6.7   Tracy's Enzyme Assay Dilutions

Sample	Dilution	A_{420}
Enzyme Sigma Grade VIII	1/10	3.2422
Enzyme Sigma Grade VIII	1/100	1.0234
Enzyme Sigma Grade VIII	1/1000	0.1026

So, there were 0.0388892 U/10 μL.

- She then calculated how many units were present on a per milliliter basis:

$$\frac{0.0388892\text{U}}{10\mu\text{L}} = \frac{?}{1000\mu\text{L}} \quad ? = 3.88892\text{U}/1000\mu\text{L} = \mathbf{3.88892\ U\ /\ mL}$$

- She multiplied by the dilution: 3.88892 U/mL × 100 ≈ 389 U/mL.
  For the 1/1000 dilution she performed these steps:
- Plugging into the Formula 6.1 provided this result:

$$\frac{0.1026(0.380)}{10\,\text{min}} = \mathbf{0.0038988\ units}$$

So, there were 0.0038988 U/10 μL.

- Tracy calculated how many units were present on a per milliliter basis:

$$\frac{0.0038988\text{U}}{10\mu\text{L}} = \frac{?}{1000\mu\text{L}} \quad ? = \mathbf{0.38988\ U\ /\ mL}$$

- Tracy multiplied by the dilution: 0.38988 U/mL × 1000 ≈ 390 U/mL.

The two dilutions, therefore, gave strikingly similar results, confirming that she was in the linear range of the assay. Neither value is as high as the manufacturer's (622 U/mL), but that is easily explained as being caused by a loss of activity during shipping and storage.

Planning Your Work: The Beta-Galactosidase Assay

1. What is the purpose of this exercise?
2. Write the assay procedure in the form of a flowchart or a sketch, whichever seems best to you. Place it in your laboratory notebook or on a separate sheet of paper, as directed by your instructor.
3. What data will you collect?
4. How will you analyze your data?
5. What will you have learned after you analyze your data?
6. What are the controls in this assay?

Safety Briefing

The buffer you will be using contains dithiothreitol (DTT), which is toxic. Avoid contact with it. You may want to wear gloves, but if you do, change them frequently. Remember, gloves used improperly provide no protection.

LABORATORY PROCEDURE

This assay method is derived from Lab #5 in Jeffrey H. Miller, *Experiments in Molecular Genetics* (New York: Cold Spring Harbor Laboratory, 1972).

Note: This procedure, as written here, is not designed for use with an older Spec-20 style spectrophotometer because the total volume is only 1.7 mL. Modifying the assay for a volume large enough to be used with a Spec 20 would require a costly amount of enzyme.

1. Obtain an ice bucket with crushed ice to keep your enzyme cool until you are ready to do the assay.
2. Make sure you have a water bath or other method of bringing your solutions to 37°C.
3. Make several dilutions of the β-galactosidase unknown provided by your instructor:
   3.1. Use Z buffer as the diluent.

3.2.   Consider making 1/10, 1/100, 1/500, and 1/1000 dilutions as a start. The total volume does not matter—you only need 10 μL for each assay.

3.3.   Be sure to mix dilutions thoroughly, either by using a vortex mixer or by covering and inverting them.

3.4.   If the amount of enzyme is limited, omit the 1/10 dilution.

3.5.   Place each dilution on ice in an ice bucket as you prepare it.

4.   Prepare a blank that contains 1.0 mL of Z buffer, pH 7.0.

5.   Label small test tubes (e.g., that can hold 5 mL or 10 mL), one for each dilution of the original sample.

6.   Place 1.0 mL of Z buffer in each labeled tube. These are your "reaction tubes."

7.   Add 10 μL of each diluted sample to the properly labeled tube.

8.   Add 10 μL of Z buffer to the blank.

9.   Pre-equilibrate the following to 37°C:

- Each enzyme solution reaction tube from step 6
- Substrate solution (ONPG at a concentration of 4 mg/mL in 0.1 M sodium phosphate buffer, pH 7.5, made fresh or thawed each day)
- Blank

10.   Add 0.2 mL of ONPG to each tube, mix briefly, and record the time.

11.   Incubate the tubes at 37°C for 10 minutes.

12.   Add 0.5 mL of 1 M Na_2CO_3 (stop solution) to terminate the reaction.

13.   Read the A_{420} of each tube and record the results.

14.   **Data analysis**

14.1.   Use Formula 6.1 to calculate the concentration of β-galactosidase in your unknown. Compare your value to the answer key provided by the instructor.

LAB MEETING/DISCUSSION QUESTIONS

1.   Compare your results for the unknown with that of your classmates and comment on the accuracy and precision of your class. Note that your instructor is basing the answer key on the manufacturer's value for the enzyme. This may not be exactly correct as the activity of an enzyme can diminish over time.

2.   Optional: Follow the instructions in Box 6.1 to calculate your own absorptivity constant for ONP. Use this value in your calculations to determine the activity of β-galactosidase in your samples. Compare the results when you use your own absorptivity constant to the results when you use the published value.

Box 6.1   Optional Method to Use Your Own Value for the Absorptivity Constant for ONP

$$U \text{ of } \beta\text{-gal} = \frac{A_{420} \times 0.380 \mu mol}{Minutes \, at \, 37°C}$$

The purpose of this formula (Formula 6.1) is to determine the number of units of β-gal in a sample. This formula is based on the published value for the absorptivity constant of ONP, which is 4500 L/mol-cm. However, you calculated your own value for this constant in Laboratory Exercise 13. You can, therefore, derive the formula for units of β-galactosidase based on your own value for the absorptivity constant. We will go through the derivation of the formula based on the published absorptivity constant. You can substitute in your value for this published value. Put your value into step 2 in the place of 4500 L/mol-cm. Then, continue with the calculations at each step, using your values. *Make sure that your value for the absorptivity constant is in units of liters/mole-cm.*

1.   Beer's law tells us the following:

$$A_{420} = \alpha b \left[ONP \right]$$

where

α = molar absorptivity constant for ONP

b = path length through the cuvette (usually 1 cm)

2. $\quad A_{420} = \dfrac{4500\ \text{L}}{\text{mol}\text{-}\cancel{\text{cm}}}\ 1\ \cancel{\text{cm}}\ [\text{ONP}]$

Substitute your absorptivity constant value for 4500 L/mol-cm.

3. $\quad \dfrac{A_{420}(\text{mol})}{4500\ \text{L}} = [\text{ONP}]$

Divide both sides of the equation by your value for the absorptivity constant.

4. $\quad \dfrac{A_{420}(\text{mol})}{\text{L}} \times \dfrac{1}{4500} = [\text{ONP}]$

Rearrange the equation with your value in place of 4500.

5. $\quad \dfrac{A_{420}(\text{mol})}{\text{L}} \times 0.000222222 = [\text{ONP}]$

The number 0.000222222 will be different for you.

6. $\quad \dfrac{A_{420}(10^6\ \mu\text{mol})}{\text{L}} \times 0.000222222 = [\text{ONP}]$

(Remember that 1 mol = 10^6 µmol.)

7. $\quad \dfrac{A_{420}(\mu\text{mol})}{1000\ \text{mL}} \times 222.222 = [\text{ONP}]$

(Remember that 1 liter = 1000 mL.)
The number 222.222 will be different for you.

8. $\quad \dfrac{A_{420}(\mu\text{mol})}{1\ \text{mL}} \times 0.222222 = [\text{ONP}]$

The number 0.222222 will be different for you.

9. $\quad \dfrac{A_{420}(\mu\text{mol})}{1\ \cancel{\text{mL}}} \times 0.222222 \times 1.7\ \cancel{\text{mL}} = \textbf{amount of ONP}$

The equation in Step 7 will give the *concentration* of ONP in the tube, but you want to know the *amount* (in units of µmol) of product. The volume in the tube is 1.7 mL. (Volume = 1 mL of reaction buffer + 0.2 mL of ONPG + 0.5 mL of stop solution.) So multiply the concentration, which is in units of µmol/mL, times 1.7 mL.
The number 0.222222 will be different for you.

10. $\quad A_{420} \times 0.3777\ (\mu\text{mol}) = \textbf{amount of ONP}$

The number 0.3777 will be different for you.

11. $\quad A_{420} \times 0.380\ (\mu\text{mol}) = \textbf{amount of ONP}$

Round your answer.

12. $\quad \dfrac{A_{420} \times 0.380\ (\mu\text{mol})}{\text{minute}} =$

$$\dfrac{\mu\text{mol of ONP formed}}{\text{minute}} = \textbf{units } \beta\text{-gal}$$

(Divide by the number of minutes the reaction proceeded to get the answer on a per minute basis.)
The number 0.380 will be different for you.

Laboratory Exercise 27: Comparing the Specific Activity
of Two Preparations of Beta-Galactosidase

OVERVIEW

Scenario

BEING A DEVELOPMENT SCIENTIST; COMPARING TWO PURIFICATION
METHODS

Eventually, your company will be manufacturing beta-galactosidase that is purified from large vats of *E. coli* bacteria that produce the enzyme. One of your tasks as a research and development scientist is to optimize the purification procedure for your product. Your task today is to compare two purification methods to see which is better. To do so, you will use both the Bradford protein assay and the β-galactosidase assay to calculate a purity value. This measure of purity is called **specific activity**. Specific activity is defined in the Background section that follows.

In this activity, you will simulate being a development scientist. Imagine that you are comparing two methods of purifying an enzyme preparation. You want the method that provides the purest enzyme. In reality, you will be comparing enzyme preparations that are purchased from a commercial supplier. The two preparations are of two different grades, a more expensive and more highly purified grade and a lower grade that is somewhat less expensive and less highly purified.

As you explore the idea of specific activity in this laboratory exercise, you will achieve the following objectives:

- Learn how to combine a protein assay and an enzyme assay to assess the purity of an enzyme preparation.
- Deepen your understanding of how to optimize accuracy and precision in assays.
- Develop an understanding of how enzymes function to catalyze reactions.
- Begin to understand the overall process of purifying a product in a research and development setting.

BACKGROUND

Many important biotechnology products are proteins. The most common biopharmaceutical products, for example, are proteins that have a beneficial effect in the body. The purity of these protein products is critical—imagine what might happen if an injected drug product were contaminated with extraneous substances. Assays are, therefore, required to test the purity of proteins at each step in their purification.

In this exercise, you will determine "specific activity," a common measure of protein purity. **Specific activity** *communicates the amount of a specific protein per milligram of total protein* (see Figure 6.10). For example, β-galactosidase is our product of interest in our scenario. β-gal is extracted and purified from bacteria in a series of steps. After each step, a protein assay can be performed on a small amount of the extract, and a β-gal enzyme assay can be performed on another small sample of the extract. The specific activity is then calculated as shown in Formula 6.2.

As a protein is progressively purified, its specific activity increases. This is because there are fewer and fewer protein impurities as the enzyme becomes purer. Examine Figure 6.11 and make sure you can see why specific activity increases as the preparation's purity increases.

Formula 6.2

Calculating Specific Activity

$$\frac{\text{Units of }\beta\text{-galactosidase activity/mL}}{\text{mg of total protein in the sample/mL}}$$

Observe that milliliters in the numerator and denominator cancel.

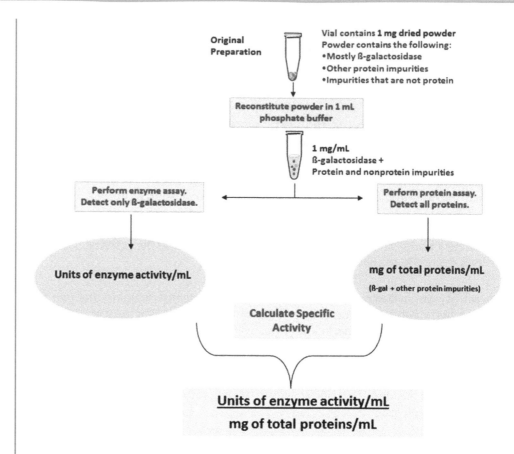

Original Preparation

Vial contains **1 mg dried powder**
Powder contains the following:
• Mostly ß-galactosidase
• Other protein impurities
• Impurities that are not protein

Reconstitute powder in 1 mL phosphate buffer

1 mg/mL
ß-galactosidase +
Protein and nonprotein impurities

Perform enzyme assay. Detect only ß-galactosidase.

Perform protein assay. Detect all proteins.

Units of enzyme activity/mL

mg of total proteins/mL
(ß-gal + other protein impurities)

Calculate Specific Activity

Units of enzyme activity/mL

mg of total proteins/mL

FIGURE 6.10 **The calculation of specific activity.** Specific activity is an indication of the purity of a protein preparation. If the protein of interest is an enzyme, as is our protein of interest, then specific activity is the units of enzyme activity/mL divided by the milligrams of total protein/mL. To calculate this value, it is necessary to perform both an enzyme assay and a protein assay on the sample.

Example 6.3

Suppose you have a sample that contains 142 U/mL of β-gal activity, based on the β-galactosidase assay. Suppose that it also contains 0.6 mg/mL of total protein based on the Bio-Rad protein assay. The specific activity is as follows:

$$142\,\text{U/mL divided by } 0.6\,\text{mg/ml} = \textbf{237U/mg}$$

Planning Your Work: Specific Activity

1. Begin by making sure you understand the concept of specific activity. Review Figure 6.10, which shows an overall flowchart for the calculation of specific activity. Then review Figure 6.11, which compares the specific activity of two preparations, like those you will evaluate in this laboratory exercise. Also review the calculation in Example 6.3.

2. This exercise requires putting together the procedures from two previous laboratory exercises. Begin by creating a flowchart for your work. You can start with the flowcharts in Figure 6.11, but add in more procedural detail.

3. Write a protocol to perform this laboratory exercise. Your supervisor will tell you whether to use a word processor or to write the protocol in your laboratory notebook. As always, be sure to distinguish between what you plan and what you actually do.

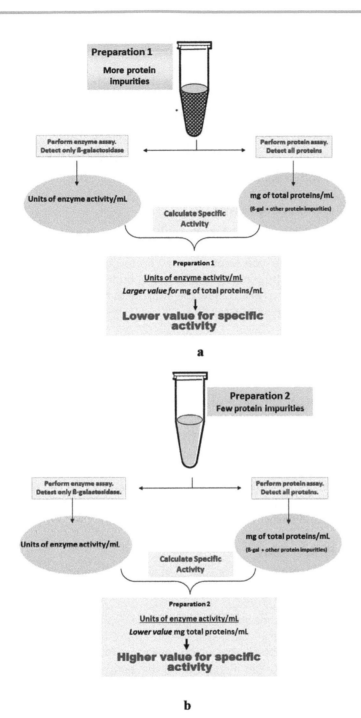

FIGURE 6.11 **Two preparations are compared.** (a) Preparation with more protein impurities. (b) Preparation with few protein impurities. Observe how this affects the values calculated for specific activity.

LABORATORY PROCEDURE

Part A: Perform Protein Assay

A.1.   Obtain from your instructor samples of the two β-gal enzyme preparations to be compared.

   A.1.1.   Place them on ice in an ice bucket.

   A.1.2.   Observe that each of the two preparations will have been dissolved for you to a concentration of 100 μg/mL in phosphate buffer.

   A.1.3.   Prepare a 1/5, 1/10, and 1/20 dilution of each of the two preparations.

A.1.3.1. Prepare the dilutions in water. Phosphate buffer interferes with the Bio-Rad assay, but once the enzyme is diluted in water, the phosphate buffer will be present at a low enough concentration that it will not interfere.

A.1.3.2. Make your dilutions so they have a final volume of 4 mL.

A.1.3.3. Place the resulting six tubes of diluted enzyme on ice.

A.2. Perform a Bio-Rad assay as you did in Laboratory Exercise 24.

A.2.1. Use BSA to make your standards.

A.2.2. Use the standard curve to estimate the concentration of total protein in both β-gal samples.

Part B: Perform β-Galactosidase Assay

B.1. Perform a β-gal enzyme assay on both samples as you did in Laboratory Exercise 26.

B.1.1. In this case, you want the enzyme to be active, so make your dilutions in Z buffer instead of water. Z buffer provides the proper ionic strength and pH to allow the enzyme to function. It will not interfere with the enzyme assay.

Part C: Compare the Specific Activities of the Two Preparations

C. **Data analysis**

C.1. For both samples, calculate the specific activity. Example 6.4 shows how to perform the calculations.

C.2. Decide which preparation is more pure.

Example 6.4

Tracy, our student intern, performed Laboratory Exercise 27 by using two different grades of β-gal, one of which was specified to be purer than the other. A brief summary of her results and analysis is shown in this example.

Part A: Perform Protein Assay

Tracy performed a protein assay on both enzyme preparations, as described in Laboratory Exercise 24:

• She began with a 1 mg/mL stock of each enzyme in phosphate buffer.
• She diluted the enzyme in water, as shown in Table 6.8.
• She made BSA standards in water at concentrations from 1 µg/mL to 25 µg/mL.
• She added 1 mL of Bio-Rad reagent to all the samples and all the standards, waited 10 minutes, and read the absorbances at 595 nm.
• She plotted a standard curve based on the absorbances of the standards (see Figure 6.12).
• Her absorbances for the samples are shown in Table 6.8.
• Based on the results for her 1/100 dilution, she calculated that she had ≈980 µg/mL total protein in enzyme 1 and ≈790 µg/mL in enzyme 2.

Part B: Perform β-Galactosidase Assay

• Each enzyme came as dried solid. Tracy dissolved each one in Z buffer to a concentration of 1 mg dried solid/mL.
• She diluted each preparation in Z buffer, as shown in Table 6.9.
• She performed a β-galactosidase assay on enzyme 1 and enzyme 2 with the results as shown in Table 6.9.

TABLE 6.8   Tracy's Protein Assay

Sample	Enzyme Stock (1 mg/mL)	Water	Absorbance in Bio-Rad Assay	Protein Concentration, Based on Standard Curve (μg/mL)
Enzyme 1 1/50 dilution	80 μL	3920 μL	0.7357	≈16
Enzyme 1 1/100 dilution	40 μL	3960 μL	0.4291	≈9.8
Enzyme 1 1/200 dilution	20 μL	3980 μL	0.2460	≈5.6
Enzyme 2 1/50 dilution	80 μL	3920 μL	0.5953	≈14
Enzyme 2 1/100 dilution	40 μL	3960 μL	0.3484	≈7.9
Enzyme 2 1/200 dilution	20 μL	3980 μL	0.1989	≈4.5

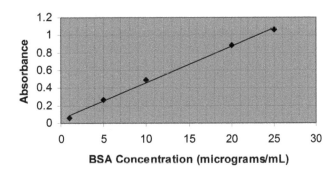

FIGURE 6.12   Bio-Rad Protein assay standard curve for Tracy's results from Laboratory Exercise 27.

TABLE 6.9   Enzyme Assay

Sample	Dilution	A_{420}
Enzyme 1	1/10	3.4247
Enzyme 1	1/100	0.8966
Enzyme 1	1/500	0.1912
Enzyme 1	1/1000	0.0744
Enzyme 2	1/10	3.3200
Enzyme 2	1/100	0.4707
Enzyme 2	1/500	0.0679
Enzyme 2	1/1000	0.0403

- She plugged the values she obtained into Formula 6.1 and calculated the enzyme activity for the dilutions that were in the linear range of the assay, that is, within 0.3–0.9 absorbance units:

Enzyme 1: 0.8966 (0.380 μmol) /10 minutes → 0.0340708 U/10 μL

$$\frac{0.0340708U}{10\mu L} = \frac{?}{1000\mu L} \quad ? \approx 3.40708U/mL$$

$$(3.40708\,\text{U/mL})(100) \approx \textbf{341\,U/mL}$$

Enzyme 2: She performed the same calculations for enzyme 2, using the 1/100 dilution, which gave a result in the linear range of the assay. The result was $\approx$ 179 U/mL.

Part C: Compare the Specific Activities

Tracy calculated the specific activities of both preparations using Formula 6.2:

Enzyme 1: 341 U/mL/0.980 mg/mL $\approx$ 348 U/mg protein
Enzyme 2: 179 U/mL/0.790 mg/mL $\approx$ 227 U/mg protein
Discussion: Enzyme 1 had a higher specific activity than did enzyme 2. This is what she expected because, in this example, the manufacturer specified that enzyme 1 was a purer grade than enzyme 2.

LAB MEETING/DISCUSSION QUESTIONS

1. Compare and discuss your values with your classmates.
2. Your instructor will have specific activity values from the manufacturer for each enzyme preparation—because this is a simulation. If this were a real research project, you would not have values to compare with your results. Compare your values and those of your classmates with the values provided by the manufacturer. Note that enzymes can lose activity over time, so the manufacturer's values may not be the same as yours.
3. What laboratory issues did you encounter when combining two procedures? Are there any problems that you would avoid in the future?
4. Explain in your own words why specific activity is calculated when purifying proteins.

Laboratory Exercise 28: Using Spectrophotometry for Quality Control: Niacin

OVERVIEW

Scenario

WORKING AS A QUALITY CONTROL ANALYST

You work in the quality control (QC) department in a nutraceutical company that is trying to compete in the lucrative vitamin market. You have been provided with a standard operating procedure (SOP) for an assay to determine the quality of niacin tablets. Your job now is to become familiar with this assay and to use it to test the niacin tablets of several of your competitors.

In this laboratory exercise, you will use UV spectrophotometry to perform QC assays on niacin tablets. You will compare the spectra of niacin from several companies with the spectrum of a niacin standard solution to see if they are the same. This comparison is a qualitative assay of identification. You will also calculate the concentration of niacin in the purchased tablets, which is a quantitative assay. In this QC situation, a form is provided for documentation purposes, and you will not use your laboratory notebook.

In this exercise you will achieve the following objectives:

- Practice following an SOP exactly.
- Learn about the use of spectrophotometry in a pharmaceutical quality control setting.
- Practice a spectrophotometric assay that is both qualitative and quantitative.
- Practice using a form for documentation purposes.

BACKGROUND

A. USP Niacin Assay

Niacin is an essential vitamin that is widely available in drug and health food stores. Niacin is sometimes prescribed in high dosages to lower cholesterol. People also take niacin supplements because they think niacin helps ease the severity of migraine headaches and alleviates gastrointestinal disturbances.

Consumers who purchase niacin supplements expect that the pills will, indeed, contain niacin in the amount shown on the label. They also expect that the tablets will not be contaminated with unexpected materials. Manufacturers need to have methods to assay their niacin tablets to be sure that they meet these requirements. In this exercise, you will use a spectrophotometric method to test niacin tablets. This assay will allow you to see if the tablets do indeed contain niacin. This is a qualitative test of *identity*. This assay will also detect whether certain contaminants are present. This is a qualitative test of *purity*. The assay will also allow you to test the concentration of niacin in the tablets. Figure 6.13 shows student results when performing this assay.

The particular method you will be using is derived from the *United States Pharmacopeia* (*USP*). The *USP* is a compilation of officially recognized standard methods relating to medicines. The methods in the *USP* are mandatory in the pharmaceutical industry and other situations where a company is compliant with current Good Manufacturing Practices (cGMP).

The *USP* describes UV assays for drugs in a general way as follows (*USP 23*, Rockville, MD: United States Pharmacopeial Convention, Inc., 1994, p. 1724):

Ultraviolet Absorption—a test solution and a standard solution are examined spectrophotometrically, in 1 cm cells [cuvettes], over the spectral range from 200 to 400 nm … Dissolve a portion of the substance under examination in the designated Medium to obtain a test solution having the concentration specified …

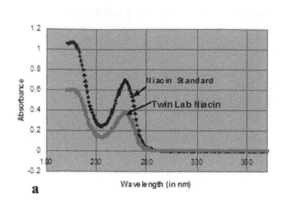
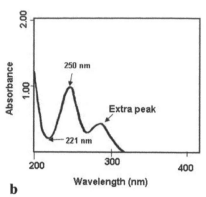

a

b

FIGURE 6.13   Niacin standard. (a) The higher spectrum is niacin reference standard freshly diluted from 0.4 mg/mL stock. The absorbance peak maximum is close to 262 nm, and the minimum UV absorbance is close to 237 nm. The second niacin preparation was purchased at the store. It has a similar pattern of absorbance to the standard, although the height of the niacin peak is somewhat lower than that of the reference standard. Both niacin preparations pass the test for niacin in the SOP. (b) Degraded niacin standard. Diluted niacin reference standard material was stored four months at a concentration of 20 µg/mL in a clear bottle. The absorbance peak maximum is at about 250 nm, and the minimum is at about 221 nm. An extra peak appears. This indicates that niacin degrades with storage.

for Solution. Similarly prepare a Standard solution containing the corresponding USP Reference Standard.

Record and compare the spectra concomitantly obtained for the test solution and the Standard solution. Calculate absorptivities and/or absorbance ratios where these criteria are included. The requirements are met if the ultraviolet absorption spectra of the test solution and the standard solution exhibit maxima and minima at the same wavelengths and/or absorbance ratios are within specified limits.

As you perform this laboratory exercise, note how the tasks you perform match this general description from the USP.

B. Stability Testing

Have you ever opened your medicine cabinet to get an aspirin or cold medication and noticed that the drug you want is expired? The FDA requires expiration dates on drug products because drugs degrade over time. Pharmaceutical manufacturers must perform **stability tests** *to establish how long a drug remains safe and effective*. Stability testing can involve simply waiting various lengths of time and then testing the performance and chemistry of the drug. It is also possible to **accelerate the stability-testing process** *by subjecting the product to heat, light, altered pH, etc.*

Your instructor may have you experiment with accelerated stability tests on niacin by using this spectrophotometric assay. After ensuring that a niacin solution has the proper spectral features, you can try subjecting it to light and/or heat to see if you can detect changes in the product. If you do detect changes, think about what this means in terms of the proper storage of the product. What happens, for example, to the product if a consumer stores it in a hot kitchen?

Planning Your Work: Niacin Quality Control

Plan your work before coming to lab by using a method that you have found to be effective.

LABORATORY PROCEDURE

Follow this SOP exactly as it is written. If you have questions, ask your supervisor (instructor) before proceeding.

Cleanest Genes Corporation page 1 of 2
SOP #S708, revision 01

SOP For the Qualitative and Quantitative Analysis of Niacin (Nicotinic Acid)*

Written by <u>Noreen Warren</u> Date <u>8/11/10</u> Approved by <u>Lisa Seidman</u> Date <u>9/2/10</u>

Approved by <u>Diana Brandner</u> Date <u>9/6/10</u>

1. **Purpose.** This SOP describes the quality control procedures to verify the identity, purity, and concentration of niacin.

2. **Scope.** This SOP applies to testing of in-process samples and final lots of niacin products.

3. **Responsibility.** Analysts in the quality control department.

4. **Reagents and materials.**

   4.1. Niacin reference standard (Sigma catalog #N5410 CAS No. 59-67-6; niacin is $C_6H_5NO_2$)

   4.2. UV spectrophotometer, properly calibrated

   4.3. Quartz cuvettes, preferably a matched pair

   4.4. Analytical balance, properly calibrated

   4.5. Ultrapurified water

   4.6 500 mL and 100 mL volumetric flasks

5. **Associated Forms.** Record results on Form F708.

6. **Procedures.**

   6.1. **Prepare the niacin reference standard (RS).**

   6.1.1. Weigh out 200 mg of niacin RS with a calibrated analytical balance.

   6.1.2. Transfer to a 500 mL volumetric flask.

   6.1.3. Dissolve in ultrapurified water.

   6.1.4. Mix thoroughly on a stir plate.

   6.1.5. Remove stir bar.

   6.1.6. Bring to volume of 500 mL (concentration is now 0.4 mg/mL).

   6.1.7. Label as "RS, 0.4 mg/mL" along with date, preparer's name, and source. This stock solution of RS should be stable for six months if stored in the dark.

   6.1.8. On the day of assay, remove exactly 5.0 mL of prepared solution and transfer to a 100 mL volumetric flask.

   6.1.9. Dilute to volume with ultrapurified water (concentration is now 20 µg/mL).

   6.1.10. Label as reference standard (RS).

   6.1.11. Check the RS.

   6.1.11.1.   Blank the spectrophotometer using ultrapurified water in a quartz cuvette.

   6.1.11.2.   Determine the absorbance at 237 nm and at 262 nm for the RS, using the same cuvette as was used for the blank or using a matched cuvette.

   6.1.11.3.   Calculate the A_{237}/A_{262} ratio for the RS.

   6.1.11.4.   The A_{237}/A_{262} ratio must be between 0.35 and 0.39; otherwise, prepare a new RS and begin again.

   6.2. **Prepare the niacin test samples (TSs) that are to be assayed to a concentration of 20 µg/mL.**

   6.2.1. Weigh out 200 mg of niacin TS. Use a calibrated analytical balance.

   6.2.2. Transfer to a 500 mL volumetric flask.

   6.2.3. Dissolve in ultrapurified water.

   6.2.4. Mix thoroughly on a stir plate.

   6.2.5. Remove stir bar.

6.2.6. Bring to volume of 500 mL (concentration is now 0.4 mg/mL).

6.2.7. Remove exactly 5.0 mL of prepared solution and transfer to a 100 mL volumetric flask.

6.2.8. Dilute to volume with ultrapurified water (concentration is now 20 µg/mL).

6.2.9. Label as a TS; include your name, date, and source of sample.

6.3. **Prepare absorbance spectra for both the RS and TSs.**

6.3.1. Using water as a blank in a quartz cuvette, blank the spectrophotometer from 200 nm to 400 nm.

6.3.2. Transfer the RS to a cuvette.

6.3.2.1. Clean and use the same cuvette as was used for the blank, or use a previously matched cuvette for the RS.

6.3.3. Prepare an absorbance spectrum for the RS from 200 nm to 400 nm.

6.3.4. Transfer the TS to a cuvette.

6.3.4.1. Clean and use the same cuvette as was used for the RS, or use a previously matched cuvette for the TS.

6.3.5. Prepare an absorbance spectrum for the TS.

6.3.6. Repeat steps 6.3.4 and 6.3.5 for each TS.

6.4. **Determine the A_{262} nm and A_{237} nm for the reference and test samples.**

6.4.1. Using ultrapurified water in a quartz cuvette, blank the spectrophotometer at 262 nm.

6.4.2. Transfer the RS to a cuvette.

6.4.2.1. Clean and use the same cuvette as was used for the blank, or use a previously matched cuvette for the RS.

6.4.3. Determine the absorbance at 262 nm and 237 nm.

6.4.4. Transfer the TS to a cuvette.

6.4.4.1. Clean and use the same cuvette as was used for the RS, or use a previously matched cuvette for the TS.

6.4.5. Determine the absorbance at 262 nm and 237 nm.

6.4.6. Repeat steps 6.4.4 and 6.4.5 for each TS.

7. **Analyze the data.**

7.1. **Perform qualitative analysis.**

7.1.1. Compare the spectra for the RS and each TS.

7.1.1.1. Determine the absorbance maxima and minima.

7.1.1.1.1. If the maxima and minima of the RS and TS are at the same wavelength ($\pm$ 2 nm) they are within specification, otherwise, reject the TS.

7.1.1.2. Determine the ratio of absorbance at 237 nm and 262 nm (A_{237}/A_{262}) for the TS.

7.1.1.2.1. If the A_{237}/A_{262} ratio for the TS is between 0.35 and 0.39, accept the sample; otherwise, reject it.

7.2. **Perform quantitative analysis.**

7.2.1. Calculate the concentration of the niacin in the TS by using the following equation:

Concentration niacin = C (A_{TS}/A_{RS}), where

C = concentration of niacin in the RS

A_{TS} = the absorbance of the niacin TS at 262 nm

A_{RS} = the absorbance of the niacin RS at 262 nm

7.2.2. Compare the value for the TS to the specifications on the product label.

*Derived from *USP/NF 2000*, monograph pp. 1080–1081 and 1724.

<table>
<tr><td align="center">**page 1 of 1**
Cleanest Genes Corporation
Form #F708, revision 01</td></tr>
</table>

FORM FOR SOP FOR THE QUALITATIVE AND QUANTITATIVE ANALYSIS OF NIACIN (NICOTINIC ACID)

Written by <u>Noreen Warren</u>　　Date <u>8/11/10</u>

Approved by <u>Lisa Seidman</u>　　Date <u>9/2/10</u>

Approved by <u>Diana Brandner</u>　　Date <u>9/6/10</u>

FORM #F708

Scope: This form is to be used to record information with SOP # S708, revision 01:

SOP for the Qualitative and Quantitative Analysis of Niacin (Nicotinic Acid)

Date_____ **Name of analyst** _____

Signature of analyst _____

Reference Standard (RS) ID number _____

Date prepared _____ Date diluted _____

RS A_{237} _____ RS A_{262} _____ RS A_{237}/A_{262} _____

RS Pass or Fail (Is it within 0.35-0.39?) _____ *(Do not continue if RS does not pass)*

RS Absorbance Maximum _____nm RS Absorbance Minimum _____nm

Test Sample (TS) ID number _____ Brand Name _____

Date prepared _____ Date diluted _____

TS A_{237} _____ TS A_{262} _____ TS A_{237}/A_{262} _____

TS Pass or Fail (Is it within 0.35–0.39?) _____

TS Absorbance Maximum _____ Pass/Fail (within ± 2 nm of RS?) _____

TS Absorbance Minimum _____ Pass/Fail (within ± 2 nm of RS?) _____

TS concentration (show calculation) _____

Test Sample (TS) ID number _____ Brand Name _____

Date prepared _____ Date diluted _____

TS A_{237} _____ TS A_{262} _____ TS A_{237}/A_{262} _____

TS Pass or Fail (Is it within 0.35-0.39?) _____

TS Absorbance Maximum _____ Pass/Fail (within ± 2 nm of RS?) _____

TS Absorbance Minimum _____ Pass/Fail (within ± 2 nm of RS?) _____

TS concentration (show calculation) _____

LAB MEETING/DISCUSSION QUESTIONS

1. Compare the shapes of the spectra for your test samples to the reference standard. Also compare the maxima and minima for each spectrum. What do you think these spectra tell you about the identity of your test samples?

2. Did the concentration of niacin in the test samples match the concentration specified on their label? Show and explain your calculations. Comment on the purity of the niacin samples.

3. What factors may have affected the accuracy of your test results? Consider, for example, the preparation of the samples and the calibration of the spectrophotometers. What might happen if your used two spectrophotometers—one for the RS and the other for the TS?

4. Report on your results if you were able to perform a stability study.

5. What is the purpose of quality control assays? How can spectrophotometry be used in a quality control setting for both qualitative and quantitative purposes?

UNIT DISCUSSION: ASSAYS

1. Consider this situation. A research and development (R&D) team in a biotechnology company is working on a new product. The product is an enzyme that will be useful for breaking down cellulose for making biofuels. The team purifies the enzyme through a series of steps. At each step, team members measure the activity of the enzyme using an enzyme assay. The enzyme has a pH optimum of 6.5, although it functions in a pH range from 6.0 to 7.3. They routinely perform the enzyme assay in a buffer at pH 6.5. Unbeknownst to the team, however, a technologist in the buffer preparation department has accidentally made one batch of the buffer that is pH 7.3 instead of 6.5. The R&D team uses the buffer, thinking it is pH 6.5. A few weeks later, the technologist in the buffer preparation department realizes the mistake and makes a new batch of buffer that is pH 6.5. The technologist gives the new batch to the R&D team but does not tell them that the old batch was pH 7.3.

   a. Speculate as to the consequences of this error.
   b. What should the technologist have done?
   c. Can you think of any systems the company might have put in place to prevent this sort of mistake from ever occurring?

2. Consider the overarching themes of good laboratory practices in the diagram that follows. Jot down notes on the diagram or on another piece of paper that describe how the ideas and skills that you practiced in this unit relate to these overarching themes.

Good Laboratory Practices
DOCUMENT
PREPARE
REDUCE VARIABILITY AND INCONSISTENCY
ANTICIPATE, PREVENT, CORRECT PROBLEMS
ANALYZE, INTERPRET RESULTS

VERIFY PROCESSES, METHODS, AND RESULTS	
WORK SAFELY	

Biological Separation Methods

UNIT VII

DOI: 10.1201/9781003360742-7

UNIT INTRODUCTION

Ignorance more frequently begets confidence than does knowledge: It is those who know little, and not those who know much, who so positively assert that this or that problem will never be solved by science.

—**Charles Darwin,** *The Descent of Man,* **1871, http://www.literature.org /authors/darwin-charles/the-descent-of-man/introduction.html**

A. SEPARATING BIOLOGICAL MATERIALS

This unit is about biological **separations**, *methods used to separate specific substances from one another.* The ability to separate materials is essential in both research and production settings.

Separations of biological materials rely on differences among them. Filtration is an example of a relatively simple and familiar separation method. The basis for separation by filtration is size difference. We use filtration in the kitchen to separate liquid coffee from the grounds. Liquid can pass through the filter while the solid coffee grounds are too big and remain on the filter surface.

Separation methods all require a force. In the case of making coffee, that force is gravity. Gravity causes the liquid to move through the filter; without this force, the separation would not occur.

Filtration, simple as it is, has important applications in the sophisticated biotechnology laboratory. You will be introduced to an application of filtration in Classroom Activity 13. You will also use filtration for sterilization purposes in Unit VIII. Filtration, however, is only one method of separation. Others that are commonly used by biologists include the following:

- **Centrifugation. Centrifuges** *accelerate the rate at which substances separate from one another by rapidly spinning the samples, thus creating a force many times that of gravity.* Centrifugation is introduced in Classroom Activity 13.
- **Chromatography. Chromatography** is a *class of techniques that separate compounds based on differences in their affinities for two phases, one of which is stationary and one of which is mobile.* Laboratory Exercise 29 explores the meaning of the term "affinities for two phases." Laboratory Exercise 30 introduces the concept of a "mobile" and a "stationary" phase. Laboratory Exercise 35 shows how one type of chromatographic method is used as part of a strategy to purify a protein.
- **Electrophoresis. Electrophoretic methods** *separate molecules with the force of an electrical field.* Electrophoresis is one of the most important technologies in molecular biology. Laboratory Exercises 31–34 deal with the performance and optimization of separations by electrophoresis.

B. PREPARATIVE VERSUS ANALYTICAL SEPARATIONS

Separations have two major goals, as summarized in Figure 7.1. The first goal is to separate a desired substance from other materials so that the substance of interest can be used in the future. A researcher, for example, might discover a new protein that must be separated and purified out of tissue for further study. The separation process functions to isolate the protein from all the other substances present in the tissue. The product of the separation is a protein, a tangible item. Another example is the purification of a biopharmaceutical drug product from the cells that produce it. **Preparative separations** *produce a tangible item for future use.*

Separations are also used for analysis, to learn something about a sample. The result of the separation is information, not a tangible item. A forensic investigator, for example, might

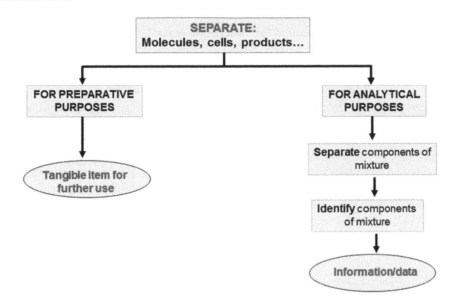

FIGURE 7.1   Separations have two key goals: preparative (to isolate a tangible product for future use) and analytical (to obtain information about a sample).

separate the components of a blood sample from one another to see if a particular drug is present in the bloodstream. **Analytical separations** *produce information.* Typically, when a separation is used for analytical purposes, the separation is the first of two steps. The second step is to identify the isolated components.

Filtration and centrifugation are routinely used for preparative purposes, as you will see when you perform Classroom Activity 13. Centrifugation can also be used for analytical purposes, although we will only talk about preparative centrifugation in this manual.

Chromatography and electrophoresis are used for both preparative and analytical purposes. Chromatography is particularly important in biomanufacturing where it is often used to purify protein products. You will practice analytical chromatography in Laboratory Exercise 30 and preparative chromatography in Laboratory Exercise 35.

Electrophoresis, unlike chromatography, is very limited in terms of the volume of product it can produce. A typical electrophoresis run will produce less than a milligram of product. Therefore, it is molecular biologists, who work with very small amounts of molecules, who are most likely to perform preparative electrophoresis. In this manual, we cover only analytical electrophoresis, as explored in Laboratory Exercises 31–34.

Classroom Activity 13: Planning for Separating Materials Using a Centrifuge

SCENARIO

Being a Research and Development Scientist

You are a research and development (R&D) scientist in a biopharmaceutical company. Your team has discovered a protein, isolated from the venom of a rare Amazonian snake, which shows promise as an anticancer agent. Everyone in your small company is excited by the potential of this protein, but you know that you cannot obtain enough protein from rare snakes to test in mice, let alone use for human patients. To obtain large amounts of the protein, you plan to take the snake gene that encodes the protein, insert the snake gene into cultured cells, and use the cells to manufacture the protein. If all goes well, this will allow you to make nearly unlimited amounts of the protein for testing and later, hopefully, for use in patients.

OVERVIEW

This activity is a classroom simulation in which you will act as a research and development scientist. In this role, you will be given portions of three procedures, each of which has a centrifugation step. You will need to consider how to accomplish the centrifugation requirements of each of the three procedures:

- In the first procedure, you must separate cells that were frozen for long-term storage from their storage solution.
- In the second procedure, you must separate large volumes of cells from their growth medium.
- In the third procedure, you must concentrate a protein product that is diluted in buffer.

Although this activity is a simulation, the procedures are real ones that are routinely used in many laboratories.

There are serious hazards associated with centrifugation that will also be described in this activity. Thus, in this activity, you will perform the following tasks:

- Explore three common applications of centrifugation in the biotechnology industry.
- Use manufacturers' technical information and online resources to design a centrifugation strategy.
- Practice planning to perform a procedure found in the scientific literature.
- Learn about safety issues involved in centrifugation.

BACKGROUND

A. What Is Centrifugation?

Consider the graduated cylinder that contains gravel, sand, and water in Figure 7.2. The substances in this graduated cylinder will tend to settle out, that is, sediment, over time. The larger gravel particles will sediment more quickly than the smaller sand particles, so there will eventually be a bottom layer of mostly gravel, a layer of mostly sand, and a relatively clear liquid layer in the upper part of the cylinder. The force that causes this separation to occur is gravity, and the basis of separation is the relative size of the substances. This process, however, is slow. The separation can be greatly accelerated if more force is applied, as is the case in centrifugation. A **centrifuge** is *a piece of equipment that accelerates the rate of sedimentation by rapidly spinning the samples, thus creating a force many times that of gravity.*

There are many applications of centrifugation in biology laboratories, for example, to rapidly perform the following functions:

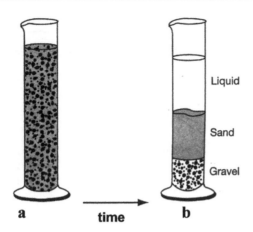

FIGURE 7.2   Separation of particles and liquid under the influence of gravity.

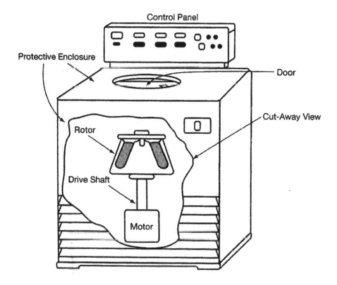

FIGURE 7.3   The basic parts of a centrifuge.

- Separate small "particles" from a liquid mixture:
  - Bacteria
  - Viruses
  - Cells
- Separate two immiscible liquids from one another. Immiscible liquids do not mix or blend with one another.
- Isolate cellular organelles from an extract of cells.
- Isolate DNA, RNA, and proteins from a mixture.

B. Basic Design and Types of Centrifuges

The basic design of a centrifuge is shown in Figure 7.3. This is a floor model; smaller centrifuges sit on a countertop. Centrifuges spin samples rapidly around a central drive shaft. The **rotor** *sits on the drive shaft and holds tubes or bottles containing the samples.* All modern centrifuges have a protective enclosure because there is a lot of force on the rotors and their contents as the centrifuge spins. The enclosure protects people and property from damage in the event a container breaks or a malfunction scatters debris out of the centrifuge.

There are many designs and models of centrifuge. One way to classify them is based on their maximum speed of rotation:

- **Desktop, or clinical centrifuges** *spin samples at rates slower than 10,000 revolutions per minute (rpm).*

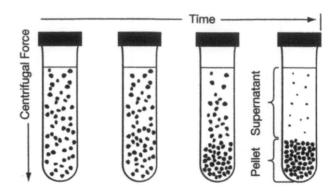

FIGURE 7.4   **Formation of a pellet during centrifugation.**

- **Superspeed centrifuges** *spin samples at speeds from 10,000 to 30,000 rpm.*
- **Ultracentrifuges** *spin samples at speeds up to and sometimes exceeding 80,000 rpm.*

There are also different sorts of rotors. The rotor in Figure 7.3 is a **fixed angle rotor**, *which means that it holds the tubes at a stable angle relative to the centerline of the centrifuge.* The tubes in a fixed angle rotor are held securely by the rotor and do not swing out as the rotor starts to spin. In contrast, a **swinging bucket rotor** *allows the tubes to swing out when the rotor spins.*

Consider separating particles from a liquid in a centrifuge. The sample is inside a centrifuge tube that is placed into the rotor. Inside the rotor, the sample spins around a great many times during a centrifuge run. After centrifugation, the particles will have separated from the liquid. The **supernatant** *is the liquid at the top of the tube;* the **pellet** *is the aggregated particles at the bottom* (see Figure 7.4).

C. How Much Force Is Generated in a Centrifuge?

The magnitude of the force that arises in centrifugation is dependent on two factors. The first is intuitively obvious; the faster the centrifuge spins, the more force there is—and the faster the separation will occur. The second factor is less obvious—unless you have played the game "crack the whip." In this game, children hold hands to form a line and spin around in a circle, perhaps on ice skates. The child who experiences the most force is the child at the outer end of the line. Similarly, in a centrifuge, the further a particle is from the center of rotation, the more force there is on it (see Figure 7.5).

Relative centrifugal force (RCF) refers to *the amount of force that the materials in a centrifuge experience.* RCF is the same as "× g," that is, "times the force of gravity." There is an equation to calculate the RCF generated in a particular situation (see Formula 7.1). As you can see in this equation, the amount of force applied to a particle in a centrifuge depends on both the speed at which the centrifuge spins the sample (in revolutions per minute) and the distance of the particle from the centerline (in cm). The equation is rearranged in Formula 7.2 for situations where you need to calculate how many RPM to use to achieve a particular force.

Formula 7.1

Calculating the Force Applied to a Particle in a Centrifuge

$$RCF = 11.17 (r)(n/1000)^2$$

where:
    RCF = relative centrifugal force
    r = radius of rotation in centimeters from centerline
    n = rotor speed in rpm

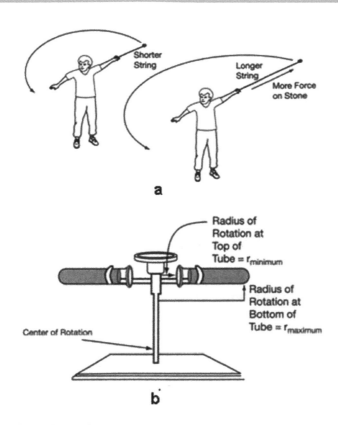

a

b

FIGURE 7.5   The radius of rotation is the distance from the centerline to a particle. (a) The longer a string that is being whirled (or the longer a line of children playing "crack the whip"), the more force there is at the end of the line. (b) In a centrifuge, at a constant rate of rotation, the further a particle is from the center of rotation, the more force it experiences. (This is a "cartoon" centrifuge, created for purposes of illustration, which is missing many essential features.)

Formula 7.2

Calculating the RPM Required to Achieve a Particular Force on a Particle in a Centrifuge

$$1000\sqrt{\frac{RCF}{11.2r}} = RPM$$

where:
  RCF = relative centrifugal force
  r = radius in centimeters from centerline
  RPM = revolutions per minute

Example 7.1

A tube is in a fixed angle rotor, as shown in Figure 7.6. Observe that the radius of rotation (r) is different along the length of the tube because of the angle. A particle in the tube, therefore, experiences a different amount of force depending on its location in the tube. Suppose the rotor is spinning at 30,000 rpm. Calculate the force on a particle at each of the following distances from the centerline:

* At the minimum radius of rotation (r_{min}): 3.84 cm
* At the average radius of rotation (r_{ave}): 6.47 cm
* At the maximum radius of rotation (r_{max}): 9.10 cm

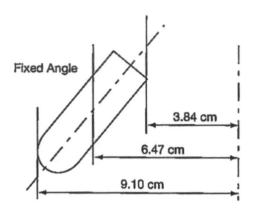

FIGURE 7.6 **Example of tube orientation in fixed angle rotor.**

Answer

- At r_{min}: RCF = 11.17 (3.84) (30,000/1000)^2 ≈ 38,603 (in this situation, most people would round for convenience to 38,600)
- At r_{ave}: RCF ≈ 65,000
- At r_{max}: RCF ≈ 91,400

Safety Briefing

Centrifuges look sturdy; floor models look a bit like a washing machine. But centrifuges are fragile and are among the most hazardous laboratory instruments used by biologists. This is because of the immense forces that arise during centrifugation. Never use an ultracentrifuge without proper instruction. Always operate centrifuges according to the manufacturer's instructions.

One of the major centrifuge safety concerns is that a rotor in an ultracentrifuge (the type of centrifuge that achieves the highest forces) might come off the shaft while spinning at high speeds. The rotor then becomes a missile, capable of penetrating cement walls, devastating a laboratory, and obviously adversely affecting any individuals in its path. Manufacturers provide protective enclosures to help prevent this from occurring, but it is essential that rotors and centrifuges are used according to the manufacturer's guidance.

The creation and distribution of aerosols during centrifugation is a more subtle, but also serious concern. Aerosol formation is inevitable during normal centrifugation, and very high levels are released if an accident occurs. Manufacturers, therefore, make special rotors, caps, bottles, and tubes that can be firmly sealed to prevent the escape of aerosols, even in the event of an accident. If you work with pathogenic or toxic substances, use these containment devices.

ACTIVITIES

You will complete a series of three activities that relate to centrifugation in your imaginary company:

- Activity Part A: Centrifugation will be used in the early stages of research to thaw cells. Your challenge in this part will be to order adapters that allow you to use the proper size of centrifuge tubes in your centrifuge.

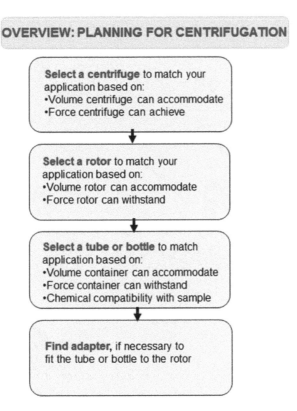

OVERVIEW: PLANNING FOR CENTRIFUGATION

Select a centrifuge to match your application based on:
• Volume centrifuge can accommodate
• Force centrifuge can achieve

Select a rotor to match your application based on:
• Volume rotor can accommodate
• Force rotor can withstand

Select a tube or bottle to match application based on:
• Volume container can accommodate
• Force container can withstand
• Chemical compatibility with sample

Find adapter, if necessary to fit the tube or bottle to the rotor

FIGURE 7.7   Planning centrifugation requires matching the centrifuge, rotor, bottles/tubes, and adapters to each other and to your application.

- Activity Part B: Centrifugation will be used to separate growing cells from their nutrient broth. Your challenge will involve matching the adapters, rotor, and centrifuge.
- Activity Part C: Centrifugation will be used to concentrate a product so that quality control tests can be run on it. Your challenge in this part will be to purchase a new centrifuge that will allow you to concentrate your product.

Figure 7.7 shows an overview of the steps required to plan a centrifugation run. Figure 7.8 shows how centrifugation is used in your imaginary company at various stages in the life cycle of your product.

Part A: Procedure 1; Early Stage of Research: Thawing the Cells That Will Be Transfected

You are working on the early steps in the process of taking the snake venom gene of interest (GOI) and inserting it into cultured cells. The cells you will use are currently in ultracold storage in liquid nitrogen. You must carefully thaw the cells and begin growing them under laboratory culture conditions. The thawing procedure that you plan to follow is a common one. Observe that there is a centrifugation step early in this procedure that will separate the stored cells from the storage solution in which they were frozen. You must be sure that you have the centrifugation equipment that will be required to complete this step. A portion of the procedure is shown in Box 7.1.

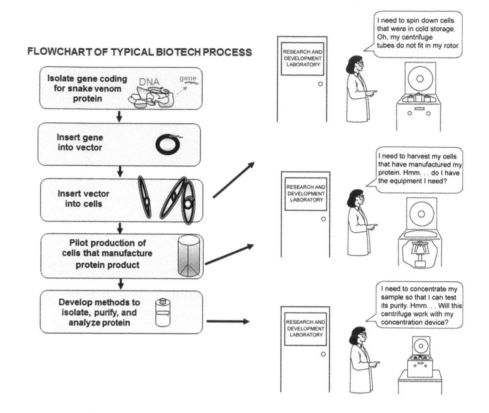

FIGURE 7.8   **Three procedures that require centrifugation during research and development of the snake venom anticancer drug.**

Box 7.1   Procedure 1; Using Centrifugation When Thawing Cells

Procedure to Thaw Cells That Have Been Stored in Liquid Nitrogen

(Only a portion of the procedure is shown here.)

1. Gently pipette cells from the freezer vial into a sterile 15 mL conical centrifuge tube using a 1 mL pipette.

2. Add 4 mL cell culture medium to the cells, slowly, drop by drop, to avoid osmotic shock. Mix gently by swirling as medium is added.

3. Gently pipette the cells up and down to release clumps.

4. Centrifuge the cells at 200 × g for 5 minutes.

5. Remove and discard the supernatant.

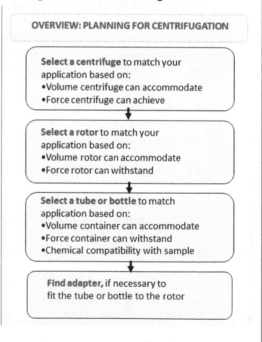

Observe that a centrifuge is used in this procedure to separate cells, which are a kind of "particle," from the liquid storage medium in which they were frozen. During centrifugation, the cells will go to the bottom of the centrifuge tube where they will form a loose pellet. You will then be able to pipette the spent storage medium, the supernatant, away

Eppendorf Centrifuge 5702

Rotor A-4-38

FIGURE 7.9   The problem. You have a suitable centrifuge, but the 15 mL conical tubes specified in the procedure are too small for the rotor sleeves.

from the pellet. The cells are to be centrifuged at 200 times gravity ($\times$ g). The cells will be damaged or killed if you spin them with more force. However, if you spin the cells with too little force, they will not form an adequate pellet.

You take an inventory of what you have on hand in the laboratory:

- For step 1 in the procedure, you find a rack of sterile 15 mL conical centrifuge tubes.
- For step 4 in the procedure, you find a tabletop centrifuge, shown in Figure 7.9. It is possible to use this centrifuge to spin the cells at 200 $\times$ g.

Unfortunately, when you try to insert the 15 mL conical tubes into the centrifuge, they clatter around in the rotor, as shown in Figure 7.9. Centrifuge tubes and other containers must fit snugly into the rotor; otherwise they can break or cause the rotor to become unbalanced. An unbalanced rotor can, at worst, fly off the driveshaft. An unbalanced rotor can also cause expensive damage to the centrifuge.

Now what? You look in the cabinet where the centrifuge was kept in the hopes of finding **adapters**, which are *devices that can be placed into the rotor sleeves so that the tubes fit snugly.* You find four adapters that fit 50 mL tubes. Unfortunately, 50 mL cell culture tubes are too big for the very small volume of cells you will be thawing.

The solution to this dilemma is to order the right adapters for the 15 mL conical tubes from the centrifuge manufacturer. This is not a problem because you have allowed ample time for planning and preparation, and you have sufficient resources to buy essential equipment. All you need to do is fill out the requisition form for the proper adapters so that the purchasing manager knows what to buy for you. Fill out the form and have your supervisor (instructor) sign it.

Your To-Do List: Procedure to Thaw Cells Stored in Liquid Nitrogen

1. Using online resources and/or references provided by your instructor, find the actual part number for the correct adapters for 15 mL conical tubes to fit the rotor and centrifuge in Figure 7.9. Alternatively your instructor may have you find adapters for a different combination of tube/rotor/centrifuge.
2. Fill in the following requisition form.
3. Have the form signed by your supervisor (instructor).

---

page 1 of 1
Cleanest Genes Corporation
Requisition Form r1200 version 1.0

Name of person requesting item _____

Department _____Date _____

Name and description of item to be purchased _____

_____

Quantity_____ Purpose of item _____

_____

Manufacturer and catalog number _____

Cost (if known, otherwise purchasing manager can get quotes) _____

Accessories, if any, required. Provide manufacturer, catalog numbers, costs

<u>ACCESSORY NAME</u>   <u>MANUFACTURER</u>   <u>CATALOG #</u>   <u>COST</u>

Supervisor's signature _____ Date _____

Part B: Procedure 2; Pilot Production

Many months have passed. There have been obstacles in the development process, but overall, your team has made good progress. You have inserted the snake gene into the cultured cells, and the cells produce the protein. You are now entering pilot production, a stage where you will transition from working in smaller culture plates to vessels that hold 5 L. You will need to separate the cells from the 5 L of medium after each run. A centrifuge will again be used to separate the cells from the cell culture medium, but the situation is a little different now. First, you will want the *supernatant*, not the pellet, because the protein is secreted by the cells into their liquid medium. Second, you have a much bigger volume than you did before, and 15 mL centrifuge tubes will be too small. You will need bigger centrifuge containers for the cells. You will also need a bigger centrifuge to hold the bigger containers. Finally, you will be spinning the cells with more force because you need to completely remove the cells from the cell culture medium and the cells need not survive centrifugation. Box 7.2 shows part of the procedure you plan to follow.

Box 7.2   Procedure 2; Using Centrifugation When Harvesting Cells

Procedure to Harvest Cultured Cells for Volumes of 1–5 Liters

(Only a portion of the procedure is shown here.)

1. Pre-weigh centrifuge bottles.
2. Place nutrient broth with cells in bottles and centrifuge at 10,000 × g for 10 minutes.
3. Remove the liquid.

OVERVIEW: PLANNING FOR CENTRIFUGATION

Select a centrifuge to match your application based on:
- Volume centrifuge can accommodate
- Force centrifuge can achieve

Select a rotor to match your application based on:
- Volume rotor can accommodate
- Force rotor can withstand

Select a tube or bottle to match application based on:
- Volume container can accommodate
- Force container can withstand
- Chemical compatibility with sample

Find adapter, if necessary to fit the tube or bottle to the rotor

One of your colleagues from another laboratory agrees to let you use her centrifuge for these pilot runs. You visit her laboratory and see the equipment in Figures 7.10 and 7.11.

- The centrifuge has a nameplate; it is made by Beckman Coulter, and is a model Avanti J-E High Performance Centrifuge.
- You find three rotors in a cabinet, which fit in the centrifuge; your colleague invites you to use any rotor you like. (The three rotors are not shown to scale, but rotor C is bigger than the other two.)
    - Rotor A has 8 spaces; each space holds a 50 mL centrifuge tube properly, without an adapter.
    - Rotor B has 18 spaces; each space holds a 10 mL centrifuge tube properly, without an adapter.
    - Rotor C has 6 spaces; each space is too big for any sort of centrifuge tube.
- You discover some centrifuge bottles in the cabinet (see Figure 7.11). You fill one of them with water and find that it holds 500 mL. This bottle fits into the spaces in rotor C perfectly without any adapter.

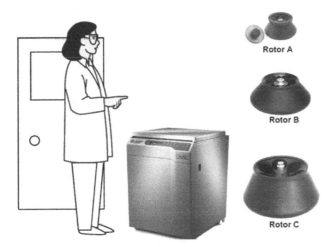

FIGURE 7.10   The centrifuge and rotors that you can use. Your colleague has three different rotors: rotor A, rotor B, and rotor C that go with the Beckman Avanti J-E High Performance Centrifuge and are available for your possible use. Note that all three are fixed angle rotors where the tubes are held at a specified angle in the rotor throughout the centrifugation run. (Photos courtesy of Beckman Coulter, Inc.)

FIGURE 7.11   The type of bottle and tubes available for your use. (Photos courtesy of Beckman Coulter, Inc.)

Your To-Do List: Pilot Production Procedure

1. **Decide first which, if any, of these rotors will work for your application.** You can start by checking their capacity (total volume they will hold) and deciding which one (if any) would be the best for your needs. In this situation, one of these rotors will work. Make a decision whether A, B, or C is preferable and record it here. _____ Once you have decided which rotor to use, you can return the other two to their storage area (or imagine returning them to the storage area).

2. **Find the model of the rotor you selected.** You chose a rotor: What is its model number? This would be easy to determine if the model number was printed on the rotor. You look hopefully all over the rotor—no number or identifying information is stamped on its surface. The Beckman Coulter centrifuge catalog will enable you to figure out which rotor you have. Find this catalog online at https://www.beckman.com/centrifuges/rotors/fixed-angle. Record here the number of the model rotor you have. _____

Note that it may not be possible to spin down all 5000 mL of your cells in a single centrifuge run.

If you need help, a senior technician offers you this advice:
   Use a search engine to find Beckman's information about the Avanti J-E centrifuge. Then find which rotors are compatible with that centrifuge. The rotor you have contains six places for centrifuge bottles. Which rotor(s) might be the one you have? _____

3. **Make sure the rotor/centrifuge is capable of providing 10,000 × g.** Check the information in the catalog for the model of rotor you will be using. Is it capable of providing at least 10,000 × g? Yes or no? _____ If the answer is no, you need to select a different rotor and perhaps even find another centrifuge to use.

At this point, you have decided to use the Avanti centrifuge, and you have selected a rotor that will work for your application. Now you need to decide if the centrifuge bottles or tubes that you found in the cabinet will work with the rotor you have chosen. You need to find out the following information:

- Is this bottle or tube recommended by the manufacturer for use with the chosen rotor?
- What is the maximum g-force that this container can withstand? (Every bottle and tube has a maximum force at which it can be used. The container is likely to break at higher forces, at best causing a mess at worst contaminating the centrifuge and the laboratory and causing the loss of a valuable sample. Therefore, you need to know what type of centrifuge container you found in the cabinet.)

It would be extremely time-consuming (perhaps impossible) to determine without doubt what part number and brand of bottle you have because bottles are seldom labeled with this information. In a working laboratory, the centrifuge bottles and tubes you find may not even be made by the same manufacturer as your centrifuge and rotor. Tubes and bottles may be purchased at a time when a particular model of centrifuge is being used, and they may remain behind when that centrifuge model has been replaced. If you are going to perform ultracentrifugation, where the forces on the tubes or bottles are extremely high, you must be certain that your centrifuge tubes or bottles are compatible with your equipment and can withstand the force you will be using. If you are not certain that your ultracentrifugation tubes or bottles are compatible, then you must order new ones from the centrifuge manufacturer. In our present scenario, however, you will be working with a centrifuge that achieves high forces, but they are not as high as those attainable by an ultracentrifuge. You, therefore, may be able to use the centrifuge bottles you found in the cabinet.

YOUR TO-DO LIST: PILOT PRODUCTION PROCEDURE (Continued)

4. **Find out if the bottles you have available in the cabinet are acceptable.** Consult the rotor information provided on the Beckman Coulter website to help determine the compatibility of the centrifuge bottles you found in the cabinet. Is there a 500 mL, wide-mouthed plastic bottle that is compatible with the rotor? To find out, check the catalog page that describes the rotor. Are there any 500 mL bottles listed as compatible with the rotor? _____ If so, what is/are the part number(s)? _____ If you find a centrifuge bottle that matches the rotor in terms of size and style, most people would use it at moderate g-forces even if they are not certain of its brand and part number. The force needed in this example, 10,000 × g, would probably be considered "moderate." If you think the bottles will work, check with your supervisor just to be sure. So, will the centrifuge bottles in the cabinet work for your purposes or do you need to order new ones? _____

5. **Find out how many rpm will give a force of 10,000 × g.** Once you have the centrifuge/rotor/bottle combination, you need to calculate how fast to run the centrifuge to obtain a force of 10,000 × g. This is because the centrifuge is adjusted according to rpm, not force, but the procedure specifies the force required. You, therefore, need to be able to convert units of force, RCF (or × g), to units of rpm. This is accomplished using Formula 7.2. Find the average radius of rotation for your rotor using the Beckman Coulter website. (Remember that r must be in units of centimeters.) Then use Formula 7.2 to calculate how fast you should set the centrifuge to get the right force. Show your calculations and answer in the following box:

Part C: Filtration and Centrifugation

It is now much later in the development process, and your team is making steady progress. The first pilot batch of protein has been purified. It is now time to check the protein preparation to see if it is pure enough to use for animal studies.

Your team plans to run many tests on the protein, one of which requires electrophoresis. If the protein is pure, it will display only a single band after electrophoresis. Impurities will show up as additional bands. (We will explore electrophoresis later in this unit. At this point, it is only important to understand that electrophoresis will be used to look for contaminants in the protein preparation.)

The purified protein is dissolved in a buffer and is fairly dilute. Electrophoresis, however, requires very small volumes of *concentrated* protein sample. Your team must, therefore, concentrate the protein, that is, remove buffer, in such a way that the protein is not harmed. You find a commercial product called a Cleanspin apparatus (an imaginary brand, but a real type of device) that is intended for exactly this purpose. The Cleanspin apparatus contains a filter with tiny pores that is held in a small, plastic centrifuge tube. Liquids can pass through the pores of the filter, but proteins are too large. You only need a few microliters of sample for electrophoresis, so the filter can be very small. One of

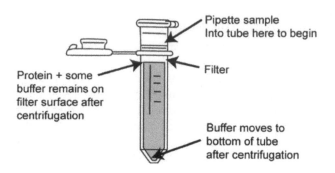

Pipette sample
Into tube here to begin

Filter

Protein + some
buffer remains on
filter surface after
centrifugation

Buffer moves to
bottom of tube
after centrifugation

FIGURE 7.12   A Cleanspin device.

the issues with this type of filter is that the pores are so fine that the liquid will not read-ily penetrate it. Therefore, more force than that of gravity is required to force the liquid through the pores. Centrifugation can provide this force.

A Cleanspin filtration device is shown in Figure 7.12. A portion of the procedure that comes from the Cleanspin manufacturer is shown next.

Box 7.3   Procedure 3; Using Centrifugation to Concentrate Protein

Procedure to Concentrate Protein Using a Cleanspin Filter

(Only a portion of the procedure is shown here.)

1. Place 100 μL of sample in the top of the Cleanspin filter apparatus.

2. Close the lid and orient the tube so the mark faces the center of the rotor. (Rotor must hold 2 mL microfuge tubes.)

3. Use an appropriate balance tube in the centrifuge opposite the position of the sample tube.

4. Centrifuge for exactly 3 minutes at 7,500 × g.

5. Open the filter unit and gently remove the protein solution from the top chamber.

OVERVIEW: PLANNING FOR CENTRIFUGATION

Select a centrifuge to match your application based on:
- Volume centrifuge can accommodate
- Force centrifuge can achieve

Select a rotor to match your application based on:
- Volume rotor can accommodate
- Force rotor can withstand

Select a tube or bottle to match application based on:
- Volume container can accommodate
- Force container can withstand
- Chemical compatibility with sample

Find adapter, if necessary to fit the tube or bottle to the rotor

This procedure requires a centrifuge that holds **microcentrifuge tubes**, which are *small plastic centrifuge tubes for small volumes, usually 2.0 mL, 1.5 mL, or 0.5 mL*. The type of centrifuge for these tubes is called a microcentrifuge or a microfuge. There are several microcentrifuges in your laboratory, one of which is shown in Figure 7.13. All the micro-centrifuges in your laboratory are the same model.

You examine this microcentrifuge and see that it does not have any way to adjust to different speeds (and therefore to obtain different g-forces). It is, therefore, not suitable for this application where a specific g-force is specified by the procedure. You check with your colleagues, but no one has a better microcentrifuge to lend you. You must purchase a new microcentrifuge. Microcentrifuges are often used in molecular biology, so you are confident that this purchase will be a good one for your laboratory.

FIGURE 7.13   A microcentrifuge. This type of centrifuge is used to centrifuge small plastic microfuge tubes. Molecular biologists often work with small volume samples, so microcentrifuges are ideal.

Your To-Do List: Microfiltration Procedure

Using the resources provided by your supervisor (online or paper catalogs), select a new microcentrifuge and rotor to purchase for your laboratory that meets the following criteria:

- Adjustable so that you can select the required g-force.
- A rotor that can hold at least eight microcentrifuge tubes at a time.
- Operational in a refrigerated mode because proteins are more stable when cold.

Provide ordering information on the requisition form that follows. Have your supervisor sign it.

---

page 1 of 1
Cleanest Genes Corporation
Requisition Form r1200 version 1.0

Name of person requesting item _____

Department _____Date _____

Name and description of item to be purchased _____

_____

Quantity _____

Purpose of item _____

Manufacturer and catalog number _____

Cost (if known; otherwise purchasing manager can get quotes) _____

Accessories, if any, required. Provide manufacturer, catalog numbers, costs.

ACCESSORY NAME   MANUFACTURER   CATALOG #   COST

Supervisor's signature _____ Date _____

Congratulations! Your team is successful.

Laboratory Exercise 29: Separation of Two Substances Based on Their Differential Affinities for Two Phases

OVERVIEW

Chromatography is a class of powerful and versatile separation techniques. This laboratory exercise is an introduction into fundamental ideas of chromatography. In this introduction, you will see how two molecules are separated from one another based on differences in how they interact with two phases. The phases in this exercise are two liquids: one is water, the other an organic solvent. The two molecules you will separate are paints that you will be able to see as they separate from one another.

You will also identify the components of a mixture of paints of two or more colors. To do so, you will need to separate the paints from one another and then compare your results to standards.

When you complete this laboratory exercise, you will have achieved the following objectives:

* Have a better understanding of the term "differential affinity for two phases."
* Have a visual image of a separation between two phases that will help you to understand chromatographic separations.
* Have used a separation method as part of a strategy to identify the components of a mixture.
* Have practiced using safe practices to handle an organic solvent.

BACKGROUND

Chromatography is a class of separation methods, all of which are based on differences between how molecules interact with two phases. Phases can be liquids, solids, or gases. The phases must be different from one another. Different molecules in a mixture interact with the two phases differently. Some chromatographic methods take advantage of differences in the sizes of molecules. (Recall that size difference is also the basis for filtration and centrifugation techniques.) Chromatography can also separate molecules based on differences in several other properties. Chromatography, for example, can separate molecules based on differences in their charge, which is useful for proteins. Yet other chromatography methods separate substances based on differences between how they dissolve in a polar, hydrophilic ("water-loving") phase and a nonpolar, hydrophobic ("water-hating") phase. This latter basis for separation is the one you will explore in this laboratory exercise.

Water and ethyl acetate are the two phases we will use in our exercise. Water is a polar solvent; ethyl acetate is a nonpolar solvent. The two substances we will separate from one another are two paints/; one water based, the other oil based. We use paints because they are easily followed visually as they partition between the two phases.

Safety Briefing

Ethyl acetate is potentially hazardous.

* Read its Safety Data Sheet (SDS).
* If possible, work in a chemical fume hood; avoid breathing fumes.
* Avoid eye contact.
* Keep away from heat, sparks, and flame.
* Keep container closed when not in use.
* Change gloves frequently—ethyl acetate may penetrate the glove plastic over time.
* Be sure your test tubes are compatible with ethyl acetate.
* Do not pour down the sink. To dispose, place in a fume hood, open top, allow to evaporate.

Planning Your Work: Separation of Two Substances

1. Read the SDS for ethyl acetate and record precautions in your laboratory notebook that you will need to follow to use this solvent safely.

2. Find out which is denser: ethyl acetate or water. Ethyl acetate and water do not mix well with one another, much like vinegar and oil do not mix in salad dressing. Ethyl acetate and water are both clear liquids, so they look the same, but the denser liquid is the one in the bottom of the tube. You can find the densities of the two materials in a *Merck Index* (see Bibliography, Appendix 3 for reference), in their SDSs, or using an online search engine. Record this information in your laboratory notebook.

LABORATORY PROCEDURE

Part A: Check Your Test Tubes to See If They Are Resistant to Ethyl Acetate

A.1.   Cover a work area in the chemical fume hood with absorbent paper.

A.2.   Make sure the tubes you plan to use have caps and are resistant to ethyl acetate.

    A.2.1.   If you plan to use glass tubes, they are resistant and do not require testing. Be careful to avoid breakage.

    A.2.2.   If you plan to use plastic tubes, find out if they are resistant to ethyl acetate.

        A.2.2.1.   You may have information from the tube manufacturer. If the manufacturer says that the tubes are resistant to ethyl acetate, it is probably safe to use them without further testing.

        A.2.2.2.   If you do not have information from the manufacturer, place a test tube into a glass beaker and place the beaker in the fume hood. The glass beaker will contain any leaks that may occur if the tube is damaged during testing.

        A.2.2.3.   Place a few milliliters of ethyl acetate into the tube.

        A.2.2.4.   Wait 45 minutes, then check to see if the tube is intact. If so, you can use this type of tube for this exercise. If the tube is damaged, consult with your instructor to find another type of tube.

Part B: Separation of Paints

B.1.   Prepare paints.

    B.1.1.   Your instructor will provide you with two or more paints whose identity is known, at least one of which is oil-based and the other water-based.

    B.1.2.   Take a pea-sized portion of each paint and mash it into 1 mL of its solvent.

        B.1.2.1.   Use water for water-soluble paints and ethyl acetate for oil-based paints.

        B.1.2.2.   You can use a weigh boat and spatula when mashing the paint.

B.2.   Combine eight drops of an oil-based paint with eight drops of a water-based paint to obtain a mixture.

B.3.   Mix the combined paint sample by vortexing it.

    B.3.1.   Describe the result.

    B.3.2.   Allow the mixture to sit undisturbed for 5 minutes and describe it again.

B.4.   Separate the components of your paint mixture from one another.

    B.4.1.   Place 7 mL of water in a test tube and add 7 mL of ethyl acetate to the same test tube.

    B.4.2.   Place your mixture from step B.3.2 into the tube with water and ethyl acetate.

    B.4.3.   Cover and mix thoroughly by vortexing it.

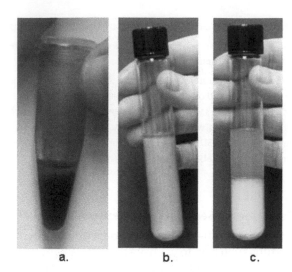

FIGURE 7.14   Separating a mixture of two paints. (a) The original sample mixture contained two different paints, mixed together in one tube. (b) The sample was placed in a tube containing ethyl acetate (nonpolar solvent) and water (polar solvent). The tube was mixed by vortex to combine the paint mixture with the two solvents. (c) In 2 minutes, the two solvents separated from one another. One of the two paints moved into the ethyl acetate phase and the other paint moved into the water phase.

        B.4.4.   Describe the appearance of the contents of the test tube immediately after mixing.

        B.4.5.   Allow the test tube to sit undisturbed for 10–15 minutes. Periodically describe the appearance of the contents. See the example in Figure 7.14.

B.5.   **Data analysis**

        B.5.1.   Which of the paints have a higher affinity for water and which for ethyl acetate?

        B.5.2.   What does the term "differential affinity for two phases" mean?

Part C: Separation and Identification of Paints

C.1.   Your instructor will provide you with an unknown consisting of two components, an oil-based and a water-based paint.

C.2.   Your instructor will also provide you with known paints to use as standards.

C.3.   Write a procedure to identify the components of your unknown.

C.4.   Identify the components of your mixture. Be sure to record the details of your procedure.

C.4.   **Data analysis**

        C.4.1.   Check your results with the answer key provided by your instructor.

LAB MEETING/DISCUSSION QUESTIONS

1.   Explain in your own words how substances can be separated based on their differential affinity for two phases.

2.   Can you imagine any practical applications of this method, that is, the separation of two or more substances by their differential affinity for two phases?

**Laboratory Exercise 30: Separation and Identification
of Dyes Using Paper Chromatography**

OVERVIEW

You separated paints based on their differential affinities for two phases in the previous laboratory exercise. The separation process that you used is related to chromatography, but it is not exactly the same. In chromatography, one of the two phases *moves*; it is **mobile**. The other phase is *fixed in place*; it is **stationary**. In this laboratory exercise, you will perform a relatively simple chromatography separation involving a stationary phase and a mobile phase. The particular type of chromatography you will use is called "paper chromatography." In **paper chromatography,** *paper fibers are the stationary phase, and a solvent is the mobile phase.*

You will again work with easily visualized dyes as your samples. You will complete the following tasks as you explore the principles of chromatography:

- Practice setting up a paper chromatography assay using the dyes in colored markers as your standards.
- Identify the dyes present in unknowns provided by your instructor.
- Watch a mobile phase interact with a stationary phase.

BACKGROUND

A. Introduction to Paper Chromatography

Paper chromatography is one of the oldest chromatographic methods but is seldom used now in the "real world" because other forms of chromatography provide much better separations. Nonetheless, paper chromatography nicely illustrates principles of chromatography.

To perform paper chromatography, a drop of the sample solution is applied to a piece of chromatography paper and allowed to dry. A liquid solvent is then placed in the bottom of a container, and the paper is suspended so that the bottom edge of the paper is in contact with the solvent. The solvent moves up through the paper by capillary action. As the solvent passes by the dried sample, the components of the sample may interact with the solvent and may move along with it. Separation will occur if the components within the sample interact to differing extents with the solvent. Compounds that interact readily with the solvent move along with it quickly. In contrast, compounds that have a higher affinity for the paper either do not move with the solvent at all or move with it more slowly.*

Some mixtures of compounds cannot be totally separated by paper chromatography with a single solvent, but they can be separated with two solvents in a two-step process. First, a mobile phase that will separate some of the components of the mixture is allowed to flow across the paper in one direction. The paper is then dried. Next, a different mobile phase is allowed to flow across the paper in a direction perpendicular to the flow of the first mobile phase. This is called "two-dimensional" chromatography. You will perform this type of separation by using isopropanol and water as the two mobile phases. Use the isopropanol first because it dries more quickly.

In all chromatographic methods, the components of the sample, once separated, are usually identified by comparison with standards. Standards are essential in chromatography; without them, chromatography cannot be used for identification. Even with standards, there is sometimes uncertainty because different substances may act identically when subjected to chromatographic conditions.

Most compounds of interest are not readily visible, so various detection methods are used to stain them or detect them. To make things simple in this exercise, however, we

---

* You may have experience using a rapid test to check whether you are infected with the virus causing COVID-19. When you perform such a test, you place a strip into liquid containing material from a nasal swab. The liquid moves up the strip by capillary action. Such rapid tests are derived from paper chromatography.

will work with colored inks. You will be provided with various inks as your known standards. Your instructor will also give you one or two unknowns to identify based on comparison with the standards.

LABORATORY PROCEDURE

1. Mark eight chromatography papers using pencil, as shown in Figure 7.15.
2. Obtain eight markers to use as standards; four water-soluble markers of different colors (e.g., Vis-à-vis) and four non-water-soluble markers of different colors (e.g., Sharpies).
3. Spot chromatography papers using your standards.
    3.1. Place one small spot of ink on each paper, as indicated on Figure 7.15. Observe that the spot is positioned 5 mm up from the first fold line you drew and 15 mm from the left side of the paper.
4. Allow the spots on the papers to dry.
5. Fill your chromatography plates to a depth of 3–4 mm with isopropanol.
6. Obtain one or two unknowns from your instructor. Your instructor will prepare the unknowns by spotting chromatography papers with one or more inks.
7. Fold each chromatography paper on the first fold line. The fold should be about 90 degrees.
8. Place the papers on plates so that the end you folded contacts the liquid and the long end of the paper is supported by the tray, as shown in Figure 7.16.
    8.1. Do not immerse the spots in the liquid.
    8.2. Fit two papers in each plate.
    8.3. It is best to obtain enough plates so that you can run all your standards and your unknown at the same time.
9. Allow the isopropanol to migrate at least 40 mm along the paper.
10. Remove the papers from the solvent and allow them to dry.
11. Flatten the dried papers and fold them at the second fold line.
12. Change the solvent in the plates to water.
13. Stand the papers in the water so that the end closest to the samples is in the solvent and the other end is on the lab bench (see Figure 7.17). Do not allow any portion of the samples to be immersed in the solvent, and do not allow the papers to touch one another.
    13.1. Allow the second solvent to move almost to the end of the papers.

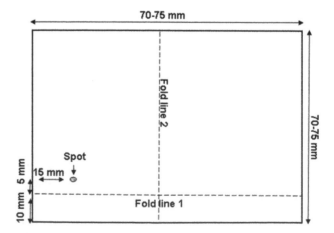

FIGURE 7.15   Marking chromatography papers.

FIGURE 7.16   Setting up paper chromatography in the first mobile phase. Eight standards and two unknowns in the first mobile phase, isopropanol.

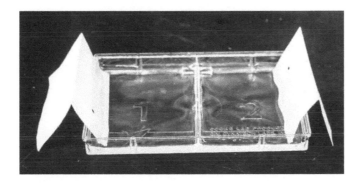

FIGURE 7.17   Chromatography papers in the second mobile phase.

14.   Remove the papers and allow them to dry.
15.   **Data analysis**
       15.1.   Identify your unknowns by comparing their appearance with the standards. Compare your answers with the answer key provided by the instructor.

LAB MEETING/DISCUSSION QUESTIONS

1.   The ink in most markers consists of one or more colored compounds that provide a certain color. How did chromatography enable you to separate the different compounds that comprise each marker?
2.   How did separating the molecules allow you to identify the compounds in your unknown?
3.   Explain the relationship between this laboratory exercise and the previous one. How are they similar? How are they different?
4.   How does the term "differential affinity for two phases" relate to this laboratory exercise?

5. In order to understand chromatography, it is necessary to understand the idea of a stationary phase and a mobile phase. What is the stationary phase in this method? What is the mobile phase? What causes molecules to separate from one another?

6. What is the force in this procedure that causes the mobile phase to move? Do you think this type of force is commonly used for chromatography? Why or why not?

7. How successful were class members in identifying their unknowns using this method?

8. A standard definition of **chromatography** is *a class of methods that separate compounds based on their differential distribution between a stationary phase and a mobile phase.* Explain this definition in your own words.

Laboratory Exercise 31: Separating Molecules by Agarose Gel Electrophoresis

OVERVIEW

The purpose of this laboratory exercise is to introduce the basic principles of agarose gel electrophoresis. **Electrophoresis** is *a class of separation methods that separate molecules under the influence of an electrical field.* The electrical field provides the force that causes molecules to move. Molecules that are positively charged move away from a positive pole toward a negative pole; molecules that are negatively charged move toward a positive pole. Therefore, differences in the charge of molecules can be used as the basis for their separation in electrophoresis. But charge is not the only factor that affects how molecules separate in electrophoresis. The samples in electrophoresis move through a matrix or gel that acts like a three-dimensional filter. Smaller molecules maneuver more readily and quickly through the matrix than larger ones. Size and charge are thus both bases for separation in electrophoresis.

Molecular biologists routinely use electrophoresis to separate DNA molecules from one another. In this exercise, however, you will use colored dyes because they will allow you to visualize the separation as it occurs. In this exercise, you will perform the following tasks:

- Look at how charge and size affect separation in electrophoresis.
- Learn how to pour a gel for electrophoresis.
- Learn how to operate an electrophoresis apparatus.
- Examine safety issues associated with electrophoresis.

BACKGROUND

Agarose gel electrophoresis is one of the most commonly used techniques in molecular biology because it separates DNA fragments based on differences in their size (see Figures 7.18 and 7.19). Agarose gel electrophoresis can be used for analytical purposes to

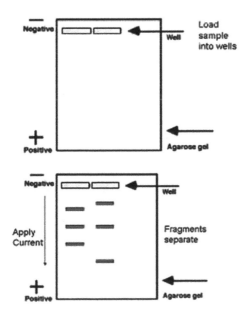

FIGURE 7.18   The principle of separation by electrophoresis. The samples are loaded into wells in a gel-like matrix, to which an electrical current is applied. DNA fragments, for example, are negatively charged and move away from the negative pole toward the positive pole. The rate of DNA migration is inversely proportional to the log of the size of the DNA fragment. The result is that DNA fragments separate from one another with the largest fragments at the top of the gel, closest to the wells, and the smaller ones lower down in the gel.

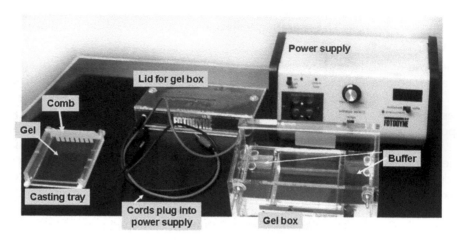

FIGURE 7.19   **The apparatus for electrophoresis.**

see which fragments of DNA are present. It can also be used for preparative purposes to isolate small amounts of a particular DNA fragment for further use.

The sample's constituents maneuver through a matrix during electrophoresis. Agarose is the matrix used most commonly to separate DNA fragments from one another. Agarose is a polysaccharide that, when warm, can be dissolved in buffer. When the solution cools, it forms a gel with a consistency somewhat like hard gelatin. You can control the shape of the gel by pouring the molten agarose into a **casting tray** that *shapes the agarose into a rectangular slab with small depressions for the samples. The depressions for the samples* are called **wells**. The agarose gel functions as a sieve. The size of the pores of the gel "sieve" can, to some extent, be controlled by adjusting the concentration of agarose used. A higher concentration of agarose creates a tighter sieve matrix than a lower agarose concentration. The buffer that you will use to dissolve the agarose contains ethylenediaminetetraacetic acid (EDTA), tris(hydroxymethyl)aminomethane (Tris), and boric acid, and is therefore called "TBE." Figures 7.20 and 7.21 show how a gel is made and how the samples are loaded into it.

To perform electrophoresis, an agarose gel is placed into an electrophoresis tank (also called "gel box") that is filled with TBE buffer. One sample is added to each well with a micropipette. An electrical current is applied across the gel. Molecules in the samples separate from one another based on their size and their charge. Positively charged molecules move through the gel toward the negative electrode. Negatively charged molecules move through the gel toward the positive electrode. A molecule that is more highly charged relative to its size moves quicker than one with less charge relative to its size. The effect of charge acts together with the effect of size; larger molecules move slower than smaller ones because larger molecules are more impeded by the gel matrix.

In this introductory electrophoresis laboratory exercise, you will not separate DNA fragments. Rather, you will work with dye molecules that have color and are, therefore, easy to follow as they move through the gel. These dyes will separate from one another based on differences in their sizes and their charges.

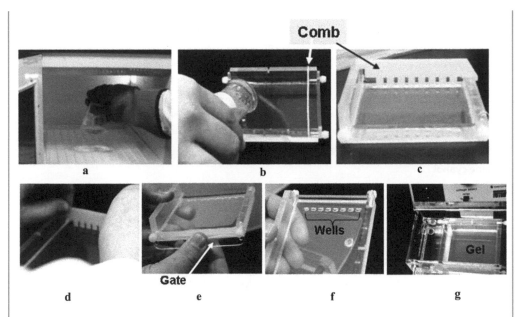

FIGURE 7.20 Preparing an agarose gel. (a) The agarose is dissolved and melted by heating in a microwave. The analyst is careful to avoid boiling the agarose or being burned by it. (b) The agarose is cooled until its flask is comfortable to hold. The liquid agarose is then poured into the casting tray in one smooth motion. Thirty milliliters of agarose makes a gel of appropriate thickness in this standard "minigel" casting tray. "Gates" at each end of the casting tray keep the gel from spilling out of the tray. (c) The gel has been poured and allowed to set for 30 minutes. It has solidified in the casting tray. The comb used with this gel has eight teeth and will provide a gel with eight wells. (d) The analyst gently removes the comb from the gel, being careful not to disturb the wells. (e) The analyst is lowering the gates. (f) The eight wells are clearly visible at the top of the gel. (g) The analyst has placed the gel in the gel box. Buffer will now be poured into the gel box until the gel is just submerged.

FIGURE 7.21 Loading the gel. The analyst is now loading the samples into the wells of the agarose gel that is resting in a plastic gel box. The wells are filled with buffer and the samples must be carefully applied into the wells. The samples sink because they contain a dense loading dye. After the samples are loaded, the box will be connected to the power supply and a current will be applied to the gel. A blue loading dye is routinely added to the samples to make them easier to see and to make them sink into the wells.

Planning Your Work: Separation of Molecules by Gel Electrophoresis

1. If your instructor has you prepare your own TBE, write a procedure in your laboratory notebook to prepare the buffer. Be sure to distinguish between what you plan to do and what you actually do when you come to lab.

2. You will separate four colored dyes from one another in this exercise. Fill in this table using internet resources or print resources provided by your instructor. Note that acidic molecules tend to give away hydrogen ions (H^+). This leaves the acid with a net negative charge.

Dye	Charge (+ or – at pH 8)	Molecular Weight
Xylene cyanol		
Bromophenol blue		
Orange G		
Safranin O		

3. Based on the information in the table, which dye(s) do you predict will move toward the positive pole during electrophoresis and which toward the negative pole?

4. Based on the information in the table, put the dyes in order based on their size.

5. Write a procedure to make 30 mL of 1% (w/v) agarose gel in TBE in a 100 mL flask (see step B.3 of the following Laboratory Procedure). Write your procedure in your laboratory notebook. Remember to distinguish between what you plan to do and what you actually do when you come to the lab.

6. What safety issues are of concern when you perform this exercise? Write safety considerations in your laboratory notebook.

Safety Briefing

TBE

- TBE can cause burns.
- Wear personal protective equipment (PPE), including gloves. Change gloves frequently. Acids and bases will penetrate plastic gloves; change them immediately in case of exposure.
- In case of a spill or accident, get assistance from your instructor immediately.

Agarose

- Hot agarose has the potential to cause burns—and to make a mess that is tedious to clean.
- Watch the flask with agarose while microwaving so that the liquid does not boil over the top of the flask.
- Agarose can become superheated; it does not boil until you move it, at which point it can boil over your hands. Swirl agarose carefully after 45 seconds of heating, wearing heat-resistant gloves to make sure it does not overheat.
- Use heat-resistant gloves to protect your hands when removing the flask from the microwave.
- Use a microwave that is not used for food.

Electrophoresis

- Electrophoresis uses potentially dangerous levels of electrical current.
- Never touch the inside of a gel box when the power supply is connected to the box.
- In the event that a crack occurs in the box, liquid may leak out of it. Do not touch the liquid to avoid exposure to the current.

LABORATORY PROCEDURE

Part A: Preparation of TBE Buffer

Your instructor may have you prepare your own TBE buffer or may have it prepared for you.

If directed to do so by your instructor, prepare TBE buffer according to these recipes:*

---

* Based on recipes in D. Micklos, G. Fryer, and D. Crofty, *DNA Science a First Course* (2nd ed.), NY: Cold Spring Harbor Press, 2003.

10X Tris/Borate/EDTA (TBE) Electrophoresis Buffer

Makes 1 L

1. Add the following dry ingredients to 700 mL of purified water in a 2 L flask:
    1 g of NaOH (molecular weight = 40.0)
    108 g of Tris base (MW = 121.10)
    55 g of boric acid (MW = 61.83)
    7.4 g of EDTA (MW disodium salt = 372.24)
2. Stir to dissolve, preferably using a magnetic stir bar and stir plate.
3. Add purified water to bring the volume to 1 L.

Note: If stored TBE comes out of solution, place the flask in a water bath at 37°C to 42°C with occasional stirring until all solid matter goes back into solution. Store at room temperature indefinitely.

1X TBE

Makes 10 L

1. Into a spigotted carboy, add 9 L of purified water to 1 L of 10X TBE.
2. Stir to mix.

Note: Store at room temperature indefinitely.

Part B: Casting an Agarose Gel

Refer to Figure 7.20.

B.1. Inspect your comb to be sure it is clean and its edges are smooth, straight, and without burrs. Place the comb Into the casting tray.
B.2. Seal the ends of the casting tray.
 B.2.1. If your casting tray has "gates," adjust the gates so that the liquid gel will be contained in the tray.
 B.2.2. If your casting tray does not have gates, seal the ends of the tray with tape so that liquid cannot run out the ends.
B.3. Make 30 mL of a 1% solution of agarose in TBE buffer according to your procedure that you wrote before coming to lab.
 B.3.1. Weigh out the amount of agarose you calculated you will need. Place it in a 100 mL flask.
 B.3.2. Add 30 mL of TBE to the flask.
 B.3.3. Microwave the flask containing the TBE/agarose mixture for about 45 seconds.
 B.3.4. Remove the flask from the microwave wearing heat-resistant gloves and gently swirl the flask. If you can still see agarose particles, return the flask to the microwave for about 15 more seconds.
 B.3.5. Repeat step B.3.4 until the agarose is all dissolved but do not let it boil.
B.4. Allow the agarose solution to cool to about 60°C.
 B.4.1. At this temperature, the mixture will not harden and it can be poured. Also, the agarose is cool enough that it will not damage the comb or casting tray.
 B.4.2. Your instructor may have you measure the temperature with a thermometer.
 B.4.3. Alternatively, many analysts rely on a sense of touch to determine when the agarose is sufficiently cooled. The bottom of the flask will feel warm to the touch but will not burn your skin.

B.5.   Pour 30 mL of agarose into the casting tray with the comb so that the teeth of the comb are just covered. Remove large bubbles with a pipette.

B.6.   Allow the gel to solidify 30 minutes; do not move the tray while this is occurring.

B.7.   Lower the gates on the casting tray or remove the tape.

    B.7.1.   Some analysts gently remove the comb at this point. Others wait until the gel is in the gel box and covered with buffer.

B.8.   Place the casting tray into the gel box with the wells closest to the negative pole.

B.9.   Fill both sides of the chamber with TBE buffer so that the buffer just covers the gel.

    B.9.1.   Look for "dimples" on the edges of the wells.

    B.9.2.   Pour in buffer until the dimples just disappear. At this point, the buffer exactly covers the gel and is the right level. If you have too much or too little buffer, current will not flow properly through the gel.

B.10.   If you have not already done so, gently remove the comb without tearing the gel. The TBE buffer helps lubricate the comb, making it easier to remove.

B.11.   You can store a gel for one or two days or use it immediately. To store the gel, keep it tightly wrapped in plastic wrap or leave it covered with TBE buffer in the gel box so it does not dry.

Part C: Familiarization with Electrophoresis Apparatus

Refer to Figure 7.19.

C.1.   Electrophoresis requires electrical current. This current is hazardous. Your apparatus is designed to help protect you from the current. Examine the apparatus.

    C.1.1.   Determine the locations of the positive and negative poles when the box is assembled.

    C.1.2.   Determine the pathway for current flow in the box.

    C.1.3.   Determine what safety features are present to help protect you from the current.

    C.1.4.   Find out if there is any way to bypass the safety features. If so, make careful note of this so you can avoid contact with the current.

Part D: Performing Electrophoresis

Refer to Figures 7.20 and 7.21.

D.1.   Your gel has a top, bottom, left, and right side. The top is the part that is nearest the negative pole. The wells are at the top.

D.2.   Load your samples into the gel from left to right, one sample in each well. The first lane on the left is Lane 1.

    D.2.1.   Load 4 µL of the following samples into the gel. Skip the first lane (well) because samples may not migrate as smoothly along the edge of the gel:

- Lane 2. Xylene cyanol
- Lane 3. Bromophenol blue
- Lane 4. Orange G
- Lane 5. Safranin O
- Lane 6. Mixture of all four dyes
- Lane 7. "Unknown" (a mixture of one or more dyes provided by your instructor)

    D.2.2.   Use a fresh tip for each sample.

    D.2.3.   It may be helpful to steady your hands by placing your elbows on the lab bench.

D.2.4.   Place the pipette tip into the well but do not puncture the bottom of the well.

D.2.5.   Gently depress the pipette plunger to release the sample into the well.

D.2.6.   Strive to ensure that the sample is entirely confined in the well.

D.3.   Cover the box and connect the leads of the electrophoresis apparatus to the power supply. The leads may be color coded: red plugs into red, black into black.

D.4.   Turn on the power and adjust the power supply to deliver 120 V.

D.4.1.   You should see bubbles appearing in the tank when the voltage is on.

D.4.2.   Read the current and voltage displayed by the power supply every 15 minutes as electrophoresis proceeds. You should see changes as the run proceeds.

D.5.   Your gel is oriented with the wells close to the negative pole. If any of the dyes are positively charged, they will run off the top of the gel into the buffer. Do you expect this to happen? Observe if it does occur.

D.6.   Run the gel until one of the dyes approaches the bottom of the gel. Then, turn off the power supply.

D.7.   Disconnect the leads and remove the casting tray and gel from the electrophoresis tank.

D.8.   Photograph the gel as directed by your instructor or draw the results.

D.9.   When you are certain you have recorded all the required information, discard your gel in the appropriate waste container.

D.10.   Label your photograph or drawing.

D.10.1.   Label the lanes with the number of the lane and the sample.

D.10.2.   Alternatively, label the lanes only with a number and refer to your laboratory notebook to see what was in each lane.

D.11.   **Data analysis**

D.11.1.   Which dyes ran to the positive and which to the negative electrode?

D.11.2.   In what order did the dyes run through the gel? Did your results match your predictions? Explain.

D.11.3.   Identify the compound or combination of compounds that make up your unknown sample. Check your answer with the answer key provided by the instructor.

LAB MEETING/DISCUSSION QUESTIONS

1.   In this exercise, did you use electrophoresis for a preparative or an analytical purpose? Explain.

2.   Compare the class results. The order in which the dyes run and their direction of migration should be the same for everyone. Is this the case? If not, discuss why not. Compare your class results with the student results shown in Figure 7.22.

3.   How successful were you in identifying your unknown? How successful were your classmates? Explain.

4.   Explain in your own words the purpose of electrical current in electrophoresis.

5.   Explain in your own words the purpose of the agarose gel.

6.   What happened to the current and voltage as the run progressed? Was either voltage or current held constant?

7.   What risks are associated with preparation of agarose gels? What precautions minimize these risks?

8.   What risks are associated with electrophoresis runs? What precautions minimize these risks?

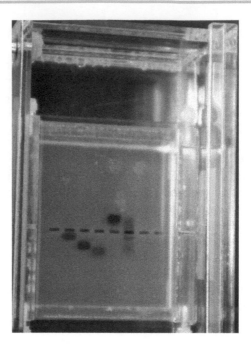

FIGURE 7.22   Student results for Laboratory Exercise 31. In this example, the comb was placed in the middle of the gel so that dyes could move in both directions without running off the top of the gel. Some electrophoresis casting trays allow this placement of the comb; others do not. Lane 1: empty. Lane 2: xylene cyanol. Lane 3: bromophenol blue. Lane 4: orange G. Lane 5: safranin O. Lane 6: mixture of all four dyes.

Laboratory Exercise 32: Using Agarose Gel Electrophoresis to Perform an Assay

OVERVIEW

The purpose of this exercise is to demonstrate how agarose gel electrophoresis is used to separate DNA fragments of different sizes. This exercise is set up as a simulation of a restriction fragment length polymorphism (RFLP) analysis.

RFLP analysis was the earliest version of "DNA fingerprinting," which is used to distinguish individuals from one another. You are likely familiar with the use of DNA fingerprinting in crime scene investigation, but DNA fingerprinting has many other applications as well. In this laboratory exercise, your mock RFLP assay will not be used in a crime scene context, but rather in a medical crisis to distinguish pathogenic viruses of two different strains.

It is possible to perform a real RFLP assay in your laboratory, and you may do so in a more advanced class. We have chosen to simulate a RFLP assay in this laboratory exercise to show you why the ability to separate DNA fragments from one another is so important.

As you perform the tasks in this laboratory exercise, you will perform the following tasks:

- Practice working with and visualizing DNA.
- Continue to improve your skills in performing electrophoresis.
- Learn about using agarose gel electrophoresis to assay DNA.

BACKGROUND

A. Agarose Gel Electrophoresis of DNA

In Laboratory Exercise 31, you learned how to perform agarose gel electrophoresis using dyes as your samples. In this exercise, you will work with samples of DNA, which are more relevant to biotechnologists. But we are not finished with the dyes xylene cyanol and bromophenol blue. These dyes are commonly added to DNA samples to monitor the progress of an electrophoresis run.

Consider what happens to DNA during agarose gel electrophoresis. The name "DNA" stands for deoxyribonucleic acid; DNA is an acid. Acids in water tend to give up hydrogen ions (H^+). At the pH at which electrophoresis is run, each subunit of DNA gives up one H^+ to become negatively charged. DNA will, therefore, migrate toward the positive electrode in electrophoresis.

Agarose gel electrophoresis is commonly used to separate mixtures of DNA fragments that are different sizes. All the DNA fragments have the same charge density, that is, one negative charge per nucleotide. Therefore, DNA fragments separate on the basis of size, with the larger fragments moving more slowly through the agarose sieve than the smaller ones. DNA fragments thus separate into "bands," with the larger fragments (bands) toward the top of the gel and the smaller ones lower down on the gel.

B. RFLP Analysis

RFLP analysis is a way to differentiate between individuals. DNA from an organism (or virus) is isolated. It is then cut into pieces with one or more restriction enzymes. **Restriction enzymes** are *a class of enzyme that cut DNA wherever a particular sequence of bases occurs.* When RFLP analysis is performed, specific restriction enzymes are chosen that produce DNA fragments that vary in number and size between individuals. In Figure 7.23, a restriction enzyme cuts a strand of DNA at two sites, resulting in three DNA fragments that vary in size. These three fragments are separated from one another by size on an agarose gel. The largest fragment (B) runs slowest, and the smallest (C) runs the fastest. The pattern of fragments viewed on the gel can vary between individuals and thus help with their identification.

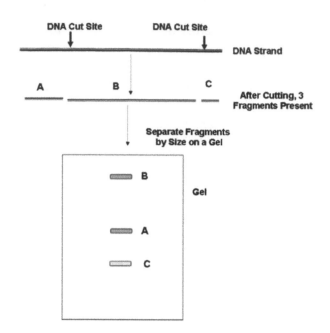

FIGURE 7.23   RFLP analysis example. This is the band pattern for one individual. Other individuals might have a different pattern of banks.

C. Ethidium Bromide Staining of DNA

In Laboratory Exercise 31, you used agarose gel electrophoresis to separate dyes from one another. Dyes have color and are easily visualized. DNA, however, is colorless and must be stained in order to see it. There are various stains used for DNA. Ethidium bromide is the stain most commonly used in the workplace because very low levels of DNA can be visualized with it. Ethidium bromide is a compound that inserts itself into the structure of DNA and fluoresces when it is activated by ultraviolet (UV) light. The DNA in an agarose gel can be stained by soaking the gel in an ethidium bromide solution. The ethidium bromide inserts itself into the DNA in the gel. The gel is then placed on a **trans-illuminator**, *a box with a UV light source.* The ethidium bromide fluoresces under the UV light, making the DNA visible. In a classroom situation, other stains may be substituted for ethidium bromide. This is because ethidium bromide is assumed to be mutagenic and carcinogenic and, thus, requires careful handling to be safely used. The directions in this laboratory exercise are for ethidium bromide, but your instructor may choose to use a less hazardous substitute.

Safety Briefing

- Ethidium bromide is presumed to be a carcinogen and mutagen.
- When working with ethidium bromide, wear gloves, safety glasses, and a laboratory coat.
- Any equipment that contacts ethidium bromide staining solution should be kept separate from other equipment to avoid spreading ethidium bromide through the laboratory.
- Gloves should be changed frequently to avoid contaminating the laboratory.
- Read the SDS for this chemical prior to use.

Scenario

(FICTITIOUS)

In March 2013, a 35-year-old man was diagnosed with "fifth disease" (a real viral disease but one that, in reality, is seldom severe) at a hospital in Albuquerque, New Mexico. He was sent home to recover, but later that evening, his symptoms rapidly worsened, and he returned to the hospital emergency room. He was admitted to the hospital, where he received 10 days of care until he was well enough to be sent home to finish his recovery. Over the course of the next two months, 3000 similar cases were diagnosed in the United States, resulting in 152 deaths.

Although the initial symptoms of this illness closely mimicked that of the virus that causes fifth disease, a common childhood illness that has few complications, the rapid progression and high death rate indicated that this was not actually a common viral illness. Scientists at the Centers for Disease Control and Prevention became involved in the study of this potential epidemic. The scientists were able to isolate the virus from patients. They used RFLP analysis to compare the DNA of the virus causing fifth disease to that of this potentially new and lethal virus. They found that they could rapidly differentiate between the fifth disease virus and this new strain of virus. When a particular restriction enzyme is used to cut the viral DNA, the pattern of DNA bands that is produced varies between the fifth disease virus and the new virus (see Figure 7.24).

In this simulation, you are a diagnostician in Albuquerque, and your job is to use the new RFLP method to test patients to see if they have fifth disease or the more serious new disease.

Planning Your Work: Using Gel Electrophoresis to Perform an Assay

Prepare for this laboratory exercise by using whatever method seems most helpful to you. You may want to create a flowchart, write out procedures, or write down the data you will be collecting and how you plan to analyze it.

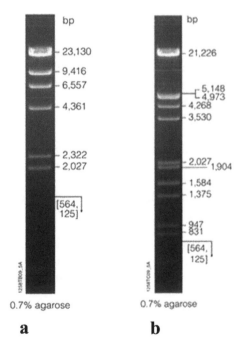

FIGURE 7.24   Comparison of RFLP patterns of fifth disease and the fictious "new" virus. (a) Fifth disease pattern after RFLP analysis. (b) New virus pattern after RFLP analysis. The units are base pairs (bp).

LABORATORY PROCEDURE

1. Cast a 0.7% agarose gel.
2. Briefly vortex the samples that you will obtain from your instructor to ensure that the samples are homogeneous. Then briefly spin them down in a micro-centrifuge to collect the sample in the bottom of the tube.
3. Load 6 μL of your samples onto the gel. You will have the following samples:
   - Lane 2: Control RFLP preparation from fifth disease virus
   - Lane 3: Control RFLP preparation from "new" virus
   - Lane 4: RFLP preparation from patient 1 with fifth disease symptoms
   - Lane 5: RFLP preparation from patient 2 with fifth disease symptoms
   - Lane 6: RFLP preparation from patient 3 with fifth disease symptoms
     - 3.1. Loading dye will have already been added to your DNA samples. Loading dye has two functions. It causes the sample to be denser than the buffer and, therefore, to sink into the wells. It also contains bromophenol blue so that you know when to stop the electrophoresis run.
     - 3.2. Remember to record which lane contains which sample.
4. Run electrophoresis at 120 V.
5. While electrophoresis is occurring, prepare an area for staining with ethidium bromide.
   - 5.1. Line a lab bench with absorbent paper.
   - 5.2. Place a sign notifying everyone that the area will be used for ethidium bromide staining.
6. When the bromophenol blue is nearly at the bottom of the gel, turn off the power supply.
7. Disconnect the leads and remove the casting tray and gel from the electrophoresis tank.
8. Put on gloves.
9. Stain the gel with ethidium bromide; remember to follow ethidium bromide safety rules as described in the Safety Briefing.
   - 9.1. Pour ethidium bromide from its stock bottle into a plastic staining tray used only for this purpose.
   - 9.2. Change your gloves.
   - 9.3. Slide the gel from the casting tray into the plastic staining tray with ethidium bromide.
   - 9.4. Do not place the casting tray or anything else on the bench reserved for ethidium bromide use.
   - 9.5. Do not touch the casting tray to the ethidium bromide staining tray.
10. Change gloves.
11. Stain for 10 minutes with gentle agitation. If you agitate the stain and gel by holding the plastic staining tray in your hand, be sure to wear gloves and change them afterward.
12. Carefully pour the ethidium bromide solution back into the stock bottle through a funnel.
13. Change gloves.
14. Rinse the gel briefly in tap water. Your instructor will tell you whether to use the regular sink for this purpose or to use an alternative method.
15. Using a spatula, remove the gel from the staining tray and place it on a UV light box.
16. Change gloves. Photograph the gel as demonstrated by your instructor.
17. Discard the gel as directed by your instructor.
18. Remove gloves and wash your hands.

19.   **Data analysis**
   19.1.   Label the lanes of your picture.
   19.2.   Compare the results of patients 1 through 3 with the two control samples.
   19.3.   Record in your notebook which of the patients is likely to be infected and with which virus.

LAB MEETING/DISCUSSION QUESTIONS

1.   Did we use electrophoresis in this exercise for preparative or analytical purposes? Explain the role of a separation technique in this laboratory exercise.
2.   Did you successfully diagnose the patients? Did the class? If not, discuss your diagnoses.
3.   Why is bromophenol blue used to determine when to end the run? Consider the relative molecular weights of bromophenol blue and DNA fragments.
4.   What are the safety issues involved in using ethidium bromide? Why change gloves so often?

Laboratory Exercise 33: Optimizing Agarose Gel Electrophoresis

OVERVIEW

Electrophoresis requires optimization to ensure trustworthy results. The purpose of this laboratory exercise is to look at factors that can affect the results of electrophoresis, which include the following:

- The amount of DNA loaded on the gel.
- The salt concentration of the buffer.
- The amount of time the gel is allowed to set before using.
- The thickness of the gel.

BACKGROUND

Planning Your Work: Optimizing Agarose Gel Electrophoresis

Write a protocol to investigate one or more of the factors previously listed that might influence the results of electrophoresis. Your instructor will tell you how many factors to investigate and might direct you to investigate a particular factor. Put your protocol into your laboratory notebook. In the laboratory, be sure to distinguish between what you planned to do and what you actually do. Include the following in your protocol:

- The factor to be investigated.
- The samples you will analyze.
- Your exact procedure. Copy this from Laboratory Exercise 31 whenever appropriate.
- How you will analyze your results.

LABORATORY PROCEDURE

In order to be able to see the effect of each factor, you will want to keep everything as consistent as possible except for the factor you are manipulating. Begin with the same conditions you used in Laboratory Exercises 31 and 32, and then vary one factor at a time. It is most efficient to work in teams, but you must be careful to be sure that everyone is doing each step the same way.

A. Amount of DNA Loaded onto the Gel

Your goal in this part of the investigation is to explore how much DNA is optimal, given a particular size of gel and a particular comb. Analyze at least three different amounts of DNA, ranging from 20 ng/well to 500 ng/well. Your results will depend on the type of DNA you analyze.

We suggest that you initially use a 0.8% agarose gel and that you analyze enzyme-digested lambda DNA where the sizes of the bands you should obtain are known. Be sure to use 30 mL of agarose so that your gels are consistent. If possible, run two gels that you expect to be identical, with the same samples and the same conditions to see if there is unwanted variability in the system.

B. Salt Concentration of the Buffer

The electrical current that flows during electrophoresis flows through the gel and the surrounding buffer. This flow of current is dependent on the presence of salt ions that carry the current. Your goal in this part of the investigation is to optimize the salt concentration in the buffer. Use three concentrations of salt: purified water (very low salt), 1X TBE (normal salt concentration), and 2X TBE (high salt concentration). If possible, run two gels that you expect to be identical, with the same samples and the same conditions to see if there is unwanted variability in the system.

C. Amount of Time to Harden the Gel

Gels do not take long to appear to be hardened, but there is evidence that it takes 30 minutes to an hour for the gel to be fully hardened. An incompletely hardened gel can result in smeary, poor bands. Check this out; try a gel that has just hardened, one that has hardened for 15 minutes, and one that has hardened for 30–45 minutes. It is desirable to run each of these three gels twice, but this may not be practical.

D. The Thickness of the Gel

The results of electrophoresis can be affected by the thickness of the gel. You might try using 25 mL, 30 mL, and 40 mL of agarose to make three gels of different thicknesses. It is desirable to run each of these three gels twice, but this may not be practical.

Lab Meeting/Discussion Questions

Prepare, as a class, a notebook of photos showing optimal and suboptimal electrophoresis conditions. Some student examples are shown in Figures 7.25–7.27. Consult your class notebook in the future as you perform additional work with agarose gel electrophoresis.

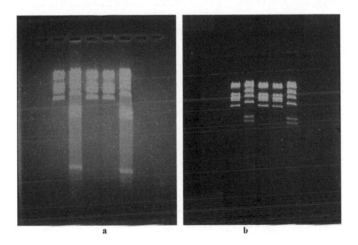

a   b

FIGURE 7.25   Staining. (a) This gel was overstained, and the bands are not clearly distinguishable. Fortunately, this problem is easily solved by destaining. (b) The same gel was allowed to soak in tap water for 10 minutes. This removes some of the stain, and the individual bands are much more clearly distinguishable.

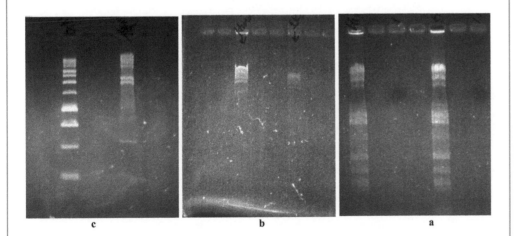

c   b   a

FIGURE 7.26   Intentional mistakes. (a) This is a reasonably good result with clearly distinguishable bands. The student, however, allowed the gel to cool too much before pouring, and you can see some swirling patterns in the lower part of the gel. (b) The student dissolved the agarose in water instead of buffer. This gel is unusable. (c) The student used a dirty comb. Note the "scalloped" appearance of the upper bands and the general smearing.

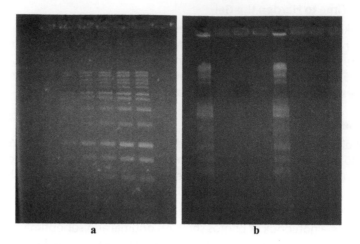

FIGURE 7.27   Amount of DNA loaded. (a) A DNA marker was used to test the effect of the amount of DNA loaded. According to the manufacturer's directions, the last lane on the right should have been overloaded, and the student expected to see poor, "smeary" bands. However, the results in the last lane are good. (b) This gel was, in fact, overloaded, resulting in "smeary" bands.

Laboratory Exercise 34: Quantification of DNA by Agarose Gel Electrophoresis

OVERVIEW

In this exercise, you will use agarose gel electrophoresis to determine the quantity of DNA in a sample. This method is based on the fact that the intensity of ethidium bromide staining is proportional to the amount of DNA present. It is, therefore, possible to quantify DNA by comparing the intensity of staining of samples to that of standards with a known concentration of DNA.

As you complete the tasks in this laboratory exercise, you will perform the following tasks:

- Practice performing an electrophoretic separation of DNA.
- Learn a method of estimating the quantity of DNA in a sample.
- Compare the electrophoretic and spectrophotometric methods of quantifying DNA.
- Consider the accuracy and precision of this analysis method.

BACKGROUND

A. Quantifying DNA

DNA that has been purified from cells or tissue must be quantified prior to use in experiments. For instance, suppose you are going to cut vector DNA with enzymes in order to insert a DNA fragment of interest into the vector. In this case, you need to be sure that you use the right amount of vector DNA, the right amount of the DNA of interest, and the proper amount of enzyme. To do this, you need to know the amounts of DNA in your preparations of the vector and the DNA of interest.

You used a spectrophotometric method of DNA quantification in Laboratory Exercise 22. The spectrophotometric method allowed you to convert an absorbance value to a DNA concentration by using the appropriate conversion factor. The spectrophotometric method has the advantage of being rapid and easy. However, it requires relatively large volumes of DNA (by molecular biology standards) and it does not provide information as to whether the DNA is intact.

In this exercise, you will use electrophoresis to estimate the amount of DNA in a sample. The gel electrophoresis method is more time intensive than the spectrophotometric method. However, electrophoresis requires only a very small volume of sample (on the order of a few microliters). Another advantage of the electrophoresis method is that, because the DNA sample is separated by size on an agarose gel, the purity and integrity of the DNA sample is easily analyzed along with its amount. At the same time, this electrophoresis method, like the spectrophotometric method, only provides an estimate of DNA quantity.

B. The Theory behind DNA Quantification Using Gel Electrophoresis

Quantifying DNA by electrophoresis depends on the fact that ethidium bromide intercalates (inserts itself) between the stacked bases in the DNA double helix. The intensity of ethidium bromide staining is proportional to the amount of DNA present. We can, therefore, use the intensity (or brightness) of ethidium bromide-stained DNA to estimate the amount of DNA present in a sample. We do this by comparing the intensity of staining of the sample DNA to the intensity of staining of a standard with a known amount of DNA. Suppose, for example, that sample band A is roughly twice as intensely stained with ethidium bromide as standard band B. This means that band A has about twice the amount of DNA as band B. If band B is known to contain 75 ng of DNA, then band A must contain approximately 150 ng of DNA. In this method, you compare only the intensity of stain. It does not matter how far the band has migrated on the gel.

C. Overview of the Electrophoresis Quantitation Procedure

Step 1: Prepare the sample DNA.

The sample is prepared for electrophoresis by combining some of it, for example, 1 μL, with loading dye and water:

1 μL DNA
8 μL water
1 μL of 10X loading dye
10 μL total

Step 2: Prepare a molecular weight standard for electrophoresis.

Various molecular weight standards can be used for this purpose. Lambda (λ) DNA digested with *Hin*dIII restriction enzyme is a commonly used standard because it is readily available and inexpensive (see Figure 7.28). The λ standard is prepared as follows for electrophoresis:

1 μL λ DNA (500 ng/μL stock concentration)
8 μL water
1 μL 10X loading dye
10 μL total

Steps 3–6: Run gel, stain with ethidium bromide, and estimate quantity of DNA.

The DNA sample and the λ standard are run on a gel and stained with ethidium bromide. The intensity of stain in the samples is compared with the intensities of the standard bands, and the amount of DNA in the sample fragments is calculated. Example 7.2 illustrates how this is done.

Example 7.2

An analyst had 100 μL of a solution containing a DNA fragment and wanted to estimate the amount and concentration of DNA in this solution by using agarose gel electrophoresis. Here are the steps performed by the analyst:

FIGURE 7.28   A DNA standard. This figure shows the sizes of the fragments in the λ standard mixture.

Step 1:

The sample was diluted 1/10.
The diluted sample was combined with loading dye and water:

4 µL DNA sample
5 µL water
<u>1 µL 10X loading dye</u>
10 µL total

Step 2:

A restriction-digested standard that contains seven fragments of known sizes was obtained. It was mixed with loading dye and water:

1 µL standard DNA (250 ng/µL stock concentration)
8 µL water
<u>1 µL 10X loading dye</u>
10 µL total

Step 3:

The standard and sample were run side by side on an agarose gel and stained with ethidium bromide. The results are shown in Figure 7.29.

Step 4:

The intensity of the sample band was compared with the intensity of each band in the standard. The analyst decided that the sample band matched the 8 kilobase (Kb) standard fragment in intensity.

Step 5:

The concentration of DNA in the sample was calculated:

- The analyst added together all the fragment sizes in the standard and obtained 54,500 bp.
- The 8 Kb standard fragment was 8,000/54,500 of the total DNA in the standard lane. So, it contained (250 ng) (8,000 bp/54,500 bp) ≈ 36.70 ng of DNA.
- This means the sample band also contained about 36.70 ng of DNA.
- That 36.70 ng was in a volume of 4 µL, so the concentration of DNA in the sample was 36.70 ng/4 µL of DNA = 9.175 ng/µL.
- The sample had been diluted 1/10, so the concentration in the undiluted sample was (9.175 ng/µL) (10) = 91.75 ng/µL.

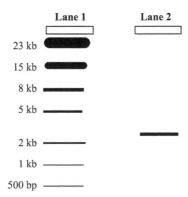

FIGURE 7.29   Hypothetical results of quantification by electrophoresis. Lane 1 is the standard. Lane 2 is the sample.

Step 6:

The total amount of DNA in the original 100 μL was calculated as follows:
$$(91.75 \text{ ng/mL}) \ (100 \text{ mL}) = \textbf{9175 ng DNA}$$

This value implies more certainty than the method provides. Most people would probably round the value, likely to 9000 ng of DNA.

LABORATORY PROCEDURE

1.   Your instructor will provide you with a DNA sample to quantify, your "unknown." Prepare three mixtures with your unknown:

Sample 1:
   1 μL unknown
   9 μL water
   <u>2 μL 6X loading dye</u>
   12 μL total

Sample 2:
   3 μL unknown
   7 μL water
   <u>2 μL 6X loading dye</u>
   12 μL total

Sample 3:
   1 μL of 1/3 dilution of unknown
   9 μL water
   <u>2 μL 6X loading dye</u>
   12 μL total

2.   Prepare one mixture with a molecular weight marker provided by your instructor:

1 μL DNA (usually 500 ng/μL stock concentration)
9 μL water
<u>2 μL 6X loading dye</u>
12 μL total

   (The manufacturer of the molecular weight marker may suggest that you heat the molecular weight marker and water mixture at 65°C for 15 minutes prior to loading it on a gel.)

3.   Electrophorese all samples on a 0.8% agarose gel.
4.   Stain the gel with ethidium bromide.
5.   Photograph the stained gel and label the lanes. Place the gel photo in your laboratory notebook.
6.   **Data analysis**
6.1.   Determine the mass of DNA present in each of the bands in your DNA standard lane.

   6.1.1.   Find a picture of λ HindIII digested standard (or the standard DNA that you actually used) in the manufacturer's catalog.
   6.1.2.   Sketch in your laboratory notebook the bands on the gel as shown in the catalog.
   6.1.3.   Record the sizes of the bands next to the bands on your sketch.
   6.1.4.   The λ genome is 48,502 bp. (If you did not use a λ standard, find the size of your DNA.)
   6.1.5.   For each band in the picture, determine the fraction of total mass per band.
      6.1.5.1.   For example, a band with a molecular weight of 3000 bp would make up 3000 bp/48,502 bp of the λ genome.
   6.1.6.   Determine the total mass of standard DNA per lane you loaded on your gel:

6.1.6.1. For example, if the DNA concentration of the standard was 500 ng/μL, and you loaded 1 μL, then you loaded 500 ng of total DNA in your lane.

6.1.7. Determine the mass of DNA in each band of the standard.

6.1.7.1. For example, from steps 6.1.5 and 6.1.6, the 3000 bp band in the standard makes up 3000/48,502 of the λ genome. There are 500 ng of total DNA in the standard lane. Therefore, in the 3000 bp band, there are

$$(3000/48502)(500\,\text{ng})\,\text{of DNA} \approx 31\,\text{ng}$$

6.1.8. Compare the intensity of the band in the DNA lanes you wish to quantify with the intensity of the bands in the standard lane. Find a band in the standard lane of equal intensity to that in the sample. Your DNA sample has about the same mass of DNA as the DNA in this standard band.

6.1.9. To calculate the concentration of DNA in your sample, divide the mass present in the lane by the volume of DNA sample you loaded into the lane. Multiply by reciprocal of dilution if sample was diluted.

LAB MEETING/DISCUSSION QUESTIONS

1. Compare the class results. How close were your estimates for the concentration of DNA? How did your estimates compare to the answer key provided by your instructor?

2. What is the purpose of running three different samples of the same DNA preparation on the quantification gel?

3. Compare this method to the spectrophotometric method in Laboratory Exercise 22. Based on class data, which method gave you more accurate results? When do you think you would choose one method over the other?

4. What factors will affect the accuracy of your estimate of DNA concentration using the electrophoresis method of quantification?

Laboratory Exercise 35: Introduction to Ion Exchange Column Chromatography

OVERVIEW

Scenario

BEING A RESEARCH AND DEVELOPMENT SCIENTIST

You are a research and development scientist in the same biopharmaceutical company as you were in Classroom Activity 13. Recall that your team has discovered a protein, isolated from the venom of a rare Amazonian snake, which shows promise as an anticancer agent. Your company has named this protein "venomX." Your team has isolated the snake gene that encodes the protein, inserted the snake gene into cultured cells, and used the cells to manufacture the protein at a pilot level. However, in the project to date, company scientists have used an expensive purification process that is not suitable for preparing larger quantities of the protein. If the protein has any chance of being developed further as a commercial product, a less expensive purification method must be found. You have been assigned to find out whether a particular chromatography method, ion exchange chromatography (IEC), can be used for this purpose. You are not yet familiar with IEC, never having performed this technique before, so you must first read about this method. You will then perform preliminary testing of two types of stationary phase to see if either is likely to work to purify the venomX protein.

Separation strategies are important in the biotechnology industry. Biotechnology products must be purified away from contaminants in order to be sold. Many of the products that are produced in modern biotechnology companies are proteins. For example, proteins may be produced to use as drugs, such as insulin or tissue plasminogen activator, or they may be used as molecular tools to enable biotechnology research, such as the enzymes used to cut and paste DNA. **Protein purification** is *the separation of a specific protein from contaminants in a manner that produces a useful end product.* Protein purification strategies exploit the fact that the structures and chemical compositions of each protein are different.

In this laboratory exercise, you will explore one type of chromatography, IEC. Protein separation by IEC takes advantage of the fact that proteins have charges on their surface, and the type of charge and its distribution on the protein's surface varies greatly among proteins. Your task is to study the behavior of venomX in an IEC system because IEC can potentially provide a relatively inexpensive method to purify large quantities of product.* Thus, as you complete the tasks in this laboratory exercise, you will perform the following tasks:

- Study the principles of ion exchange chromatography.
- Learn how to set up a chromatography apparatus.
- Test your protein, venomX, using two different stationary phase materials.

---

* This is a simulation. In reality, there is no anticancer protein called "venomX." The protein you will use is really ß-galactosidase.

BACKGROUND

A. An Illustrated Introduction to the Use of Ion Exchange Chromatography to Purify a Protein

Column Chromatography Begins with a Column

(See Figure 7.30a.) You will perform ion exchange column chromatography. The **column** *holds the stationary phase* (see Figure 7.30b). The stationary phase is a slurry of tiny beads made of resin that are suspended in buffer. The stationary phase is poured into the column where the beads settle to form an even, moist bed of resin. The beads are prevented from falling out of the column by a screen. **Packing the column** is *the process of pouring the beads into a column and allowing them to settle.*

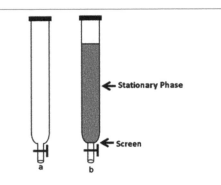

FIGURE 7.30 The column. (a) Empty column. (b) Column with stationary phase.

The Mobile Phase is a Liquid

You will use Tris buffer as the mobile phase. *The mobile phase is contained* in a **reservoir**, which can be as simple as a beaker sitting on a shelf. The reservoir is connected to the top of the column by a tube. During chromatography, the mobile phase runs from the reservoir into the column and then through the stationary phase (see Figure 7.31).

As the mobile phase flows through the column, drops come out the bottom. These drops are captured in tubes that are positioned beneath the column. A certain amount of liquid is collected in a tube; then that tube is moved away and another tube is placed beneath the column. *These tubes, each of which contains a certain volume of liquid that has run through the column,* are called **fractions**.

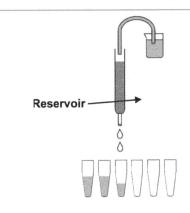

FIGURE 7.31 Mobile phase runs through the column and is collected as fractions.

Separation Is Caused by the Mixture Components' Differential Affinities for the Mobile and Stationary Phases

You might be wondering how any separation of molecules occurs in column chromatography. As with all chromatography, the answer is that the components of a mixture interact differently with the stationary and mobile phases. Some substances interact more with the mobile phase and move quicker through the column, while others interact more with the stationary phase and are slowed.

To understand how you might separate venomX from contaminants, you must know that proteins all have a complex structure and every protein has a unique structure. The structure of a given protein is somewhat dependent on the pH of its liquid environment. At a given pH, some proteins have a net positive charge, others have a net negative charge, and some are neutral. Some positively charged proteins have a large number of positive charges on their surface, while other positively charged proteins have fewer positive charges. Similarly, negatively charged proteins may have many negative charges or may be only moderately negatively charged. Determine the net charge on each of the five proteins the artist has illustrated in Figure 7.32.

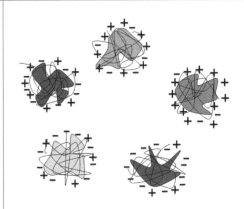

FIGURE 7.32 Proteins vary from one another.

Stationary Phases Consist of Beads with a Positive or Negative Charge

The beads that constitute the stationary phase are manufactured to have specific chemical characteristics. You will work with two kinds of beads. One kind, **diethylaminoethyl cellulose (DEAE) Sepharose**, *is modified so that the surface of every bead is positively charged*. The other type, **carboxymethyl (CM) Sepharose**, *is negatively charged* (see Figure 7.33). If you select CM Sepharose for your stationary phase, then positively charged proteins will stick to the beads and will not move quickly through the column. The more positively charged the protein, the more tightly it will interact with the stationary phase. Conversely, if you select DEAE Sepharose, then negatively charged proteins will be retarded in their movement through the column.

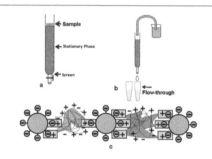

FIGURE 7.33   DEAE and CM Sepharose beads are charged.

The Sample Is Applied to the Column; Its Components Interact with the Stationary and Mobile Phases

In Figure 7.34a, there is a column that was packed with CM Sepharose, a negatively charged bead. A sample solution of the mixture of proteins shown in Figure 7.32 was applied to the top of the column. The mobile phase was then allowed to begin flowing through the column. As the mobile phase flowed through the column (Figure 7.34b–c), the positively charged proteins bound to the beads, while those that were neutral or negatively charged ran right through to be collected in the early fractions. *The proteins that ran straight through the stationary phase without being retarded are said be in the* **flow-through volume**.

FIGURE 7.34   The sample is applied to the column. (a) The sample is gently pipetted onto the column surface. (b) The mobile phase is allowed to flow through the column. (c) Artist's vision of the inside of the column, greatly magnified.

Proteins Bound to the Stationary Phase Must Be Eluted

At this point, some of the proteins from the sample are located in the flow-through volume, and some are bound to the stationary phase. But this is not particularly useful if we want to harvest a protein that is now stuck to the beads in the column. It is necessary to **elute** the rest of the proteins, that is, *to make them flow through the column with the mobile phase*. In the type of chromatography you will perform, the proteins are eluted by changing the mobile phase to one that includes a higher concentration of salt (NaCl). The salt will dissociate to form Na^+ and Cl^-. Na^+ ions compete with positive charges on proteins for sites on the beads. This causes the proteins to move into the mobile phase. Some proteins elute from the column at one concentration of salt (see Figure 7.35a), while other proteins elute when the salt concentration is higher (see Figure 7.35b). Thus, in our example the following are true:

- The neutral and negatively charged proteins are in the tubes containing the flow-through volume.

- Eluting the positively charged proteins with different salt concentrations allows separation of individual proteins.

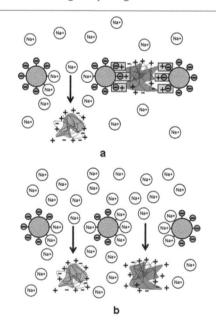

FIGURE 7.35   Elution. (a) Low salt in the mobile phase. (b) High salt in the mobile phase.

B. Testing venomX in Ion Exchange Chromatography

Your task in this laboratory exercise is to determine whether IEC shows promise as a method to purify venomX protein. In order to be successful, you need to find a system in which venomX sticks to a column, while contaminants are washed through it. If venomX does not stick to the column, then it will simply flow through together with lots of other contaminants. Once you find a column to which the protein sticks, it is necessary to make sure that the venomX can be eluted from the column. Obviously, if the protein remains bound to the column forever, it will not be of any use to your company.

Your supervisor has provided you with an initial protocol to guide you, as shown in the following Laboratory Protocol. This protocol is designed to determine whether venomX sticks to either CM Sepharose or DEAE Sepharose, both of which are readily available ion exchange resins. If the protein sticks to one or the other, you will then test whether it can be eluted with a mobile phase that includes a relatively high concentration of salt. This is as far as you will go in your initial experiments. If you have promising results, your supervisor (instructor) may ask you to develop the method further by adding a step in which another concentration of salt is used. Also, your supervisor might have you try to separate a mixture of proteins from one another.

An overview of the steps you will perform as you carry out the IEC protocol is shown in Figure 7.36. Note that there is one additional step that we have not yet discussed, that is, equilibrating the column. The purpose of equilibrating the column is *to adjust the pH of the liquid surrounding the beads throughout the column and to stabilize the charge on the beads' surface.* Recall that the nature and degree of charge on the surface of any protein depends on the pH of its aqueous environment. Also, the ability of proteins to stick to the column requires that the beads have a stable charge. Therefore, equilibrating the column is critical for the success of the separation. To equilibrate the column, you will use an equilibration buffer that contains a high concentration of Tris buffer.

Planning Your Work: Ion Exchange Chromatography

Read this laboratory exercise thoroughly and draw in your laboratory notebook a sketch of the steps in the procedure as you plan to perform it. Locate any equipment or supplies that you may need and note any questions that you have.

It is important to note that we are using three separate buffers in this procedure, and it is very important not to confuse them.

LABORATORY PROTOCOL

Because you will be comparing the performance of CM and DEAE columns, each of the following steps must be performed for each column. Our suggestion is that one team member operates the DEAE column while another works on the CM column. It is difficult for one individual to operate two columns at the same time so if you work alone, the columns should be managed sequentially. Remember that the columns should never be allowed to dry out.

1. Set up column and equipment.
   1.1. Attach each column to a vertical support (e.g., a ring stand).
   1.2. If using a syringe or pipette column, place glass wool in the bottom of the column to form a screen to retain the beads. (If using a commercial column, it will probably have a screen.)
   1.3. Attach a piece of tubing to fit over the bottom of the column. Use a clamp on this tubing to open and close the column. (If using a commercial column, you will probably have a stopcock to open and close it.)
   1.4. Find a container to use as a reservoir and locate tubing that will connect the reservoir to the top of the column. It is also possible to use a funnel as a reservoir if it is secured to the top of the column.
   1.5. Label the column as either "DEAE" or "CM," according to the solid phase you will be using in that column. Close off the column so that liquid will not pass through the column.

Step 1: Set Up Two Columns
- Place each column on a vertical support.
- Devise method of stopping and starting flow through columns.
- Add glass wool screen to bottom of columns.
- Set up buffer reservoirs above columns.
- Close columns.

Step 2: Pour Columns
- Gently swirl bead suspensions to ensure homogeneity.
  - One team member uses CM Sepharose.
  - Another team member uses DEAE Sepharose.
- Smoothly pour CM Sepharose into one column and DEAE Sepharose into the other.
- Allow stationary phases to settle.
- Set up two sets of fraction collection tubes; each set consists of 10 tubes, labeled 1-10.
- Mark each tube with a line indicating 1 mL.

Step 3: Pack and Equilibrate Columns
- Run high concentration (10X) of Tris buffer through each column to stabilize the pH and the charge on beads.
- Equilibrate each column until pH of buffer running out of column = pH of buffer going into column.
- After equilibration, remove excess Tris from the column by washing it with low salt buffer with 1X Tris.

Step 4: Apply Sample to Column and Run Low-Salt Mobile Phase
- Close columns. Pipette 0.5 mL of sample solution onto the top of each column.
- Open each column; allow sample to flow into column; do not allow the top of column to dry.
- Allow the low-salt mobile phase to run through each column.
- Collect fractions in tubes: 1 mL in each fraction.
- Collect 10 fractions for each column.
- Close columns.

Step 5: Test Flow-Through Fractions to See if They Contain VenomX
- Test column fractions by adding a drop of detection solution to a drop of each fraction.
- Record the presence or absence of yellow color.

FIGURE 7.36 **Overview of the IEC procedure you will perform.**

2. Pour beads into column.
   2.1. Make sure column is closed.
   2.2. Obtain an aliquot of column beads from your instructor. Gently swirl the tube containing the beads so that there is a homogeneous slurry of beads and liquid. In one continuous motion, taking care to avoid splashes, gently pour the bead slurry into the column.
   2.3. Allow the beads to settle in the column while you do other tasks. Check for leaks; liquid should not be coming out of the column.
   2.4. Label two sets of numbered (1–10) fraction collection tubes. Fill an unlabeled tube with 1 mL of water and note the height of the liquid; use this tube as a standard to mark each of the numbered tubes.

Step 6: Elution: Change Mobile Phase to High-Salt Buffer
- Set up two more sets of fraction collection tubes.
  - ○ Label each set of 10 tubes 1-10.
  - ○ Mark each tube with a line indicating 1 mL.
- Allow a high-salt mobile phase to run through each column.
- Collect fractions in tubes: 1 mL in each fraction.
- Collect 10 fractions for each column.

Step 7: Test High Salt Fractions to See if They Contain VenomX
- Test column fractions by adding a drop of detection solution to a drop of each fraction.
- Record the presence or absence of yellow color.

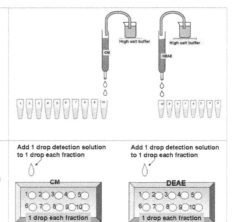

FIGURE 7.36   (Continued)

3. Pack and equilibrate column.
   3.1. Remove the excess liquid from the top of the column. Do not disturb the flat, level surface of beads at the top of the column, and do not allow the column resin to dry out. Add equilibration buffer (0.2 M NTM with 10X Tris) to the top of the column.
   3.2. Place a waste beaker (50 ml) under the tubing at the bottom of the column to collect buffer coming through the column.
   3.3. Connect a reservoir filled with equilibration buffer to the top of the column. Open the column and allow the beads to be "packed" under the pressure of the liquid flowing through it. If you are not using a reservoir, continue to add equilibration buffer to the column *without disturbing the flat, level surface of the top of the column.*
   3.4. After two to three column volumes have run through the column, test the pH of a drop of the liquid coming out of the bottom of the column with pH paper and compare it to the pH of the "equilibration buffer." Continue to run equilibration buffer through the column until the pH of the buffer coming out of the column is the same as that entering the top of the column.
   3.5. When the pH of the column has stabilized, wash the column with two column volumes of "low-salt buffer" (0.2 M NTM) in order to remove the 10X Tris.
   3.6. Stop the column flow. The column is now ready for you to apply the protein sample.
       Stop Point: If necessary, the columns can be left overnight at this point. If the procedure cannot be resumed the next day, the columns should be stored in the refrigerator. If storing the columns, it is essential to make sure that liquid does not leak or evaporate. (Parafilm can be used to seal the column.)
4. Apply the sample to the column and run "low-salt" mobile phase.
   4.1. Position your first fraction collection tubes at the bottom of the column.
   4.2. Remove any excess buffer at the top of the column but do not allow the beads to dry out.
   4.3. Using a pipette, transfer the entire protein sample (0.5 mL) to the top of the column. Open the column and watch as the sample flows into the stationary phase.
   4.4. After the sample has migrated into the stationary phase, add low-salt buffer (0.2 M NTM) to the top of the column. Continue to add low-salt buffer to the column as needed or connect a buffer reservoir. You

will need to run approximately 10 mL of buffer through the column before doing the first analysis.

    4.5. Collect 10 fractions, moving each collection tube under the column as needed. Note that the column will flow faster when there is a greater volume of liquid on the top of the column, but it will slow down as the liquid level falls. This will not affect the results for this procedure. After you have collected 10 fractions, stop the column flow.

5. Test flow-through fractions to see if they contain venomX.

    5.1. Use a ceramic spot plate or a microscope slide over a sheet of white paper to test the fractions.

    5.2. Add a drop of detection solution to a drop from each fraction and observe.

    5.3. Record the presence or absence of yellow color in Table 7.1. If yellow color is present, give it a + rating, from one (+) to four (++++), depending on how yellow it is.

6. **Data analysis**

    6.1. Which fractions contain venomX?

    6.2. Did venomX come off in the flow-through volume for either of the columns?

    6.3. Did venomX interact with the column beads for either one of the columns? Explain.

    6.4. Based on your results, do you think that venomX is negatively or positively charged?

    6.5. At this point, do you predict that DEAE or CM would be a better choice for your separation?

7. Elute the protein. If you did not see color in any of the low-salt buffer fractions from one of the columns, venomX is likely to be bound to that column. In order to show that it did bind, you now need to show that it can be released from the column by increasing the salt concentration. Use the elution buffer, 0.5 M NTM.

    7.1. With the column flow stopped, remove excess low-salt buffer from the top of the column, being careful not to disturb the top surface of the beads in the column. Replace the liquid with elution buffer (0.5 M NTM).

    7.2. Position a numbered fraction tube under the bottom of the column. Turn on the column flow and begin collecting 1 mL fractions.

TABLE 7.1   Low-Salt Column Fractions

DEAE Fraction Number	Detector Reaction + to ++++	CM Fraction Number 0.2 M NTM	Detector Reaction + to ++++
1		1	
2		2	
3		3	
4		4	
5		5	
6		6	
7		7	
8		8	
9		9	
10		10	

TABLE 7.2   Elution Buffer Column Fractions

DEAE Fraction Number	Detector Reaction + to ++++	CM Fraction Number 0.2 M NTM	Detector Reaction + to ++++
1		1	
2		2	
3		3	
4		4	
5		5	
6		6	
7		7	
8		8	
9		9	
10		10	

7.3.   Connect a reservoir or continue to add elution buffer to the top of the column, as needed.

7.4.   Stop the column after 10 fractions have been collected.

7.5.   Test the column fractions as in step 5.2 by adding a drop of detection solution to a drop of each fraction.

7.6.   Record the presence or absence of yellow color in Table 7.2. If yellow color is present, give it a + rating, from one (+) to four (++++), depending on how yellow it is.

LAB MEETING/DISCUSSION QUESTIONS

1.   Summarize your results. Were you able to determine if DEAE or CM would be useful as the stationary phase for an IEC separation of your target protein? What would you recommend to your development group about a possible purification method?

2.   Compare your results to those of the other teams in your class. Did you get similar results?

3.   If you did not have a detector solution, can you think of any other method that could be used to analyze the fractions?

4.   Does this analysis tell you if any other proteins are present? What sort of assay could you use to determine purity?

5.   What are some factors that could contribute to the success or failure of this method?

6.   What would happen if you made a mistake and applied the protein sample to the column in elution buffer (0.5 M NTM)? Predict how that might have changed your results.

UNIT DISCUSSION: BIOLOGICAL SEPARATION METHODS

1. Look through one or two biotechnology journals, such as *BioTechniques* or *Nature Biotechnology*. Describe three examples of articles that involve a separation method. You should discover many examples of centrifugation, chromatography, and electrophoresis.
2. Consider the overarching themes of good laboratory practices in the diagram that follows. Jot down notes on the diagram or on another piece of paper that describe how the ideas and skills that you practiced in this unit relate to these overarching themes.

Good laboratory practices
DOCUMENT
PREPARE
REDUCE VARIABILITY AND INCONSISTENCY
ANTICIPATE, PREVENT, CORRECT PROBLEMS
ANALYZE, INTERPRET RESULTS
VERIFY PROCESSES, METHODS, AND RESULTS

Growing Cells

UNIT VIII

DOI: 10.1201/9781003360742-8

UNIT INTRODUCTION

For over two billion years, through the apparent fancy of her endless differentiations and metamorphosis, the Cell, as regards its basic physiological mechanisms, has remained one and the same. It is life itself, and our true and distant ancestor.

—Albert Claude, winner of the Nobel Prize for Physiology in Medicine, 1974, http://nobelprize.org/nobel_prizes/ medicine/laureates/1974/claude-lecture.html

This unit introduces fundamental skills relating to maintaining cultured cells in the laboratory. **Cultured cells** are *living cells maintained inside plates, flasks, vials, dishes, bioreactors, or fermenters.*

You will work with two kinds of cells in culture: *Escherichia coli* (*E. coli*), to model a bacterial cell type, and Chinese hamster ovary (CHO) cells, to model a mammalian cell type. *E. coli* is a type of bacterium that is a normal resident in the gastrointestinal tract of humans and other animals. Most strains of *E. coli* are harmless, but a few cause serious disease. In the laboratory, you will use only **nonpathogenic**, *non-disease-causing* strains. CHO cells descend from cells originally cultured from the ovary of the Chinese hamster. CHO cells are often used in biological and medical research and commercially in the production of biopharmaceuticals.

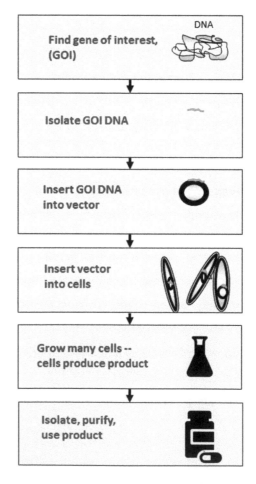

FIGURE 8.1 **Using microbial cells to produce a product.** Biotechnologists sometimes genetically modify microbial cells, as shown in this diagram, so that the cells manufacture products. However, microbes not genetically modified have long been used in "traditional" biotechnology for brewing, cheese making, antibiotic production, etc.

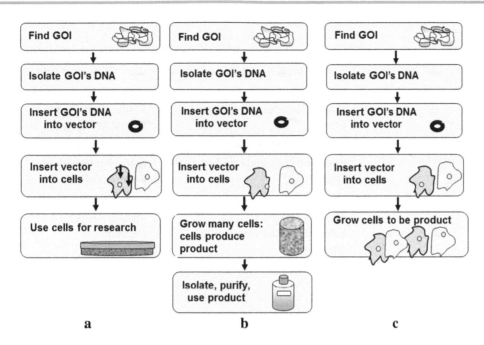

FIGURE 8.2   Examples of how mammalian cells are used. (a) Cultured cells are often used for studying cellular processes. The cells in this illustration have been genetically modified, but unmodified cultured cells have also been used for many years to study cellular processes. (b) Biotechnologists use genetically modified cells as "factories" to manufacture a protein product. (c) The cells themselves are sometimes the product (e.g., skin cells for burn victims). Cells for such uses may or may not be genetically modified.

There are many reasons to culture cells. First, consider bacterial cells. Food safety analysts, for example, might test foods for contaminating bacteria. Plant scientists might study bacteria that affect the properties of soils. Ecologists might study the roles of bacteria in nutrient cycling throughout an ecosystem. Environmental scientists might use bacteria to remediate soil that is contaminated with hazardous waste. Molecular biologists might grow bacteria for genetic transformation procedures in which they insert a gene of interest into the cells. Transformed bacteria have many uses in research and can also be used to manufacture a protein product of interest (see Figure 8.1).

What about mammalian cells? There are also many reasons to culture mammalian cells (see Figure 8.2). Researchers might use cultured mammalian cells to try to answer questions. How do cells communicate? How do disease agents damage cells? What are the mechanisms by which drugs act on cells? Cultured cells are used in pharmaceutical testing to study the effects and possible toxicity of new drugs, cosmetics, and chemicals. In a biotechnology company setting, genetically modified cells are used as "factories" to manufacture biopharmaceuticals and industrial enzymes. Human and animal tissues are grown outside the body for therapeutic purposes, such as providing skin grafts for burn victims.

Whatever the purpose of cultured cells, they need to survive, grow, and reproduce while in an artificial, aqueous, "culture" environment. Cell culturists must therefore have particular skills to accomplish the following:

- Avoid contaminating their cultures with external organisms from the environment.
- Provide a suitable liquid environment for the cells.
- Provide a suitable physical environment for the cells (e.g., a proper container, sufficient gas exchange).
- Understand cell growth and reproduction.
- Characterize and assess the condition of the cells.

These skills relate to all cells, regardless of type. There are, however, significant differences between microbial cells and "higher-order" (e.g., mammalian) cells, which influence how

TABLE 8.1   Working with Bacterial and Mammalian Cells

Purpose	Laboratory Exercise—Bacterial Cells	Laboratory Exercise—Mammalian Cells
Avoid contaminating cultures.	**Laboratory Exercise 37** **Aseptic Technique on an Open Lab Bench**	**Laboratory Exercise 43** **Aseptic Technique in a Biological Safety Cabinet**
Provide a suitable liquid environment for the cells.	**Laboratory Exercise 40** **Preparing Phosphate-Buffered Saline**	**Laboratory Exercise 40** **Preparing Phosphate-Buffered Saline**
	Laboratory Exercise 38 **Working with Bacteria on an Agar Substrate: Isolating Individual Colonies**	**Laboratory Exercise 44** **Making Ham's F-12 Medium**
		Laboratory Exercise 45 **Examining, Photographing, and Feeding Cells**
		Laboratory Exercise 47 **Subculturing CHO Cells**
Provide a suitable physical environment.	**Laboratory Exercise 38** **Working with Bacteria on an Agar Substrate: Isolating Individual Colonies**	
Monitor cell growth and reproduction.	**Laboratory Exercise 41** **The Aerobic Spread-Plate Method**	**Laboratory Exercise 46** **Counting Cells Using a Hemacytometer**
	Laboratory Exercise 42 **Preparing a Growth Curve for *E. coli***	**Laboratory Exercise 48** **Preparing a Growth Curve for CHO Cells**
Characterize and assess cells' condition.	**Laboratory Exercise 36** **Using a Compound Light Microscope** **Laboratory Exercise 39** **Gram Staining**	**Laboratory Exercise 45** **Examining, Photographing, and Feeding CHO Cells**

they are handled. One difference is that mammalian cells have a much slower rate of division than bacterial cells. This means that mammalian cell cultures can be overrun by contaminating cells much more easily than bacterial cultures. Mammalian cell culturists, therefore, learn to be extremely vigilant to avoid contamination. Another difference is that bacterial cells are unicellular; each cell normally survives alone in the external environment. In contrast, mammalian cells evolved to survive in a multicellular body. Mammalian cells in the body are constantly bathed by nutrients; oxygen is brought to them via the bloodstream; waste products are removed; and temperature and pH are constant. Mammalian cell culture, therefore, requires simulating a stringently controlled environment. You will see that there are thus many parallel techniques for growing bacterial and mammalian cells; yet there are significant differences between the two kinds of culture. Table 8.1 shows the purposes of the techniques you will learn as you complete this unit and the relevant laboratory exercises for bacterial and mammalian cells. As you perform these exercises, look for the similarities and the differences between bacterial and mammalian cell culture.

Laboratory Exercise 36: Using a Compound Light Microscope

OVERVIEW

The purpose of this exercise is to introduce the proper use of a light microscope. You have two main goals when using a microscope. The first is to take care of the microscope. Microscopes are expensive, fragile, precision instruments that can easily be damaged. The second goal is to obtain an image that reveals as much information as possible about your sample. Both goals are considered in this laboratory exercise in which you will perform the following tasks:

- Study the parts of your microscope and their functions.
- Practice using the microscope by viewing prepared slides provided by your instructor.

BACKGROUND

The invention of the light microscope was one of the most important technological advances in the history of biology. A light microscope uses **glass lenses** *that bend light in specific ways to create a magnified image of a specimen*. **Compound light microscopes**, the type routinely used in biology laboratories, *combine two lenses in succession to magnify the image more than either one could alone*.

In a compound light microscope, light is directed through a specimen that is affixed to a glass slide. The light then passes through an objective lens and finally through an ocular lens. The human viewer looks through the ocular lens to see the magnified image of the specimen. Figure 8.3 shows the parts of a typical compound light microscope.

Do Not Damage the Microscopes

Advice from the senior technician

- Always carry the microscope with two hands.
- Never touch a lens with anything other than lens paper—never touch it with your fingers, cloth, Kimwipes, facial tissues, etc.
- Do not clean lenses with solvents unless told to do so by the manufacturer. Solvents can dissolve the glue that holds the lens in place.
- Never crash an objective into a slide. To avoid doing so, never use the coarse focus knob with the 40× or 100× objectives.
- Look at the slide as you move the higher-power objectives into place.
- Never dust a lens by blowing on it. Saliva will be deposited on lenses and is harmful.
- When using the oil immersion objective, be sure oil does not get on the other objectives.
- Never take apart any of the parts of a microscope unless told to do so in the manufacturer's directions. Microscopes should be serviced by trained service technicians.

LABORATORY PROCEDURE

These directions are for a "typical" microscope. Your instructor will indicate to you any issues relating to your style of microscope.

1. Remove your microscope from storage and carry it to your lab bench, holding one hand underneath the base and your other hand under the arm of the microscope. *Never carry a microscope with just one hand.*
2. Remove the dust cover; plug in the microscope.
3. Check that the light is turned to its lowest setting. Turn on the light. (Turning on the light at the lowest setting helps extend the life of the bulb.)
4. Identify on your microscope all the parts that are labeled in Figure 8.3.

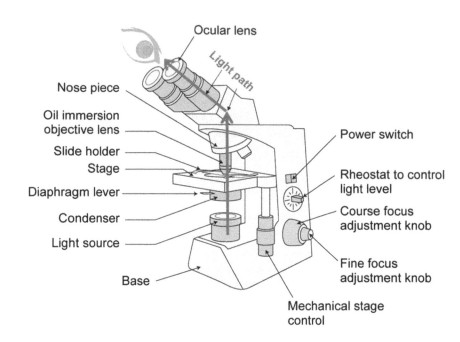

FIGURE 8.3 **A compound light microscope.** See the laboratory procedure that follows for an explanation of the parts of the microscope. This is a binocular microscope because there are two oculars present.

4.1. The **oculars** are *the eyepieces. Oculars magnify the image 4× or 10×.*
    4.1.1. You can usually adjust the distance between the two oculars to make them more comfortable.

4.2. The **objectives** are *lenses used to magnify the image.* Most student microscopes have three or four objectives arranged on a ring:
- A **scanning objective** *with a magnification of 4× provides the least magnification and is useful for quickly surveying the whole slide.*
- A **low-power objective** *is usually present with a magnification of 10×.*
- A **high-power objective** *with a magnification of 40× is usually present.*
- An **oil-immersion lens** *is often present; the magnification of this lens is 100×.*

4.3. To determine the total magnification of the sample image, multiply the magnification of the ocular times that of the objective. For example, if you use the high-power objective, the total magnification would be as follows:

$$\left(\text{Magnification of the ocular lens}\right)\left(\text{Magnification of the objective lens}\right)$$

$$= \left(10\text{X}\right)\left(40\text{X}\right) = 400\text{X}$$

4.4. The lamp emits light in all directions. Light must be directed through the specimen on the slide. The **condenser lens** *helps to focus the light from the lamp onto the slide.* Observe that this is another lens, but it is not one that provides magnification. The height of this lens may be adjustable to help optimize the image. Check to see if the condenser lens can be moved on your microscope.

4.5. The **iris diaphragm** *forms a circle of light that illuminates the sample.* The width of this circle can sometimes be adjusted by the user to optimize the image. Check to see if the iris diaphragm can be adjusted on your microscope.

4.6.   The **stage** *holds the slide*. Most microscopes have knobs to move the stage in two directions.

4.7.   The *image is focused* with the **fine and coarse focus knobs**. These knobs function to bring the objective closer to the slide or to move the objective further from the slide.

4.8.   The *intensity of the light can be adjusted* with the **light intensity knob**. You generally need more light intensity as you increase the magnification.

5.   Practice using the microscope with a prepared slide provided by your instructor:

5.1.   Gently clean the ocular and objective lenses with lens paper. Never use ordinary fabric or tissue to clean a lens as it might be scratched, or fibers and oils might be deposited on the lens.

5.1.1.   If you have a rubber bulb, you can use it to blow dust away from the lens. Do not blow with your mouth, as saliva will damage the lens.

5.1.2.   You can gently wipe the lenses with a piece of crumpled lens paper moistened with a drop of purified water. Blot the lens tissue to avoid applying too much liquid to the lens. Do not allow liquid to seep into the lens housing.

5.1.2.1.   Do not touch with your fingers the part of the tissue that touches the lenses.

5.1.3.   If a lens is especially dirty, your instructor might have you use commercial lens cleaner. If so, apply a small amount of lens cleaning solution to the crumpled lens paper and blot the lens paper to avoid adding too much liquid. (Sparkle Glass Cleaner has been recommended for this purpose.)

5.1.3.1.   Rinse with a light amount of water on crumpled lens paper and gently dry the lens with lens paper.

5.2.   Be sure that the prepared slide is clean. Place it on the stage with the specimen you will view in the center of the opening in the stage.

5.3.   Ensure that the 4× objective lens is in place (or the 10× if no 4× is present). It not, rotate the ring until the 4× objective is over the slide.

5.4.   Watch the objective and lower it towards the slide as low as it will go.

5.5.   Look through the oculars and slowly raise the objective with the coarse focus knob until the sample comes into focus. Use the fine focus knob to fine-tune the image.

5.5.1.   You do not need to wear glasses for microscope work if your glasses correct only nearsightedness or farsightedness. If you wear glasses to correct a severe astigmatism, you may be better off wearing your glasses. Special attachments for the oculars are available if you must wear glasses.

5.6.   Focus the two oculars independently of one another to obtain a sharper view of the sample. This compensates for any difference in vision between your two eyes:

5.6.1.   The focus of one of the two oculars will be adjustable by turning a ring. Cover up the ocular whose focus is adjustable. (You can place a card over the ocular.)

5.6.2.   With both eyes open, bring the specimen into focus for the other eye with the fine focus knob.

Keep your vision relaxed by looking up frequently to distant objects. This helps prevent eyestrain.

5.6.3.   When the ocular is focused for your first eye, switch the card to cover the other ocular.

5.6.4.   This time use the focusing ring on the ocular to bring the same point on the slide into focus with the other eye. Follow the same procedure for relaxed viewing.

5.6.5.   Some people see a double image of the sample when first beginning to use a binocular microscope. We do not know why this happens, but it usually goes away with time and practice. If you see two images, keep both eyes open anyway so your eyes and brain can get used to the effect. Also, it may help to make sure that the distance between the oculars is optimized.

5.7.   Raise the condenser to its highest position and fully close the iris diaphragm (if adjustable).

5.8.   Looking through the oculars, slowly lower the condenser just until the graininess disappears. Slowly open the iris diaphragm just until the entire field of view is illuminated. This should provide an image with suitable contrast.

5.9.   Move the stage around at this point to explore the slide.

5.9.1.   Structures are usually obvious when you view commercially stained slides. When you view slides that you have made yourself, particularly if the specimen is unstained, it is often difficult to tell when something is meaningful and when it is not. If you move the slide around and the object moves, then you know the object is really on the slide and is not a piece of debris on one of the lenses. An object on the slide still may not be meaningful—it might be a bubble or debris—but at least you are focused on the slide.

5.10.   When you find something on the slide that you want to study in more detail, rotate the 10X objective into place *without adjusting the focus knobs*. The image should be mostly in focus.

5.11.   Use the fine focus knob to sharpen the image. You should not need to use the coarse focus knob.

5.12.   If you want to increase the magnification to 40×, center the sample in the middle of the field of view. Rotate the ring until the 40× objective is in place; do not touch the focus knobs. The 40× objective will be close to the slide, but if you have focused properly before, it will not touch the slide. If the objective begins to touch the slide for any reason, immediately move the objective out of the way. The objective can be damaged if it crashes into the slide and breaks the coverslip.

5.13.   Carefully focus with the fine adjustment knob, but do not use the coarse adjustment knob.

5.14.   Increase the light intensity, if necessary, as you increase the magnification. Also check the iris diaphragm setting.

5.15.   Notice that as the magnification increases, the **field of view**, *the portion of the sample that you see*, becomes smaller.

5.16.   Use the 100× objective to view bacteria:

5.16.1.   Center the specimen in the field of view and then rotate the ring halfway between the 40× and 100× objective lenses.

5.16.2.   Place a small drop of immersion oil directly over the center of the slide.

5.16.3.   Rotate the 100× objective into position while watching the objective to be sure that the lens does not crash into the slide.

5.16.4.   Make sure that the immersion oil does not get on the other lenses.

5.16.5.   Note the thin film of oil that forms between the slide and the lens.

      5.16.6.   After the 100× objective is in place, focus using the fine-adjustment knob. Never focus with the coarse-adjustment knob when you are using the oil immersion objective.

      5.16.7.   Increase the light intensity and adjust the iris diaphragm, if necessary.

      5.16.8.   Scan the sample at this magnification. Record what you see.

      5.16.9.   When you are finished with this lens, move it out of place and carefully remove the oil from the lens as directed in Step 5.1.

      5.16.10. Clean the oil from the slide. You can use a Kimwipe on the slide—but not on the lens.

6.   Before you put the microscope away and finish the exercise, be sure you can answer these questions:

    6.1.   What happens to light intensity as you increase magnification?

    6.2.   What happens to the field of view as the magnification increases? In other words, can you see as large an area of the sample with high magnification as with low magnification?

    6.3.   What happens to the working distance as magnification increases? The **working distance** is *the distance between the objective lens and the specimen.*

    6.4.   What happens to the depth of field when you increase magnification? The **depth of field** is *the thickness of the specimen that is in focus.*

7.   When you are finished with the microscope, follow these steps:

    7.1.   Be sure there is no oil on the oil-immersion lens. Be sure no oil has gotten onto any other objective.

    7.2.   Turn the light intensity down all the way.

    7.3.   Turn off the light.

    7.4.   Move the 4× objective (or 10× if there is no 4×) into place. Your instructor may also have you lower the stage. This is to ensure that the objectives and stage are safely separated from one another when the next person uses the microscope.

    7.5.   Wrap the cord around the microscope or put the cord away in its storage location.

    7.6.   Put the dust cover back on the microscope.

    7.7.   Return the microscope to its storage location, carrying the microscope with two hands.

Laboratory Exercise 37: Aseptic Technique on an Open Lab Bench

OVERVIEW

Growing cells in culture requires that the cell population of interest remains free of contamination by external microorganisms or foreign cell types. **Aseptic technique** *is a system of laboratory practices that minimizes the risk of contamination in cultured cells. The first step in learning to work with cells is, therefore, to learn aseptic technique. In this laboratory exercise, you will practice manipulating pipettes and culture medium using aseptic technique—but without any actual cells involved. You will perform the following activities:*

- Arrange your workspace.
- Perform manipulations aseptically.
- Clean the work environment when activities are complete.
- Check to see if you were successful in your manipulations.

BACKGROUND

A. Aseptic Technique Protects Your Cells, Yourself, and the Environment

The air, surfaces of objects, skin, and breath all harbor billions of potential contaminants. These contaminants include bacteria, yeast, molds, and viruses, all of which have the potential to destroy a culture. The most essential tool in aseptic technique is constant *awareness*—an understanding that we are always surrounded by contaminants with the potential to destroy our cultures.

The guiding principles of aseptic technique are the same, regardless of the type of cells with which you work (see Box 8.1). The specific practices for handling mammalian cells, however, are much more stringent than those for bacterial cells. This is because a single contaminating bacterial cell can divide and multiply quickly to take over a culture of much slower-growing mammalian cells. Bacterial and fungal cultures are less susceptible to major contamination by a few stray organisms because they are typically inoculated with millions of rapidly dividing cells. This does not mean that one should be careless when handling bacterial cultures, but it does mean that nonpathogenic bacteria can frequently be handled on an open lab bench if proper aseptic technique is used. In contrast, it is routine to use *protective enclosures*, **laminar flow cabinets**, also called **biological safety cabinets (BSCs)**, when manipulating mammalian cells and other slow-growing cultures. You will practice aseptic technique on an open lab bench in this laboratory exercise, and you will learn aseptic technique in a laminar flow cabinet in Laboratory Exercise 43.

Cross-contamination is another concern when working with cell culture. **Cross-contamination** is *where cells from one culture accidentally enter another culture.* Cross-contamination usually results in loss of the cultures and invalidated research results or useless products. Moreover, cross-contamination can create unknown hazards for lab workers who are unaware of their exposure to contaminating organisms. The basic principle of avoiding cross-contamination is separation of different cells types and all the equipment, media, and supplies that are used for each. In this course, you will work with bacterial cells and then with mammalian cells. These cell types must be strictly isolated from one another. In a workplace, bacteria and mammalian cells are generally handled in different laboratories, each with their own equipment, supplies, pipettes, etc. This separation may be impossible in an academic setting. If there is overlap in the spaces or equipment used for bacteria and mammalian cells, then thoroughly clean all surfaces, water baths, equipment, etc., with 10% bleach or another disinfectant before beginning mammalian cell culture work.

In addition to protecting your cultures from contamination, aseptic technique has the added purpose of preventing cultured cells from harming humans in the laboratory or escaping to the environment. We work with cell types that are not known to be pathogenic to humans in the laboratory exercises in this manual. Nonetheless, you should

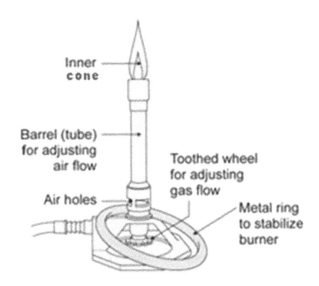

Inner cone

Barrel (tube) for adjusting air flow

Toothed wheel for adjusting gas flow

Air holes

Metal ring to stabilize burner

FIGURE 8.4   **A Bunsen burner.**

avoid exposing yourself to these cells by strictly following aseptic technique. You may encounter cells that have the potential to cause harm when you enter the workplace, and you will want to follow additional precautions, as directed by the institution's safety personnel.

B. The Use of Flame in Aseptic Technique

Flame, normally supplied by a Bunsen burner, is used routinely as part of aseptic technique when working on an open lab bench with bacteria and other microorganisms, see Figure 8.4. (Flame is not recommended in a BSC.) Flame plays several roles:

- A flame can be used to kill the microbes on a device. For example, a metal loop can be used to transfer microbes from one place to another. The loop is first sterilized by flame.
- A flame can be used to create air currents that move upward and help keep dust and particles from falling onto a work surface.
- The tops of bottles are passed for a second or two through a flame to create air currents that keep microbes from falling into the bottles when they are opened and to keep microbes on the lips of the bottle from falling into them.

Flame introduces safety issues. Carefully read the safety briefing regarding Bunsen burners before coming to lab.

Safety Briefing

In this unit, you will begin to work with living cells. Cells have the potential to be biohazardous agents, as was discussed in Unit I of this manual. The agents with which you will work are categorized as biosafety level 1 agents, which will not cause disease in healthy people. Nonetheless, it is important to treat these agents with caution and to strictly follow all Standard Microbiological Practices shown in Box 1.5.

In addition, your instructor will direct you in special procedures for disposing of contaminated waste— waste that contains cells. These procedures vary among institutions, but every institution has a method of ensuring that cells are killed before disposal.

Safety Briefing

Bunsen burners provide an open flame that can be dangerous. Follow these rules:

- Never leave a Bunsen burner unattended when it is lit.
- Check the tubing connecting the burner to the gas valve for cracks before use.
- Make sure the burner is securely anchored.
- Never pass your hands or sleeves over a burner.

- Tie back hair and loose clothing.
- Be prepared to shut off the gas in case of emergency.
- Make sure that no flammable solvents are in use anywhere in the laboratory. Vapors from solvents can move invisibly in the room and ignite at a distance from their source.

Box 8.1   General Principles of Aseptic Technique for Bacterial and Mammalian Cell Work

- An object, surface, or solution is sterile only if it contains no living organisms. If any organisms are present, the item is nonsterile.
- Objects or solutions are not sterile unless they have been treated to eliminate microorganisms (e.g., autoclaved, heat sterilized, irradiated). Items used in the culture of living cells must be initially sterilized (e.g., nutrient medium, test tubes, pipettes). Some items are sterilized by the user, commonly by autoclaving or filtration. Alternatively, disposable sterile supplies and culture media may be purchased from suppliers.
- A sterile object, surface, or solution that comes in contact with a nonsterile item is no longer sterile. Contaminated items must be replaced with sterile items if the procedure is not complete.
- Air is not sterile unless it is sterilized in a closed container. Contaminating substances can fall into any container or object that is open to the air. When materials (e.g., nutrient medium, bacteria, mammalian cells) are transferred from one location to another, special practices are used to minimize exposure to air and to avoid contact with nonsterile surfaces and objects.
- Human skin and breath are rich sources of contaminating microorganisms. Proper aseptic technique protects cultures from human contact.
- Humans must be protected from the cells with which they are working. Assume that all cells are potentially hazardous and avoid contact with them. The same practices that protect the cells from contamination will also protect you from exposure to the cells.

Using Aseptic Technique on an Open Lab Bench

Advice from the senior technician

- Watch the tip of your pipette. If it touches any surface or object, assume it is contaminated and discard it.
- Discard every pipette and tip after one use. Do not put the same pipette or tip twice into a culture or media bottle.

- Avoid putting caps and lids down on an open bench top. Hold a cap in the crook of your left hand with your little finger wrapped around it, if right-handed and vice versa. Never touch the inside of a cap.
- When opening Petri dishes, hold the lid partway over the bottom of the plate with the open end away from your face to protect the plate from airborne particles.
- Open tubes, bottles, and flasks at a slight angle away from your face.
- Flame the neck of a bottle or flask for a couple of seconds to create upward air currents.
- Immediately cover flasks, bottles, plates, boxes of sterile tips, and tubes after use.
- Do not allow drops of liquid to get on the neck of a container underneath the cap or lid.
- Do not pass your hands, arms, or sleeves over any open plate, flask, bottle, or tube.

Planning Your Work: Aseptic Technique on an Open Lab Bench

1. Prepare to perform aseptic work so that cultures and plates are exposed to the air for the shortest possible time. Review the laboratory exercise and the list of supplies required. Sketch in your laboratory notebook how you will organize all your materials on your lab bench to perform this exercise.

2. On your sketch, star the items that will need to be flamed.

3. Review the safety briefings and the advice from the senior lab technician carefully. Note in your lab manual each step in the procedure where one of more of the tips from the senior technician applies.

4. Rehearse in your mind how you will perform each manipulation.

LABORATORY PROCEDURE

You will be aseptically transferring 10 mL of sterile nutrient medium into each of the following:

- Two sterile glass bottles (e.g., 50 or 100 mL)
- Two sterile Petri dishes
- Two sterile tubes

You will check each of these for contamination after one, two, and three days. None of these items will be contaminated if your aseptic technique is effective.

Refer to Figure 8.5 for illustrations of aseptic transfer on an open lab bench.

1. Prepare to perform work aseptically.
   1.1. Tie back long hair. Avoid dangling clothing.
   1.2. Clear all materials from your work surface. Move your laboratory manual and lab notebook to the side. Remove all other papers, books, and objects.
   1.3. Wash your hands.
   1.4. Spread disinfectant across the entire work area. A number of commercial disinfectants are available. Alternatively, use 10% bleach or 70% alcohol. Spread the disinfectant with a clean paper towel, and allow excess disinfectant to evaporate.
2. Prepare a **control plate** *to determine the level of contamination that occurs because of exposure to the air and surroundings.* Obtain a previously prepared agar plate from your instructor and leave it open to the air throughout the lab period.
3. Label your storage bottles, Petri dishes, and tubes with your name, date, and type of nutrient medium you are using.
4. Gather your supplies:
   - Bunsen burner and striker (or matches) for bacteria
   - Sterile pipettes (10 mL size)
   - Stock bottle of nutrient medium
   - Two sterile, empty glass bottles with caps
   - Two sterile Petri dishes
   - One pipette bulb or aid
   - Two sterile, test tubes with rack
   - Disposal container for used pipettes
   - Squirt bottle of disinfectant (for small spills)
5. Prepare your Bunsen burner and place it in a location where you will not pass your hands or sleeves over it. (Refer to the diagram of a "typical" Bunsen burner in Figure 8.4.)
   5.1. Connect the burner to the source of gas using a thick-walled tube. Be sure the burner is secure and cannot fall over; anchor it if necessary.
   5.2. Turn on the gas and light the flame.
      5.2.1. Hold a striker just above and slightly to the side of the top of the burner. Squeeze the striker until sparks ignite the burner.
      5.2.2. Alternatively, use a match to light the burner.
   5.3. Adjust the burner to provide the optimal mixture of air and gas.

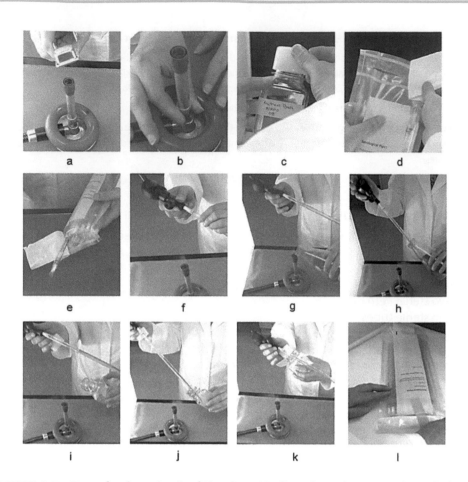

FIGURE 8.5 **Transferring 10 mL of Nutrient Medium by Using Aseptic Technique.** Refer to steps 5–18 in the Laboratory Procedure. (a and b) The Bunsen burner is lit, and its flame is adjusted. (c) The caps of the flask containing nutrient broth and the receiving bottle are loosened, but not removed. (d) A bag of 10 mL sterile pipettes is unsealed. (e and f) A single pipette is shaken out without placing hands inside the bag. A pipette aid is placed on the pipette. (g) The top of the flask that contains sterile nutrient medium is quickly passed over the flame. The culturist is being careful not to allow the pipette tip to touch any surface. Observe that the culturist is holding the cap of the flask in the crook of her right hand. (h) Next, 10 mL of sterile nutrient medium is withdrawn from the flask. The cap is placed back on the flask (not shown). (i) The top of the sterile receiving bottle is briefly flamed. Note its cap in the crook of the right hand of the culturist. (j) The sterile nutrient medium is dispensed into the receiving bottle. (k) The top of the bottle is quickly flamed again (not shown), and it is recapped. (l) The bag of sterile pipettes is resealed.

5.3.1. When the air gas mixture is optimal, the flame will consist of a cone of blue light about 4 cm high, surrounded by a translucent, violet-colored outer cone. The hottest spot is just above the top of the inner cone. A diffuse red or orange flame is not ideal.

5.3.2. You may need to adjust the gas flow.

5.3.2.1. Turn the wheel on the bottom of the burner to adjust gas flow. Clockwise provides more gas; counterclockwise provides less gas.

5.3.2.2. Use more gas if the flame is too small. If the flame is too high, reduce the gas flow.

5.3.3. You may need to adjust the air supply.

5.3.3.1. Turn the barrel of the burner to open and close holes that supply air.

5.3.3.2. If the flame is red or orange, increase the airflow by opening the holes.

5.3.3.3. If you do not see an inner cone, try decreasing the amount of air.

6. Arrange the bottle of nutrient medium and the empty sterile bottles in such a way that you will not need to reach over any open bottles.

7. Loosen, but do not remove, the caps of the three bottles.

8. Prepare a 10 mL sterile pipette.

8.1. If you have a bag of sterile pipettes, open the top of the bag (the end away from the tips).

8.2. If you have individually wrapped pipettes, open the paper wrapper by forcing the top of the top of the pipette through the wrapper.

8.3. Open the can if you have a can of sterile pipettes.

8.4. Do not fully remove the pipette from the bag, wrapper, or can. You want just the top accessible.

9. Hold the pipette near its top and insert it into the pipette aid.

10. Remove the pipette completely from the sterile packaging. Be careful not to touch the sterile tip end of the pipette to anything. If you do, throw it away and get another pipette.

11. Quickly pass the tip of the pipette through the flame, if directed to do so by your instructor. Some people flame the tips of their sterile pipettes; others do not.

12. With your left hand (if right-handed), remove the cap of the nutrient medium stock bottle and hold the cap in the crook of your hand, with your little finger wrapped around it. Avoid putting the top down on the lab bench or touching the lip of the cap with your hand.

13. Run the top of the bottle through the flame for a second or two.

14. Hold the container of sterile nutrient medium at a slight slant away from your face so that dust and particles will not readily settle into the bottle. Pipette out 10 mL of nutrient medium.

15. Immediately reflame the top of the nutrient medium stock container and loosely replace the cap.

16. Remove the cap of one of the storage bottles, again holding the cap in your little finger.

17. Pass the top of the storage bottle through the flame for a second or two.

18. Expel the medium into the bottle.

19. Quickly reflame the top of the bottle and replace the cap.

20. Discard the pipette in the proper receptacle.

21. Repeat steps 8–20 with the second storage bottle, the sterile tubes, and the two Petri dishes:

21.1. When you open a Petri dish, hold the cover above the dish to protect it from particles in the air.

22. Place your vessels with nutrient medium in a 37°C incubator. Remember to cover the control plate and put it into the incubator also.

23. Clean your work area. Dispose of all items according to your instructor's directions. Swab the lab bench with 70% ethanol or other disinfectant.

24. Wash your hands before leaving the laboratory.

25. Check the results after one, two, and three days of incubation.

25.1. Check the nutrient medium in all six vessels. There should be no cloudiness or turbidity in any bottle or plate. If there is, then your aseptic technique failed and contaminants entered. Record all information in your laboratory notebook.

25.2.   Check the air control plate each day. Draw and describe or photograph what you see on the control plate.

25.3.   If you see turbidity in any vessel, observe a drop of the liquid under the microscope at high power, using oil immersion, so you can see what the contaminant looks like.

LAB MEETING/DISCUSSION QUESTIONS

1.   Compare the class results for the control plates. The results for the entire class should give you an idea of the potential for contamination by airborne particles.

2.   How did the class fare in terms of avoiding contamination? If some people had contamination and some did not, try to analyze the differences in procedures. Individuals who had contamination may want to repeat this exercise to improve their aseptic technique.

Laboratory Exercise 38: Working with Bacteria on an Agar Substrate: Isolating Individual Colonies

OVERVIEW

This laboratory exercise has two parts. The first part is to prepare agar plates that provide a suitable nutritional and physical environment for growing bacterial colonies. The second part is to use some of your agar plates to isolate individual colonies of *E. coli*. In this exercise, you will therefore perform the following tasks:

- Mix and autoclave a nutrient medium.
- Check to see that the autoclave reached the required temperature.
- Pour agar plates for this laboratory exercise and also for Laboratory Exercise 41.
- Check your plates to see that they are not contaminated.
- Streak plates to isolate individual colonies.
- Save your streaked plates with isolated colonies for Laboratory Exercise 39.

BACKGROUND

A. Agar Plates for Bacterial Culture

In Unit V, you learned how to prepare buffered solutions to support biological macromolecules, such as DNA, RNA, and proteins. Cultured cells similarly require a buffered aqueous environment that contains specific dissolved solutes in the proper concentrations. Working with living cells, however, introduces some additional solution-related issues. The most obvious is that cells require nutrients if they are maintained in culture for more than a few hours. Another feature of intact cells is that they are surrounded by a cell membrane that is permeable to the flow of water and some, but not all, solutes. Intact cells, therefore, require an osmotically balanced environment; their environment must be isotonic. When cells are placed in **isotonic solutions**, *there is no net flow of water across the cell membrane*. As a result, the cells will neither shrink nor swell and burst (see Figure 8.6). Living cells also require a culture environment that is sterile except for the cells of interest. In this laboratory exercise, you will practice making a **culture medium** (also called **nutrient medium** or just **medium**), *a liquid environment that is suitable for*

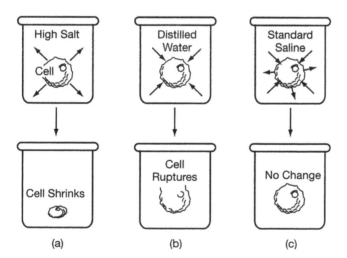

FIGURE 8.6   The flow of water across a cell membrane. (a) If there is a higher solute concentration outside the cell than inside, then water flows out of the cell and the cell shrinks. (b) If there is a lower solute concentration outside the cell, water flows into the cell and it may burst. (c) An isotonic solution is in equilibrium with the interior of the cell.

the growth and reproduction of cells. The particular type of medium you will make in this exercise contains agar and is used to support the growth of bacterial colonies.

The techniques you will use in this laboratory exercise have been used by scientists for more than a hundred years. In the 1800s, pioneering microbiologists, including Louis Pasteur and Robert Koch, studied the role of bacteria in causing disease. They discovered that many types of bacteria could be grown in nutritional, water-based, liquid media (often termed "broths") containing extracts from such sources as animal liver, heart, and brain; and yeast. Bacteria grow suspended throughout a liquid broth. Broths are generally used today for propagating large numbers of bacteria, either in the laboratory or for industrial-scale fermentation.

One of the advances that emerged from Koch's laboratory was the discovery that agar, a substance derived from seaweed, could be used to harden nutrient broths. The hardened medium creates a solid substrate on top of which bacteria readily grow, extracting the nutrients from the medium on which they sit. The use of an agar surface enabled Koch and other scientists to isolate individual colonies of specific disease-causing organisms. Agar plates are still routinely used as a substrate for bacterial growth.

You will use agar plates in this exercise to obtain colonies of *E. coli* bacteria that you can observe, describe, and store for later use. You will apply the bacteria to the surface of the plates and then "streak" them in such a way that individual cells are physically separated from one another. These individual cells each divide to form colonies; each colony is presumed to be derived from one parent cell. This means that each colony is likely to be genetically homogeneous. Isolated colonies are used by molecular biologists when genetic manipulations are performed on bacteria. If, for example, a biotechnologist intends to use cells to manufacture a protein of interest, it is important that all the cells that are grown are homogeneous and contain the gene of interest. If this is not the case, the culture might be contaminated by unwanted cells, or the cells might produce heterogeneous protein products.

B. Using an Autoclave to Sterilize a Solution

An **autoclave** is *a pressure cooker in which substances are sterilized by exposure to steam under high pressure.* Autoclaving is the method of choice for sterilization of substances that can withstand the heat and pressure.

An autoclave generates high-pressure steam and high temperatures; it, therefore, must be operated properly to avoid injury. Proper use of an autoclave requires the following:

- Knowledge of how the autoclave works and its safety features.
- Indicators for each run to show that the autoclave is functioning properly.
- Procedures for properly preparing bottles, plates, and supplies to be autoclaved.
- Procedures for loading and unloading the autoclave correctly.

Your instructor will demonstrate the particular model of autoclave in your laboratory. Read the safety briefing carefully before working with an autoclave.

Planning Your Work: Working with Bacteria on an Agar Substrate

1. Draw a sketch in your laboratory notebook of how you will organize your materials on your lab bench to perform Part A and then Part B of this exercise.

2. What items will you flame? When?

3. Review the Safety Briefings regarding Bunsen burners, autoclaves, and *E. coli*. Be sure you are aware of all the safety considerations described.

Safety Briefing

Autoclaves *use high-pressure steam to sterilize items.*

- Learn how to operate your autoclave properly and never bypass its safety features.
- Items that have been autoclaved are hot; handle them with heat-resistant gloves.
- Put liquids in containers that will hold twice their volume to prevent liquids from boiling over.
- Loosen the covers of vessels containing liquids so that pressure can equilibrate between the inside and outside of the vessels.
- Use a slow exhaust setting when autoclaving liquids to allow the temperature and pressure to slowly decrease after an autoclave run. This helps prevent liquids form boiling over. (You can use a fast exhaust setting if the run does not include any liquids.)
- When the run is over, slowly open the autoclave door an inch or so to allow the steam to escape.
- Wait 10 minutes before removing liquids from the autoclave because liquids can become superheated and can boil over if agitated.

Safety Briefing

You will be working with a nonpathogenic strain of *E. coli*, a bacterium that is a normal part of the flora of the human gut. In the laboratory, you will use special strains of *E. coli* that lack the ability to survive in the gut or the environment. Although these bacteria are regarded as safe, you should still follow Standard Microbiological Practices shown in Box 1.5 and also follow these precautions:

- Keep your face away from your work as you pipette and handle bacteria.
- Sterilize spreaders and loops after use to be sure no living bacteria remain on them.
- Avoid opening plates that contain bacterial colonies; look through the cover whenever possible.
- For small spills of bacterial cultures, for example, a few drops, immediately cover the spill with 70% ethanol or another disinfectant that is used in your classroom. Allow the disinfectant to sit for 15 minutes and then, while wearing gloves, clean up the spill with paper towels. Discard the paper towels according to the disposal method used in your institution. Discard the gloves and wash your hands.
- For larger spills of bacterial cultures, inform your instructor immediately. Stay away from the contaminated area and instruct your classmates to do so also.
- Properly dispose of all materials that might have been contaminated by bacteria. The method of disposal may vary in your institution:
  - In many institutions, contaminated solid waste is placed in heavy-gauge bags intended for this purpose and autoclaved for 45 minutes at 121°C.
  - In some institutions, contaminated materials are soaked in 10% bleach solution, allowing at least 15 minutes for bacteria to be killed.
- Wipe down the lab bench with 70% alcohol or other disinfectant at the end of every lab period.
- Wash your hands before leaving the laboratory.

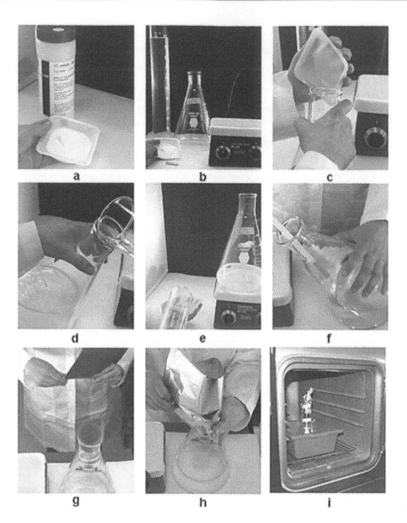

FIGURE 8.7 **Preparing LB agar medium.** (a and b) The materials were assembled. In this case, a premixed commercial agar was used, so the individual components did not need to be weighed out. The premixed agar medium was weighed according to the manufacturer's recipe. One liter of purified water was obtained. (The manufacturer does not specify that the mixture should be brought to volume; rather, the powdered medium is dissolved in 1 L of water.) (c) The LB powder was poured into a 2 L flask. (d) Purified water was added to the powder. (e) The weigh boat was rinsed with a small amount of water and the residue was added to the flask. A stir bar was added to the flask, and the flask was placed on a magnetic stir plate to dissolve the powder. (f) The stir bar was removed from the flask after the powder had dissolved. (g–i) The flask was covered first with a sponge plug. Then an aluminum foil "cap" was made to cover the plug. The aluminum foil cap prevents spills if the plug is ejected during autoclaving. Autoclave tape was used to attach the aluminum foil cap. A temperature-indicating strip was also attached to the flask (not shown). The indicator strip will indicate the temperature attained during autoclaving. The flask was placed in an autoclave-safe plastic basin to protect the chamber in the event of a spill. Note that most plastics melt at high temperature; never place a plastic basin in the autoclave unless you know it is heat resistant. Trying to clean a molten mass of plastic from the internal structures of an autoclave is an unpleasant job.

LABORATORY PROCEDURE

PART A: Make LB Agar Plates

Refer to Figure 8.7.

A.1.   Weigh out the ingredients of lysogeny broth (LB) agar:*

> **LB Agar**
>
> (per liter)
>
> 10 g tryptone
>
> 5 g yeast extract
>
> 10 g NaCl (MW = 58.44)
>
> 15 g agar
>
> 0.5 mL of 4N NaOH

A.2.   Add all ingredients except NaOH to a 2 L flask that has been rinsed with purified water.

A.3.   Add 1 L purified water.

A.4.   Add 0.5 mL of 4 N NaOH. Handle NaOH with caution.

A.5.   Stir to dissolve dry ingredients on a stirring hotplate. If a small amount of undissolved material remains, it will dissolve during autoclaving.

A.6.   Cover the flask with cotton or sponge plug. (The plug will prevent microbes from entering the flask but will still allow gas exchange.) Cover the plug loosely with a large, square piece of aluminum foil. Tape foil to the flask with autoclave tape.

A.7.   Autoclave for 15 minutes at 121°C.

    A.7.1.   Allow an hour for the entire autoclave run, including time for the autoclave to heat up and cool down.

    A.7.2.   Wear a heat-protective glove when removing the flask from the autoclave.

    A.7.3.   Determine if the autoclave has reached sterilizing temperature for a sufficient length of time. We suggest using autoclave tape because it is a convenient and inexpensive heat indicator that can be routinely applied to materials before autoclaving. The tape changes color to indicate that the proper sterilizing temperature was reached. After items are removed from the autoclave, the presence of autoclave tape indicates to laboratory personnel that a particular item was autoclaved. Autoclave tape can be used to seal an item so that when the tape is broken, it means the autoclaved item has been opened. Use autoclave tape or a similar product.

Part B: Pour Agar Plates

B.8.   While the agar is in the autoclave, prepare to perform work aseptically.

    B.8.1.   Tie back long hair. Avoid dangling clothing.

    B.8.2.   Clear all materials from your work surface. Move your laboratory manual and lab notebook to the side. Remove all other papers, books, and objects.

    B.8.3.   Wash your hands.

    B.8.4.   Spread disinfectant across the entire work area. Spread the disinfectant with a clean paper towel and allow excess to evaporate.

    B.8.5.   Place a Bunsen burner where you will not pass your hands or sleeves over the flame.

---

* Recipe from D. Micklos, G. Fryer, and D. Crofty, *DNA Science: A First Course*, 2nd ed., Cold Spring Harbor Press, 2003.

B.8.6. Arrange your bottles, tubes, and other supplies where they are easy to reach without reaching over the flame.

B.8.7. Label Petri plates and other items before beginning.

B.9. Allow the flask to cool after autoclaving just until the bottom of the flask can be held in bare hands (about 55°C–60°C). If the agar begins to solidify, you can heat the flask in a microwave until it melts.

B.10. Remove sterile Petri plates from their plastic sleeve. Be careful that they do not open. Save the plastic sleeve for later.

B.11. While the agar is cooling, label the bottoms of a stack of plastic Petri dishes with your name(s), the date, and the type of medium. (Label the bottoms of the plates because the tops can get mixed up accidentally.)

B.12. Pour your plates (see Figure 8.8).

B.12.1. Light a Bunsen burner. Arrange your materials to be easily accessible so you do not reach across the Bunsen burner.

B.12.2. When the flask is just cool enough to hold, remove the plug covering the flask. It is easiest to have a partner hold the plug, if you are working in teams. Flame the top of the flask for a couple of seconds.

B.12.3. While holding the flask in one hand, use the other hand to lift the lid of a Petri dish just enough to pour in the agar.

B.12.4. Quickly pour in agar to just cover the bottom of the plate, about 25–30 mL.

B.12.5. Replace the plate's lid. Swirl plates to spread the agar evenly across the plate.

B.12.6. Continue to pour agar into the rest of the plates. Flame the top of the flask occasionally.

B.12.7. Allow the agar to harden. Do not move the plates until the agar is firm.

B.13. Incubate the plates upside down overnight at 37°C. This dries the agar and helps reduce condensation later.

B.14. Wash your hands before leaving the laboratory.

B.15. Check every plate the next day to make sure there are no contaminating colonies. Discard any plate with any growth on it. You should not see contamination on any plate if your technique is good.

FIGURE 8.8   Pouring agar plates. (a) The lab bench is set up for pouring agar plates. (Bunsen burner is not shown.) (b) An experienced worker is shown pouring four plates in rapid succession so that neither the flask nor any plate is open for more than a few seconds. (Red dye has been added to agar for illustration purposes.)

B.16. Return the plates to their original plastic sleeve for storage in the refrigerator. Plates should be good for 3 months.

Part C: Streaking Plates for Individual Colonies

C.1. Prepare to work aseptically as you did in Part B of this exercise.

C.2. Remove two agar plates from their plastic sleeve. Label the bottoms of each plate with your name, the date, and the organism with which you are working.

C.3. Remove a colony from a plate (see Figure 8.9).

    C.3.1. Hold the handle of your inoculating loop and move the tip into the Bunsen burner flame. Wait until the tip glows red and then slowly pass the next couple of inches of the shaft through the flame.

    C.3.2. While holding the sterilized loop so it does not contact any surface, lift the cover of the Petri dish provided by the instructor that contains colonies of bacterial cells. Use one hand to hold the sterile loop and the other to open the dish.

    C.3.3. Keep the lid poised above the plate to protect the plate from dust and particles in the air.

    C.3.4. Cool the inoculating loop by touching it twice to the side of the agar where there are no colonies.

    C.3.5. Gently scrape up a colony or group of cells onto the loop; do not scrape up any agar.

C.4. Streak your plate (see Figure 8.10).

    C.4.1. Open a fresh, empty agar plate, again holding the lid so that it protects the plate. Gently pass the loop with cells on it in a streaking pattern, as shown in Figure 8.10a, being careful not to scratch the agar surface. Close the lid.

    C.4.2. Reflame the loop to sterilize it.

    C.4.3. Cool the loop on a part of the agar plate that contains no cells.

    C.4.4. Pass the loop through the end of the streak and streak a fresh area of the plate, as shown in Figure 8.10b. Close the lid.

    C.4.5. Reflame and cool the loop as described previously.

    C.4.6. Pass the loop through the end of the second streak and streak back and forth one more time (Figure 8.10c).

    C.4.7. Reflame the loop.

 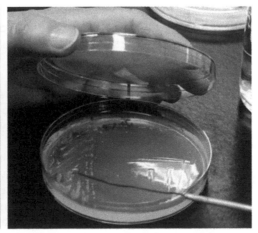

a   b

FIGURE 8.9   **Removing a colony.** (a) A loop was sterilized by passing it through a flame, as described in step C.3.1. (b) The loop was cooled on the side of the agar. A part of a single colony was gently scraped from the plate surface.

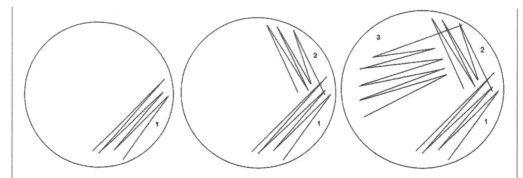

FIGURE 8.10   Streaking a plate. (a) Apply cells to a sterile loop. Without lifting the loop (and without gouging the agar) run your loop back and forth on a third of the Petri plate to form streaks. This set of streaks will have the densest cells. It is unlikely that you will see individual colonies here. (b) Sterilize and cool the loop. Pass it through the end of the first set of streaks to pick up some, but not too many, bacteria. Streak another third of the plate. (c) Repeat as before to form a third set of streaks.

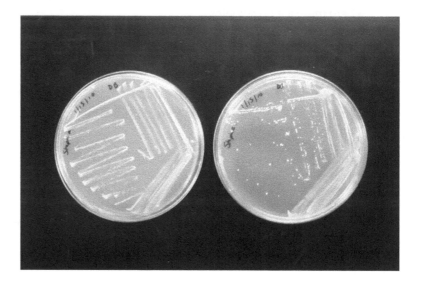

FIGURE 8.11   Results for two plates the next day. (a) Observe that individual colonies on this plate are present only at the end of the third set of steaks. This is not an ideal result. (b) This is another plate with clearly defined individual colonies that appear at the end of the second set of streaks.

C.5.   Repeat streaking with a second agar plate to practice the technique again.
C.6.   Place the two plates in a 37°C incubator overnight, upside down. This prevents condensation from the lid from falling onto the colonies and smearing them.
C.7.   Clean your work area.
    C.7.1.   Discard bacteria-contaminated items in proper bags for autoclaving.
    C.7.2.   If waste items cannot be autoclaved, disinfect them for at least 15 minutes in 10% bleach and then discard them in the regular trash.
    C.7.3.   It is good practice to wipe down racks and other items with disinfectant in case any unnoticed contamination occurred.
    C.7.4.   Swab disinfectant across your work area and let it evaporate.
C.8.   Wash your hands before leaving the laboratory.
C.9.   View your plates the next day. You are successful if you see isolated, individual colonies. The colonies should all have the same color and shape, although some will be larger than others (Figure 8.11). Is this the case? Draw and describe the appearance of the plates. Photograph the plate, if possible.

C.10. Save your best plate for Laboratory Exercises 39 and 41. Seal the plate with Parafilm and store it at 4°C.

LAB MEETING/DISCUSSION QUESTIONS

1. What is a bacterial colony?
2. You begin this exercise with a stock of bacteria that we will presume to be an uncontaminated culture of the proper strain. The colonies will have the proper appearance for this strain on this medium. Do the colonies on everyone's plates look the same? Familiarize yourself with this appearance so you recognize it in the future. Note if you see any colonies of different appearance as these are likely to be contaminants.
3. Did everyone get isolated colonies? If not, can you determine why not? What is important in obtaining isolated colonies?

Laboratory Exercise 39: Gram Staining

OVERVIEW

In this laboratory exercise, you will learn a common method of staining bacteria, called Gram staining. You will use Gram staining to visualize bacteria from one of your agar plates from Laboratory Exercise 38. **Gram staining** *is traditionally used to classify bacteria into one of two groups, as the first step in their identification.* The majority of bacteria can be classified as being either Gram positive or Gram negative, depending on how their cell walls interact with the Gram stains. Staining methods for bacterial identification, such as the Gram stain, are being replaced in many laboratories by molecular biology methods that are far more powerful. However, the Gram stain method is still sometimes performed, and it will enable you to more easily see your bacterial cells under the microscope than if they are unstained. In this exercise, you will thus accomplish the following objectives:

- Learn to perform a Gram stain.
- Become familiar with the appearance of *E. coli*, which are Gram negative and are shaped like rods.
- Check the quality of your cultures.

BACKGROUND

Bacteria (unlike mammalian cells) have a cell wall that surrounds their plasma membrane. The cell wall of Gram-positive bacteria contains a thicker layer of peptidoglycan (a mesh-like polymer of sugar and amino acids) compared to Gram-negative bacteria. The thickness of this peptidoglycan layer affects the staining properties of the cells.

To perform a Gram stain, the bacterial cells are dropped onto a glass microscope slide and affixed there by using heat. Crystal violet, the "primary" stain, is dropped onto the slide. Crystal violet penetrates the cell wall of both Gram-positive and Gram-negative bacteria and stains the cells purple. Gram's iodine is then added to the slide where it forms complexes of crystal violet-iodine. Alcohol or acetone is next applied to the slide. Gram-negative bacteria lose the crystal violet stain fairly quickly in the solvent. In contrast, Gram-positive cells do not lose the crystal violet as quickly, presumably because the stain has interacted with the thick peptidoglycan layer in their cell wall. A second stain, called the "counterstain," is next added to the slide. We will use safranin as the counterstain. The safranin gives the decolorized Gram-negative bacteria a pink or red color. In the end, Gram-positive cells should be purple; Gram-negative cells should be pink or red.

Planning for Lab: Gram Staining

1. Measure the length of a human cell and several bacterial cells in Figure 8.12. How much larger are human cells than bacterial cells?
2. You can see the nucleus clearly in the human cells. Can you see any internal structures in the Gram- stained bacterial cells in Figure 8.12? If not, why do you think you cannot see internal structures?
3. Do bacteria have nuclei?

LABORATORY PROCEDURE

Refer to Figure 8.13 to see the steps in Gram staining.

1. Prepare a slide of cells from your mouth.*
   1.1. With a toothpick, extract a small amount of material from the space between your teeth.

---

* The idea for this slide is from the University of Pennsylvania Health System Education and Training website, http://www.uphs.upenn.edu/bugdrug/antibiotic_manual/Gram2.htm.

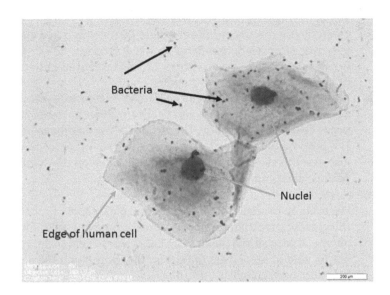

FIGURE 8.12   Gram-stained slide of sample from mouth. This sample from the mouth contains both human epithelial cells that have sloughed from the skin surface and a normal assortment of bacteria. The large human cells have easily seen nuclei. You can get a sense of the different appearance and relative sizes of the two types of cells in this image.

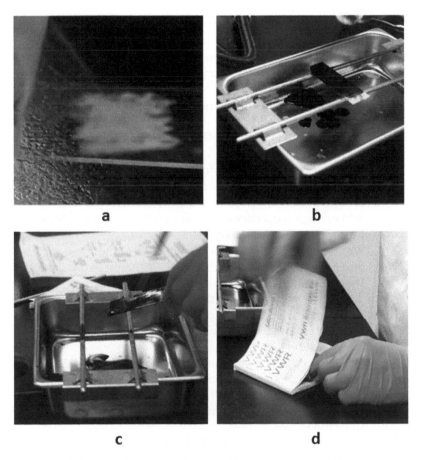

FIGURE 8.13   Preparing a Gram-stained slide. (a) The cells are smeared and dried on a slide. (b) Each reagent is successively dropped onto the slide, which is positioned above a staining tray. (c) Each reagent is rinsed from the slide. (d) The slide is placed between pieces of bibulous paper to dry it. It is not rubbed because this may damage the cells' smear.

1.2.   Smear the material into a drop of sterile water on a slide.

1.3.   Carefully dry the smear onto the slide by moving it in a circular pattern over a Bunsen burner flame for a few seconds.

1.4.   Gram stain this slide along with your *E. coli* sample (as described in the Gram staining procedure in step 3).

1.5.   After Gram staining the slide of cells from your mouth, you should see a mixture of Gram-positive and Gram-negative bacteria, along with larger cells from your body. The larger cells should have pink nuclei. The purpose of this slide is to be sure the staining procedure worked. Also, use this slide to learn to distinguish between the color of Gram-positive and Gram-negative bacteria.

2.   Prepare a slide of *E. coli*.

2.1.   Place a drop of sterile water on a second glass slide.

2.2.   Using aseptic technique, remove a small amount of cells from a colony on one of your stored plates from Laboratory Exercise 38. You should not be able to see a mass of cells on the loop; if you see a mass, you have too many cells.

2.2.1.   You can use an inoculating loop to remove the cells, or an autoclaved toothpick.

2.3.   Spread the cell smear evenly on the slide in a circle about 1.5 cm in diameter, which is about the size of a dime.

2.4.   Carefully dry the smear onto the slide by moving it in a circular pattern over a Bunsen burner flame for a few seconds.

3.   Gram stain both slides.

3.1.   Add enough primary stain, crystal violet, to cover the slides and let it sit for 1 minute.

3.2.   Rinse with a gentle stream of water for 4–5 seconds to remove unbound crystal violet. The water at this point does not need to be sterile.

3.3.   Add enough Gram's iodine reagent to cover the slides and lit it sit for 1 minute.

3.4.   Pour off the iodine and briefly rinse with water; shake off extra water droplets.

3.5.   Rinse the slide with alcohol for 3 seconds followed by a gentle stream of water. Do not use alcohol for more than 3 seconds as it may begin to decolorize Gram-positive as well as Gram-negative cells.

3.6.   Add enough secondary stain, safranin, to cover the slides and let it sit for 1 minute.

3.7.   Wash with a gentle stream of water for a maximum of 5 seconds. Blot the edges of the slide dry. You can also place the slides between sheets of bibulous paper to dry them.

4.   View the two slides, beginning with the slide of cells from your mouth.

4.1.   Begin with the scanning or low-power objective. Then move to the 40× objective. You should be able to see nucleated cells at this magnification.

4.2.   Find an area where you think there are bacteria. Move to 100× (oil immersion) unless your instructor tells you not to use this objective. See if you can distinguish Gram-positive from Gram-negative cells.

4.3.   On the slide of cells from your mouth, find differently shaped bacteria. Draw the shapes you see.

4.4.   Compare the size of the nucleated cells to the bacteria. Bacteria are roughly 1.5 µm in length while mammalian cells can be as much as 100 µm in diameter.

4.5.   Next, look at your bacterial colony slide. Begin again at a low power and then move to 40× and oil immersion. Describe the appearance of

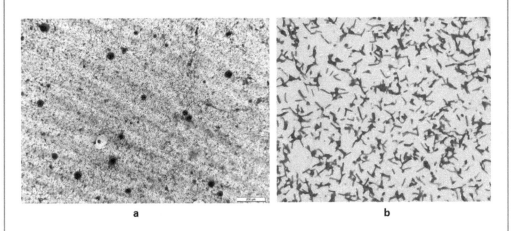

FIGURE 8.14   A Gram-stained slide. (a) This photo was taken with a 20× objective. Little detail is apparent, but a purple smear is visible. At this magnification, the purple "blobs" are not bacteria but are just clumps of stain. (b) This photo was taken with the 100×, oil-immersion objective. Note the color and characteristic rod shape of Gram stained *E. coli*.

the colonies on the agar plate and the corresponding appearance of the cells on the microscope slide. What color are the colonies? What shape? Draw or photograph the cells as they appear on the agar plate and under the microscope.

4.6.   Under the microscope, the cells should be shaped like rods (see Figure 8.14). They should also be pink in color. If you see heterogeneity in the appearance of the cells, it is possibly an indication of contamination.

4.6.1.   Even if you see all Gram-negative rod-shaped bacteria, it does not necessarily mean that you have exclusively *E. coli* of the proper strain because many other bacteria look the same. If you do not see Gram-negative rods, there is clearly a problem in the culture.

Laboratory Exercise 40: Preparing Phosphate-Buffered Saline

OVERVIEW

You will make sterile phosphate-buffered saline (PBS) for use in Laboratory Exercises 41, 46, 47, and 48. As you perform this laboratory exercise you will:

- Practice preparing a solution.
- Practice autoclaving a solution that needs to be sterile.

BACKGROUND

When cells are placed in isotonic solutions, there is no net flow of water across the cell membrane (refer to Figure 8.6). PBS is an isotonic salt solution commonly used for short-term maintenance of cells. PBS is also buffered to help prevent pH fluctuations. There are various recipes for PBS, but they all use phosphate-containing compounds to provide buffering. PBS does not contain any nutrients and so cells that are kept in PBS for longer than the length of a specified procedure will become unhealthy and eventually die.

All solutions that will be used with cultured cells must be sterilized before use. PBS does not contain heat-sensitive components and, therefore, it can be sterilized by autoclaving.

Planning Your Work: Making Phosphate-Buffered Saline

Write a procedure in your laboratory notebook to make 200 mL of PBS. When you come to lab, be sure to distinguish between what you actually do and what you planned to do.

LABORATORY PROCEDURE

1. Make 200 mL of PBS for your own personal use; adjust the following recipe accordingly. Use a source of purified water that is of suitable quality for mammalian cell culture if you plan to use this PBS for later laboratory exercises involving mammalian cells. (The water quality requirements for bacterial culture are less stringent than for mammalian cell culture.)

> **Phosphate-Buffered Saline, pH 7.4**
> Dissolve the following in 800 mL of purified water:
> 8 g of NaCl
> 0.2 g of KCl
> 1.44 g of Na_2HPO_4
> 0.24 g of KH_2PO_4
> **Adjust pH to 7.4.**
> Bring to a final volume of 1 L with purified water.

2. Check the pH and conductivity of your solution.
3. Prepare a series of dilution blanks for Laboratory Exercise 41.
   3.1. Pipette 9.0 mL of PBS into five capped test tubes.
   3.2. Pipette 9.9 mL of PBS into five capped test tubes.
4. Loosely screw the caps onto the test tubes. Do not screw the caps on tightly or the tubes may explode when autoclaved; instead screw the caps on lightly so that steam can get in and out of them.

5. Put the rest of your PBS into a loosely capped bottle. Autoclave and save this bottle for mammalian cell culture later.

6. Use autoclave tape to loosely attach the caps to the tubes and bottle.

    6.1.   Autoclave tape is a convenient heat indicator that changes color to indicate that the proper sterilizing temperature was reached. After items are removed from the autoclave, the presence of autoclave tape on them serves to indicate to laboratory personnel that a particular item was autoclaved. Autoclave tape can be used to seal an item so that, when the tape is broken, it means the autoclaved item has been opened.

7. Each model of autoclave is operated differently. Your instructor will therefore demonstrate to you the use of the autoclave in your laboratory, along with specific safety directions.

LAB MEETING/DISCUSSION QUESTIONS

Compare the conductivity and pH of everyone's solutions as a quality control check. Your instructor may provide a correctly made solution for comparison. Discard your solution if, based on these comparisons, you think its conductivity is not correct; it will not be isotonic. If you have made a mistake, try to figure out from your calculations and notes where the mistake was made.

Laboratory Exercise 41: The Aerobic Spread-Plate Method of Enumerating Colony-Forming Units

OVERVIEW

It is often important to know how many bacteria are present in a culture or a sample. It is obvious, for example, how this might be important in a food-testing laboratory where the safety of foods is being analyzed. Product safety testing in a pharmaceutical setting similarly requires a method of determining the levels of microorganisms present. Environmental scientists are concerned with microorganism counts because of the vital role microorganisms play in the environment. Researchers may want to begin an experiment with a particular number or concentration of cells. Estimating the concentration of cells present in a culture is thus a basic skill for a cell culturist.

It is, however, challenging to count bacterial cells. First, bacterial cells are minuscule and, therefore, difficult to count visually, even with an excellent microscope. Second, bacterial cells are often present in extremely high concentrations, for example, 10^7 cells/mL. Obviously, no one is going to peer through a microscope and count 10 million cells in a single milliliter of sample. Microbiologists have therefore devised assay methods to estimate the concentration of bacterial cells in a sample. You will use one of these methods, the aerobic spread-plate method, in this exercise.

The spread-plate method requires that you first dilute the sample so that the concentration of cells is reduced to a workable value. You then apply the bacteria to an agar plate and spread them around so that the plate is evenly coated with the cell suspension. The plate is incubated. During incubation, individual cells that were present in the suspension divide and reproduce to form colonies of many cells. Over time, these colonies will grow large enough to be visible without using a microscope. Given a reasonable number of colonies, for example, fewer than 250 on a plate, the colonies can be readily counted. Counting the number of colonies present on the plate allows you to calculate the concentration of bacteria in the original culture.

While it is likely that each colony on an agar plate is derived from a single parent cell, there is no way to be certain that this is true. Each colony, therefore, is said to be derived from a single "colony-forming unit" (CFU), which may or may not have been a single cell.

The spread-plate method that we will use in this exercise is commonly used in industry to enumerate bacteria in food samples. It is called "aerobic" because the plates are exposed to air throughout the process. Note that there are some bacteria that do not thrive in air; these are not counted by this method.

In this exercise you will perform the following tasks:

- Prepare an overnight culture of *E. coli*.
- Use the spread-plate method to estimate the concentration of CFU in the overnight culture.
- Explore the variability of this enumeration method.
- Explore potential sources of error in this enumeration method.

BACKGROUND

A. Preparing an Overnight Culture of *E. coli*

In this exercise, you will enumerate bacteria in an overnight culture of *E. coli*. To prepare an overnight culture, begin with an *E. coli* colony from one of your plates from Laboratory Exercise 38. Remove a colony, or portion of a colony, from the plate, using aseptic technique, and place it into sterile broth. Incubate the broth overnight, preferably with shaking. The purpose of shaking is to help provide a constant supply of air, and therefore oxygen, to the cells. Shaking also allows all the cells to be exposed to nutrients and facilitates the dilution of metabolic wastes. The cells actively divide and reproduce in the fresh aerated medium, resulting in a high density of cells. This is your overnight culture. There are many molecular biology procedures that begin with an overnight culture of cells. In this laboratory exercise, you will be using the overnight culture to practice the spread-plate counting method.

B. Diluting the Sample

In the aerobic spread-plate method, 0.1 mL of a liquid bacterial culture is evenly spread onto the surface of an agar plate. Imagine that this culture contains a great many bacterial cells, perhaps 10^7 cells/mL. This is a lot of cells, but bacterial cell densities can easily get this high or higher under good growth conditions. In this situation, the 0.1 mL of culture on the agar plate would contain about 1 million cells. This is clearly far too many cells on one plate. In a day or so, the entire plate would be covered with a lawn of bacteria; no individual colonies would be visible. For purposes of counting bacteria, you need a more reasonable number of colonies on a plate, enough to be statistically meaningful and few enough to be able to count accurately. The accepted range is 25–250 colonies for a meaningful plate count. In order to achieve a plate with 25–250 colonies, the original sample must be diluted. In this example, you would dilute the original culture from 10^6 cells/0.1 mL to, let's say, 10^2 cells/0.1 mL. A conventional way to accomplish this dilution is in a series of steps, such as the following:

From 10^6 cells/0.1 mL to 10^5 cells/0.1 mL.
Then from 10^5 cells/0.1 mL to 10^4 cells/0.1 mL.
Then from 10^4 cells/0.1 mL to 10^3 cells/0.1 mL.
Then from 10^3 cells/0.1 mL to 10^2 cells/0.1 mL.

Figure 8.15 illustrates how these four dilutions might be prepared.

The example in Figure 8.15 is simplified in that we know how many bacteria per milliliter were present in the original sample. We normally do not know this value—the point of plate counting is to find out the concentration of cells in the original sample. We,

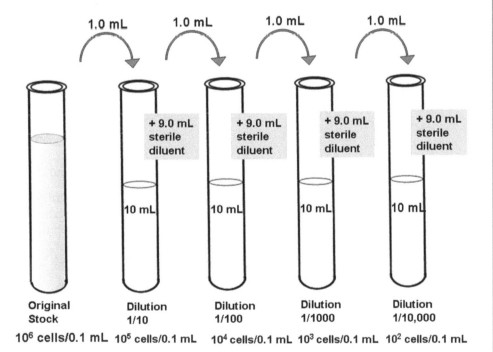

| 1.0 mL | 1.0 mL | 1.0 mL | 1.0 mL |

| | + 9.0 mL sterile diluent | + 9.0 mL sterile diluent | + 9.0 mL sterile diluent | + 9.0 mL sterile diluent |

| | 10 mL | 10 mL | 10 mL | 10 mL |

| Original Stock | Dilution 1/10 | Dilution 1/100 | Dilution 1/1000 | Dilution 1/10,000 |
| 10^6 cells/0.1 mL | 10^5 cells/0.1 mL | 10^4 cells/0.1 mL | 10^3 cells/0.1 mL | 10^2 cells/0.1 mL |

FIGURE 8.15   A simplified example: A series of dilutions, each 1/10. In this simplified example, you know that there are 10^7 cells/mL in the original stock of cells, which means there are 10^6 cells in 0.1 mL. The stock is diluted in a series of steps until, in the final dilution tube, there are 10^2 cells/0.1 mL. Each dilution combines 1 mL of cells with 9 mL of diluent to result in 10 mL of a 1/10 dilution.

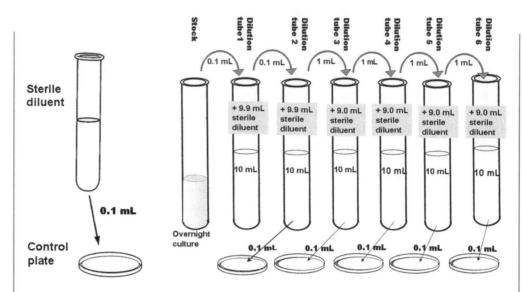

FIGURE 8.16 Dilution strategy for Laboratory Exercise 41.

therefore, usually plate two or more dilutions of the original sample with the expectation that one or more of the plates will have too few colonies, one or more of the plates will have too many colonies, but one plate will be just right with between 25 and 250 colonies on it. Figure 8.16 shows the strategy you will use in this exercise. Observe that you will plate cells from five different dilutions in order to have a high probability of obtaining one countable plate.

Safety Briefing

In this laboratory exercise, you will be placing a spreading device into a jar containing alcohol; the alcohol sterilizes the spreader. You will then put the spreader into a flame so that the alcohol burns off of it.

Do not put a hot spreader back into the alcohol because it can set the alcohol on fire.

Keep the lid to the alcohol jar close by so that if you do by chance set the alcohol on fire, you can quickly put the lid on top of it. The fire will be extinguished when the lid is put on the jar.

Planning Your Work: Spread-Plate Method of Enumerating Colony-Forming Units

1. Examine Figure 8.16. This figure shows the dilutions you will make in this exercise. On this drawing, label each tube with the total dilution in that tube. For example, the first dilution tube is a 1/100 dilution. The second dilution tube is (1/100)(1/100) = 1/10,000. Continue and label each tube with its dilution.*
2. Sketch how you will arrange your lab bench for both Part A and Part B of this exercise.
3. Review the safety briefing.

LABORATORY PROCEDURE

You will practice the plate count assay by using a colony known to contain *E. coli*. Once you have learned how to perform the assay, your instructor may direct you to repeat the assay with samples from foods or the environment. If you do so, then a known *E. coli* sample should be used as a positive control.

---

* For more information about dilution calculations, see L. Seidman, *Basic Laboratory Calculations for Biotechnology*, CRC Press, 2022.

Part A: Prepare an Overnight Culture of *E. coli*

A.1.   Prepare to perform work aseptically as described in Laboratory Exercise 37.

A.2.   Label a sterile 50 mL tube with your name, date, and procedure.

A.3.   Review in your mind the procedure you will be following so that you can perform it quickly and efficiently without making mistakes.

A.4.   Turn on your Bunsen burner.

A.5.   Loosen the cap of a container of sterile LB broth provided by your instructor.

A.6.   Loosen the cap of a sterile 50 mL tube.

A.7.   Using aseptic technique, transfer 5 mL of broth into the 50 mL sterile tube.

    A.7.1.   Attach a sterile 5 or 10 mL pipette to a pipette aid.

    A.7.2.   Optional: Run the tip of the pipette through the flame.

    A.7.3.   Remove the cap of the LB broth. Hold the cap in the crook of your hand. Flame the top of the container.

    A.7.4.   Remove 5 mL of broth, reflame the container, and replace the cap.

    A.7.5.   Remove the cap of the sterile 50 mL tube. Hold the cap in the crook of your hand and flame the top of the container.

    A.7.6.   Expel the 5 mL of broth into the tube.

    A.7.7.   Reflame the top of the tube. Replace the cap.

A.8.   Take out your agar plate with bacterial colonies from Laboratory Exercise 38.

A.9.   Sterilize your loop by placing it in the Bunsen burner flame until it is red hot. Then flame the next inch or two of the loop's shaft.

A.10.   Cool the loop by touching it twice to your agar plate, carefully avoiding touching any colonies. Remember to hold the lid of the plate over it to protect the plate from particles in the air.

A.11.   Select a well-defined colony, 1–4 mm in diameter. Gently scrape the single colony off the plate onto the loop.

A.12.   Using aseptic technique, transfer the colony to the tube with LB broth.

    A.12.1.   Remove the cap of the tube and hold the cap in the crook of your hand.

    A.12.2.   Run the top of the tube through the flame.

    A.12.3.   Immerse the tip of the loop in the broth in the tube and agitate the loop to remove the cell mass. You do not need to transfer all visible cells.

    A.12.4.   Briefly reflame the top of the tube and replace the cap.

    A.12.5.   Sterilize the loop.

A.13.   Put the cap back on the tube loosely so that air exchange can occur under the cap but particulates and microbes cannot enter.

    A.13.1.   Optional: Put a piece of tape lightly over the cap to keep it from falling off during incubation.

A.14.   Incubate the tube at 37°C with shaking. If shaking is not possible, allow at least 24 hours for the cells to multiply.*

Part B: Prepare Dilutions of your Overnight Culture

Refer to Figure 8.16 for the dilution and plating strategy. What follows are the details of each step.

B.1.   Remove your overnight culture tube from the incubator. It should be cloudy. If the broth is clear, no cells grew. In that case, share a tube with another student and review your notes to try and determine why the cells did not grow.

B.2.   Obtain two of your sterile, capped tubes with 9.9 mL of sterile PBS.

B.3.   Obtain four of your sterile, capped tubes with 9.0 mL of sterile PBS.

---

* Based on a procedure from D. Micklos, G. Fryer, and D. Crofty, *DNA Science: A First Course*, 2nd ed., Cold Spring Harbor Press, 2003.

B.4.   Line up the tubes in the front row of your test-tube rack following the illustration in Figure 8.16: from left to right, a tube with 9.9 mL sterile diluent, 9.9 mL, 9.0 mL, 9.0 mL, 9.0 mL, and 9.0 mL.

B.5.   Label each tube with the dilution it will contain.

B.6.   Label the bottoms of five agar plates from Laboratory Exercise 38 with your name or initials, date, sample, and the dilution that will be plated.

B.7.   Turn on your Bunsen burner.

B.8.   Mix your overnight culture tube.

    B.8.1.   Shake 25 times by moving the tube back and forth in a 30 cm (1 ft) arc within 7 seconds.*

    B.8.2.   Alternatively, mix it with a vortex mixer.

B.9.   Before the cells can settle, remove 0.1 mL of bacteria aseptically with a 1 mL sterile pipette. Alternatively, you can prepare your dilutions using a micropipette with sterile tips.

B.10.   Open and flame the first tube from the front left of your rack; transfer in the 0.1 mL of bacteria.

    B.10.1.   Discard the used pipette or the used tip in a container designated for this purpose.

B.11.   Close the tube and mix the solution as in step B.8.

B.12.   Remove 0.1 mL of solution from the first dilution tube and transfer it to the second dilution tube, using sterile technique.

B.13.   Mix and remove 1.0 mL from the second dilution tube and transfer to the third dilution tube.

B.14.   Mix and remove 1.0 mL from the third dilution tube and transfer to the fourth tube.

B.15.   Mix and remove 1.0 mL from the fourth dilution tube and transfer to the fifth dilution tube.

B.16.   Mix and remove 1.0 mL from the fifth dilution tube and transfer to the sixth dilution tube.

B.17.   Mix the sixth dilution tube and with a clean, sterile pipette or tip, transfer 0.1 mL of the sixth dilution to a labeled agar plate.

B.18.   Spread the 0.1 mL cell suspension evenly around the agar plate (see Figure 8.17).

    B.18.1.   Dip the spreader into the alcohol jar.

    B.18.2.   Place the spreader in the Bunsen burner flame. The alcohol will burn quickly. Do not hold the spreader in the flame. You just want to burn off the alcohol.

    B.18.3.   Cool the spreader by touching it to surface of the control agar plate, being careful not to touch the suspension.

    B.18.4.   Spread the suspension evenly across the surface of the plate. Avoid the edges of the plate because it is difficult to count colonies along the edge.

B.19.   Mix the fifth dilution tube and with a clean, sterile pipette or tip, transfer 0.1 mL to another labeled agar plate. Spread the 0.1 mL cell suspension evenly around the agar plate, as described in steps 18.1–18.4.

B.20.   Mix the fourth dilution tube and with a clean, sterile pipette or tip, transfer 0.1 mL to another labeled agar plate. Spread the 0.1 mL cell suspension evenly around the agar plate, as described in steps 18.1–18.4.

B.21.   Mix the third dilution tube and with a clean, sterile pipette or tip, transfer 0.1 mL to another labeled agar plate. Spread the 0.1 mL cell suspension evenly around the agar plate, as described in steps 18.1–18.4.

---

* This is the method recommended in Larry Maturin and James T. Peeler, "Aerobic Plate Count" in *Bacteriological Analytical Manual*, 8th ed., Chapter 3, 1998, http://www.fda.gov/Food/ScienceResearch/LaboratoryMethods/BacteriologicalAnalyticalManualBAM/ucm063346.htm.

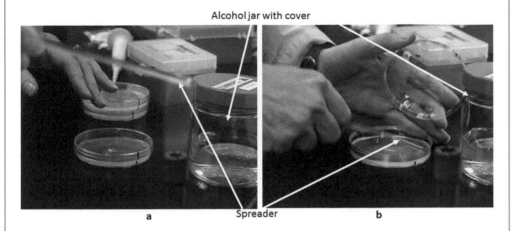

Alcohol jar with cover

a   Spreader   b

FIGURE 8.17   Spreading plates. (a) The workspace was arranged with a Bunsen burner and jar of 95% ethanol with a lid. The culturist has dipped the glass spreader into the alcohol jar and is now holding the spreader briefly in the flame of the Bunsen burner. The alcohol will quickly burn off the spreader. (b) The culturist cooled the spreader by touching it to the agar briefly. She is now spinning the agar plate with one hand as she uses the spreader to evenly distribute the cell suspension across the agar surface.

TABLE 8.2   Aerobic Spread-Plate Results

Dilution on Plate	Control	$1/10^4$	$1/10^5$	$1/10^6$	$1/10^7$	$1/10^8$
Colony count						

B.22. Mix the second dilution tube and with a clean, sterile pipette or tip, transfer 0.1 mL to another labeled agar plate. Spread the 0.1 mL cell suspension evenly around the agar plate, as described in steps 18.1–18.4.

B.23. With a clean, sterile pipette or tip, transfer 0.1 mL of sterile PBS to another labeled agar plate; this is the negative control plate. Spread the 0.1 mL cell suspension evenly around the agar plate, as described in steps 18.1–18.4

B.24. Allow the plates to sit undisturbed on the lab bench for 15–20 minutes to allow the liquid to seep into the agar.

B.25. Invert the plates and incubate overnight at 37°C.

B.26. Count the colonies.

　　B.26.1.　Copy Table 8.2 into your laboratory notebook and fill it in. If there are too many colonies to count, write "TNTC," too numerous to count. If there are no colonies on a plate, put a dash in the grid. If a plate is contaminated or otherwise unsatisfactory, label it as "LA," laboratory accident. You should have one plate with between 25 and 250 colonies. Consult your instructor if none of your plate counts fall in that range. With the plate covered, count the colonies on any plate with 25–250 colonies, including those of pinpoint size. You may want to use a marker on the cover to indicate which colonies have been counted so you do not accidentally count the same one twice.

B.27. Have another person count the colonies on the same plate. Compare your answers. The counts should be within 5% of each other. If not, consult with your instructor.

B.28. Calculate the number of CFUs per milliliter in the original overnight culture of bacteria.

TABLE 8.3   Rounding Results

Examples	
Calculated Count	**Round to**
12,700	13,000
12,400	12,000
15,500	16,000
14,500	14,000

Source: Larry Maturin and James T. Peeler, "Aerobic Plate Count" in Bacteriological Analytical Manual, 8th ed., Chapter 3, 1998, http://www.fda.gov/Food/ScienceResearch/LaboratoryMethods/BacteriologicalAnalyticalManualBAM/ucm063346.htm.

B.28.1. Select a plate with between 25 and 250 colonies. If none of your plates fall in this range, consult your instructor. If more than one plate is in this range, average the final calculated results.

B.28.2. Multiply the number of colonies counted on the plate by 10. This step is necessary because you plated only 0.1 mL and you want your answer on a per milliliter basis.

B.28.3. Multiply the value from step B.28.2 by the reciprocal of the dilution on that plate. For example, if the plate was in the $1/10^5$ column in Table 8.2, then multiply the number of colonies by 10^5. The answer is in CFUs/mL.

B.28.4. Round the final answer to two significant figures to avoid creating a misleading impression of accuracy and precision. Table 8.3 provides examples of the recommended method of rounding. (Your instructor will tell you how to round when the digit to be rounded is a 5.)

LAB MEETING/DISCUSSION QUESTIONS

1. Compare the results for the class. How much variability is there between values? What are possible causes of variability?

2. Your negative control plate should have had no colonies. Discuss the implications if any of the negative control plates in the class had cells growing on them.

3. Do you think each colony on your plate is likely to have arisen from a single cell?

4. Is it possible that the bacterial cells in a single colony are not genetically homogeneous? Explain.

5. What types of bacteria are counted by this method? What types are not? (First, consider that only living cells are counted. Also, consider the growth and nutritional conditions.)

6. The plate-count method is an assay. It is used to measure a property of a sample, that is, the cell concentration. As with all assays, there are factors that affect the accuracy and precision of the plate-count method.
    a.   What are causes of imprecision and variability in this method?
    b.   What are causes of inaccuracy in this method?

7. Optional: Use the aerobic plate-count method to enumerate the bacteria in a sample of food or a sample from the environment. When thinking about sample preparation, you may find it helpful to consult the article "Aerobic Plate Count," as referenced in the previous footnote.

Laboratory Exercise 42: Preparing a Growth curve for *E. coli*

OVERVIEW

The purpose of this laboratory exercise is to explore the growth characteristics of a culture of E. coli bacteria. In this exercise, you will generate a growth curve for a particular strain of E. coli. To prepare this growth curve, you will perform the following tasks:

- Use a spectrophotometric method to monitor the growth of an *E. coli* culture.
- Graph the growth of the culture.
- Identify the different stages of growth.
- Determine population doubling time for this culture.
- Compare spectrophotometric and plate-count methods of enumerating bacteria (optional).

BACKGROUND

Cells divide and reproduce in culture. As they do so, they metabolize the nutrients in their medium and produce waste. If nothing is added to the culture after it is inoculated and nothing is removed from it, then four stages of growth will occur in a bacterial culture (see Figure 8.18). The $\log_{10}$ of cell concentration has been plotted on the Y-axis with time on the X-axis. Observe the following four phases:

- **Lag phase.** *Immediately after the bacteria are inoculated into fresh medium, there is a phase with little cell division while the cells adapt.*
- **Exponential phase.** *A period of rapid growth follows the lag phase.* During this phase, cells are utilizing nutrients and excreting waste at maximum rates. This phase appears linear on the graph because cells *are doubling at a fixed interval, called the* **population doubling time**. You can determine the population doubling time by noting the amount of time it takes for a culture to double during the exponential phase (see Figure 8.19).

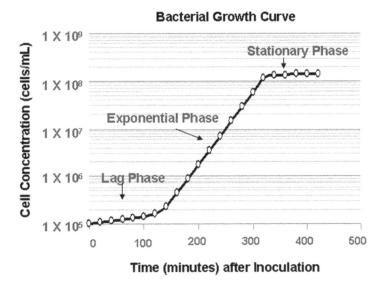

Bacterial Growth Curve

FIGURE 8.18   A typical bacterial growth curve. This graph was prepared on semilog paper, a type of graph paper that substitutes for calculating logs. The divisions on the Y-axis are initially widely spaced and then become narrower as one moves toward powers of 10. This spacing takes the place of calculating logs. This graph is on "semilog" paper where only the Y-axis has logarithmic spacing; there are normal, linear subdivisions on the X-axis. The graph thus actually portrays the $\log_{10}$ (cell concentration) versus time. Exponential phase is the part of the growth curve that appears "linear" on this semilog plot.

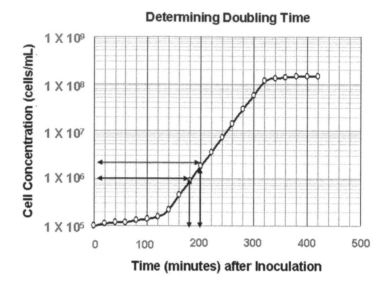

FIGURE 8.19   Determination of population doubling time. Two points are marked, each of them within the exponential phase, such that the cell concentration of one point is double that of the other point. The points chosen in this example correspond to cell concentrations of 1×10^6 cells/mL and 2×10^6 cells/mL. On the X-axis, this corresponds to 20 minutes. Therefore, the population doubling time for this culture is 20 minutes.

- **Stationary phase.** *Growth levels off as a result of nutrient depletion and a buildup of waste in the medium.* Cells remain metabolically active, but they double at an increasingly longer time interval. Eventually, no net cell division occurs. On the growth curve graph, this phase appears as a plateau.
- **Death phase.** *Eventually, the accumulation of waste and depletion of nutrients leads to a phase where more cells die than reproduce, so the population declines.* This phase is not shown in Figure 8.18 because the cells will experience neither high enough levels of waste nor low enough concentrations of nutrients to enter this phase over the course of a typical laboratory class.

In this laboratory exercise, you will explore the growth of your *E. coli* strain by performing the following tasks:

- Inoculating fresh nutrient medium with bacteria.
- Incubating the culture.
- Removing samples of culture at fixed time intervals.
- Measuring sample absorbances.
- Converting absorbance values to cell concentration.
- Graphing $\log_{10}$ (cell concentration) versus time.
- Interpreting the graph.

Observe that you will use a spectrophotometric method to estimate cell concentration. This method is based on the fact that the more bacteria present in a culture, the more turbid it will be. In a spectrophotometer, sample turbidity causes light to be scattered away from the detector and thus increases absorbance readings. The more bacteria present in a sample, the higher its apparent absorbance will be.

The spectrophotometric enumeration method is convenient and quick but has some inherent inaccuracies. In particular, the conversion factor you will use to convert absorbance to cell concentration is an estimate based on a typical bacterial strain in a typical culture medium.

Planning Your Work: Generation of an *E. coli* Growth Curve

1. Draw a sketch of how you will organize your materials on your lab bench to perform this exercise.

2. What numerical data will you collect? How will you analyze your data?

3. Optional: Write a protocol to compare the numbers of cells estimated by using the spectrophotometric method versus the number of CFUs determined by the plate-count method.

LABORATORY PROCEDURE

You will inoculate fresh bacterial medium with a small volume of stationary phase E. coli from an overnight culture. You will sample this freshly inoculated culture at 20-minute intervals throughout the remainder of the lab period. You will use your data to generate a growth curve and determine population doubling time.

1. The day before this exercise, prepare an overnight culture of *E. coli* according to the procedure in Laboratory Exercise 41.

2. Mix the overnight culture and inoculate 1 mL of it into 100 mL of LB broth in a 250 mL culture flask. The LB broth should be prewarmed to 37°C.

3. Immediately remove 1 mL of culture using aseptic technique. This is your "time = 0" sample. (If you are using a spectrophotometer that requires more than 1 mL of volume, remove the minimum amount required.)

4. Place the 250 mL culture flask into a 37°C incubator with agitation.

5. Record the A_{600} of the 1 mL of culture, using LB medium as a blank.

6. Every 20 minutes, aseptically remove another 1 mL sample of culture (or whatever volume is required to make a reading) until the end of your laboratory period. Read the sample's absorbance.

7. Clean your lab space, as directed by your instructor.

8. **Data analysis**

    8.1. Convert each A_{600} reading to cells/mL using this conversion factor:

    $$1 \text{ absorbance unit at } 600\,\text{nm} = 8 \times 10^8 \text{ cells/mL}$$

    8.2. On semilog graph paper, plot cells/mL (Y-axis) versus time in minutes (X-axis); refer to Figure 8.19. You may plot this by hand or electronically using a graphing program, such as Excel.

    8.3. Identify the exponential phase as those points that appear linear on this growth curve.

    8.4. Determine population doubling time by choosing a timepoint in the exponential phase and finding the timepoint at which the cell density is twice the value of the initial point. Then determine the difference in value between the X-coordinates of these two points. This difference will be the population doubling time; refer to Figure 8.19.

9. Optional: At two convenient time points, prepare plate counts of your cells using the directions in Laboratory Exercise 41. Compare the results of the spread-plate enumeration method to the results of the spectrophotometric method.

LAB MEETING/DISCUSSION QUESTIONS

1. Examine your growth curve. How much time did your culture spend in lag phase? How could you shorten the amount of time the culture spends in lag phase?

2. What factors could affect the population doubling time that you observe for a particular *E. coli* strain?

3.   Compare the class results for doubling time. In principle, the values should be the same.

4.   Imagine that you work for a company that sells bacterial cultures for commercial use. Your company grows large volumes of culture and then harvests the cells and freezes them in a concentrated pellet for sale. You have two strains of *E. coli* that work equally well in most applications. Strain A has a population doubling time of 45 minutes, and Strain B has a population doubling time of 20 minutes. Which strain will you choose to market and why?

5.   Optional: If you compared the spectrophotometric and plate-count methods, what were the results? Are the two methods equivalent? The plate-count method only counts viable cells. What about the spectrophotometric method? What are the assumptions for each method? In what situations would you select the spectrophotometric assay? In what situations would the plate-count method be preferable?

Laboratory Exercise 43: Aseptic Technique in a Biological Safety Cabinet

OVERVIEW

This exercise begins the section on working with mammalian cells. As you did when learning to work with bacterial cells, you will begin by practicing aseptic technique. However, you will now begin performing aseptic technique in a biological safety cabinet (BSC), also called a laminar flow hood, rather than on an open lab bench. In this laboratory exercise, you will perform the following tasks:

- Prepare a BSC for use.
- Arrange equipment and supplies appropriately inside the hood.
- Perform manipulations aseptically.
- Monitor your success at avoiding contamination.
- Clean the work environment when activities are complete.

BACKGROUND

In this laboratory exercise, you will begin working with mammalian cells. You will find both similarities and differences between the techniques used for bacterial and mammalian cells. Perhaps the most important difference between culturing mammalian cells and microbial techniques is that **tissue culture**, *working with cells from multicellular organisms*, demands much more stringent attention to aseptic technique. This is because bacterial cells may have a population doubling time of as little as 20 minutes; mammalian cell lines may take a day or more to double. Therefore, even the most seemingly insignificant microbial contamination can wreak havoc upon a slower-growing mammalian cell culture. Consider the following scenario:

1. You transfer 10 mL of mammalian cell culture at a concentration of 10^4 cells/mL into a flask. You are thus starting with a total of 10^5 animal cells.
2. Your aseptic technique is nearly perfect—but you somehow also transfer a single bacterial cell. Just *one*.
3. Now, assume that your bacterium is a strain that doubles every 20 minutes, while your animal cell line has a population doubling time of 18 hours.
4. After 18 hours, your animal cell culture ideally will have doubled, leaving you with 2×10^5 cells. But what about that single bacterium? In the worst-case scenario, your contaminating bacterial population will have gone through 54 generation times. With infinite exposure to nutrients, oxygen, etc., you could now have 2^{54} bacteria present—or about 2×10^{16} cells! In reality, your bacterial population would have been limited by exhaustion of nutrients and buildup of waste. Still, the end result is a bacterial culture contaminated with mammalian cells—not the desired result at all.

Figure 8.20 shows a mammalian cell culture plate that has been contaminated in just this way. This plate is worse than useless; it also has the potential to spread contamination to other plates.

We use BSCs for tissue culture to minimize the likelihood of contamination. When used properly, BSCs provide effective containment and a sterile air supply to the working surface (see Figure 8.21).

Work in a BSC requires special technique. You must work in such a way as to minimize disruptions in the airflow through the cabinet. Move your hands and other objects slowly into and out of the cabinet interior. Strong air currents created outside the cabinet by people moving rapidly or laboratory doors opening and closing may also disrupt the

FIGURE 8.20 **Bacterial contamination destroys a culture plate.** The plate on the right was contaminated by bacterial cells. Even though individual bacteria are too small to be visible to the naked eye, there are so many bacteria thriving in this culture that the plate is covered by a visible sludge or haze. The healthy plate on the left is included for comparison; the culture medium is transparent on the healthy plate.

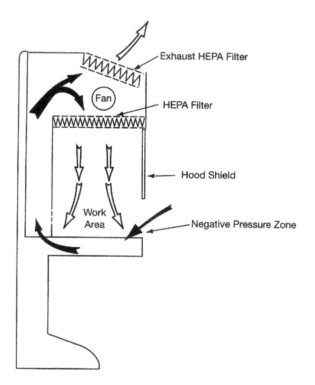

FIGURE 8.21 **Airflow in a biological safety cabinet.** Controlled airflow, combined with a HEPA filter, helps protect the work surface in the cabinet. Nonsterile air from the room (dark arrows) passes through the HEPA filter that removes all or nearly all contaminating microorganisms. Filtered air (light arrows) flows over the work surface. The air that is vented to the room is also filtered to protect the worker and the environment from the cells.

airflow and reduce the effectiveness of a BSC. Avoid Bunsen burners and open flames. Heat produces unwanted air currents that disrupt the laminar airflow in the cabinet. Also, heat may damage the cabinet's high-efficiency particulate air (HEPA) filter.

Safety Briefing

You will be working with cultured Chinese hamster ovary (CHO) cells. Although these cells are regarded as generally safe, all cultured cells should be treated as possible hazards. You should follow Standard Microbiological Practices (see Box 1.5) and also use the following precautions:

- Keep your face away from your work as you pipette and handle cells.
- Avoid contact between your skin and the cells.

- For small spills of cell cultures, for example, a few drops, immediately cover the spill with 70% ethanol or another disinfectant that is used in your classroom. Allow the disinfectant to sit for 15 minutes and then, while wearing gloves, clean it with paper towels. Discard the paper towels according to the disposal method used in your institution. Discard the gloves and wash your hands.
- For larger spills, inform your instructor immediately. Stay away from the contaminated area and instruct your classmates to do so also.
- Properly dispose of all materials that might have been contaminated by cells. The method of disposal may vary in your institution:
  - In many institutions, contaminated solid waste is placed in heavy-gauge bags intended for this purpose and autoclaved for 15 minutes at 121°C.
  - In some institutions, contaminated materials are soaked in 10% bleach solution, allowing at least 15 minutes for cells to be killed.
- Wipe down surfaces with 70% alcohol or other disinfectant at the end of every lab period.

Safety Briefing

Ultraviolet (UV) germicidal lamps *are used to kill microorganisms inside laminar flow cabinets*. Serious eye damage and burns can result if a person is accidentally exposed to the germicidal lamp. Leave the room when the UV lamp is on. While the hood window provides protection from the lamp, there is often a little space at the bottom where light can leak out of it. The lamp *must be turned off when the cabinet is in use*.

Avoid Contamination

Use these precautions to help protect your cells from contamination:

- Always work with mammalian cells in the BSC. The cabinet provides far more protection for your cells than working on an open lab bench. In addition, a BSC protects you and the environment from the cells with which you are working. The BSC is only for mammalian cell cultures; never use it for any other purpose.
- Swab any surfaces that might be contaminated with alcohol, including hood surfaces, media bottles, and hands prior to commencing work.
- Open packages of items for cell culture inside the protected environment of the laminar flow hood.
- Plan all manipulations so that nothing passes over the openings of bottles, culture vessels, and the like.
- Use the germicidal UV lamp to sterilize the interior hood surfaces before and after use. Never stay in the cell culture room while the UV lamp is on.
- Do not use the same materials, incubators, or equipment for bacterial and mammalian cell cultures.

Planning Your Work: Aseptic Technique in a Laminar Flow Hood

1. Draw a sketch of how you will organize your materials in your BSC to perform this exercise.

2. Rehearse in your mind the steps you will take in this exercise.

LABORATORY PROCEDURE

You will be aseptically transferring 10 mL of sterile cell culture medium into each of the following:

- Two sterile glass storage bottles
- Two sterile cell culture dishes
- Two sterile centrifuge tubes

You will check each of these for contamination after 1, 2, and 3 days. None of these items will be contaminated if your aseptic technique is effective.

1. At least 25 minutes before you are ready to begin work, turn on the germicidal lamp in the BSC.
   1.1. Make sure the hood's sash (window) is down.
   1.2. Leave the room.
   1.3. At the end of 15 minutes, turn off the germicidal lamp and turn on the fluorescent light and the blower.
   1.4. Raise the hood sash to the proper level. This is usually marked on the hood.
   1.5. Swab the interior hood surfaces with alcohol.
   1.6. Allow 10 more minutes before beginning work in the hood.
2. Obtain a new lab coat that you wear only for cell culture work. A lab coat that you wear when performing microbiological work can be contaminated with microbes. Avoid sleeves that can dangle over your work area.
3. Wash your hands.
4. Put on gloves.
5. Rinse your gloved hands with 70% ethanol.
6. Obtain a bottle of cell culture medium from your instructor and place it into a 37°C bath to warm it.
7. Gather your other supplies:
   - One package sterile plastic disposable pipettes (10 mL size)
   - Two sterile capped empty glass storage bottles reserved for cell culture use
   - Two sterile cell culture dishes in the original packaging*
   - Two sterile centrifuge tubes (15 mL)
   - Rack for the tubes
   - A pipette aid
   - A squirt bottle of 70% alcohol
8. Rinse and wipe the outside of the warmed bottle of cell culture medium with alcohol and place it in the hood.
   8.1. Be sure to clean the bottle surface next to the cap where microbes from the water bath can accumulate.
   8.2. Observe the medium so you know how it looks when it is uncontaminated. Record your observations.
9. Arrange your supplies inside the BSC.
   9.1. Swab all supplies with alcohol before placing them inside the hood. Move slowly to avoid creating air currents.
   9.2. Do not place any items on top of the front grating of the hood as this will disrupt the airflow.
   9.3. Place all items at least 4 inches back from the front grate.

---

* The plates that are used for mammalian cell culture look much like the plastic Petri dishes that are used for microbial culture. They are not, however, the same and should never be confused. The plastic plates, flasks, and other containers that are used for culturing mammalian cells are specially treated to facilitate growth and adherence of these cells. Bacterial Petri dishes are not treated in this way and so should never be accidentally used for mammalian cells. Mammalian cell culture plates are significantly more expensive than Petri dishes for bacterial culture and, therefore, should not be wasted on bacterial cells that do not require them.

10. Once the packages of cell culture dishes and centrifuge tubes are inside the hood, open the packages and remove two of each type of vessel.
11. Reseal the packages and remove the remaining cell culture dishes and tubes from the hood.
12. Label your two storage bottles, two plates, and two tubes.
13. Do a test run without opening any containers to be certain that you will not be reaching over the top of any open bottle, tube, or plate.
14. Transfer 10 mL of cell culture medium to the two cell culture plates (see Figure 8.22).
    14.1. Move the glass storage bottles and centrifuge tubes out of the way.
    14.2. Arrange the cell culture medium bottle and the cell culture plates so that you will not need to reach over them when they are open.
    14.3. Loosen, but do not remove, the cap of the cell culture medium bottle.
    14.4. Open the packaging of the pipette and fold it back a few inches, but do not fully remove the pipette.
    14.5. Insert the pipette into your pipette aid.
    14.6. When you are ready to transfer the medium, remove the pipette completely from the sterile packaging. Be careful not to touch the sterile end of the pipette to anything in the hood. (If you do, get another pipette.)
    14.7. Aseptically transfer 10 mL of medium to one of the cell culture plates.
    14.8. Dispose of your pipette properly, as directed by your instructor.
    14.9. With a new pipette, add cell culture medium into the second cell culture plate.
15. With two new pipettes, transfer 10 mL of cell culture medium to each of the centrifuge tubes by using aseptic technique.
16. With two new pipettes, transfer 10 mL of cell culture medium to each of the two storage bottles by using aseptic technique.
17. Place all six of your labeled vessels containing 10 mL of medium into a 37°C cell culture incubator. Also place the remainder of the bottle of cell culture medium into the incubator.
18. Remove all other items from the laminar flow hood.
19. Swab the interior of the hood with 70% ethanol.

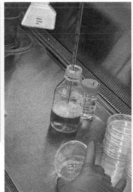

a   b   c

FIGURE 8.22   Transferring 10 mL of cell culture medium into a cell culture dish. (a) The culturist has broken open the wrapper of a sterile pipette and is placing the pipette aid onto the pipette without fully removing the wrapper. The culturist will pull the wrapper off the pipette, being careful not to touch the bottom part of the pipette. (b and c) He withdraws cell culture medium from the bottle and transfers it to the cell culture dish. Even though he is working in a BSC, the culturist is using good technique and is holding the lid over the plate to protect it from contaminants.

20. If you are the last person to use the hood for the day, turn off the blower and fluorescent light. Close the sash and turn on the UV lamp for 15 minutes. Leave the room while the UV lamp is on. Turn it off after 15 minutes.
21. Wash your hands.
22. Check your media over the next 3 days:
    22.1. Do you notice any turbidity, an indication of contamination? Do you see any change in color of the media? This might indicate a problem. Consult your instructor if you see a color change.
    22.2. Record in your laboratory notebook the appearance of all tubes and cell culture dishes at 24, 48, and 72 hours following transfer.
    22.3. If you see turbidity, examine your cell culture medium under a microscope. Draw or photograph what you observe.

LAB MEETING/DISCUSSION QUESTIONS

1. List the various precautions that you took in this laboratory procedure to minimize the risk of introducing contaminants into your culture vessels.
2. What did you learn in this lab about the importance of planning cell culture manipulations carefully prior to beginning work in the hood? Think about the planning of equipment needed for the procedure as well as the placement of materials within the hood.
3. What reasons can you provide for the need to practice extra caution when working with mammalian cell cultures relative to microbial cultures?
4. Compare the results of the whole class.

Laboratory Exercise 44: Making Ham's F-12 Medium from Dehydrated Powder

OVERVIEW

In this exercise, you will prepare cell culture medium for growing Chinese hamster ovary (CHO) cells. To make this medium, you will perform the following tasks:

- Mix dehydrated, powdered medium with water.
- Sterilize dissolved medium by filtration.
- Supplement medium with serum.*
- Perform a sterility test.

BACKGROUND

Mammalian cells evolved to live in a controlled environment in which they are in close contact with other cells. Within an animal, cells are constantly bathed in nutrients delivered by the bloodstream as their wastes are carried away. Cell culture removes tissue from this natural environment in which the cells normally flourish. The tissue is disrupted to produce individual cells that must somehow survive in what is an extremely foreign environment—a culture dish or flask. Not surprisingly, the **liquid environments that we create for cells in the laboratory**, *cell culture media*, are carefully designed and are complex in order to simulate the cells' natural environment.

In this laboratory activity, you will prepare Ham's F-12, a medium designed to allow CHO cells to grow well in a culture dish. As you prepare this medium, remember that even the slightest level of microbial contamination will quickly destroy your cell culture.

Ham's F-12 is a **defined medium**, *meaning that the quantities and identities of all of its components are known.* We purchase Ham's F-12 as a dehydrated powder that contains the following:

- Inorganic salt mixture.
    - Salts function to make the medium isotonic.
    - Salts provide ions that are required by certain cellular processes, such as membrane transport.
- Glucose as a carbon and energy source.
- Pyruvate as a source of carbon backbones necessary to synthesize other molecules.
- l-glutamine as an additional energy source to foster growth.
- Amino acids.
- Vitamins.
- Phenol red as a pH indicator to allow you to rapidly evaluate the condition of the medium. A buildup of waste products will cause the pH to drop, changing the color of the phenol red to orange and then yellow.

Powdered Ham's F-12 is dissolved in purified water and is then sterilized. Mammalian cell culture media contain components that are heat sensitive and cannot be autoclaved. The media, therefore, are sterilized by filtration. Special filtration units can be purchased for this purpose.

Because of the drastic consequences of contamination, it is good practice to perform a sterility test to ensure that freshly prepared medium is not contaminated. The sterility

---

* The use of fetal calf serum in mammalian cell culture has been standard practice for many years, and we include its use in this manual because most standard procedures require it. However, the use of fetal calf serum is problematic for a variety of reasons, and many scientists and manufacturers are trying to reduce or eliminate its use. First, there is an ethical/animal rights issue regarding this use of unborn calves. Second, fetal calf serum from different lots varies, and this variability can be a significant source of irreproducibility. On a similar note, calf serum is undefined and therefore adds unknown ingredients into the culture medium. Third, the fatal illness, bovine-spongiform encephalitis, raises concerns about the use of bovine-derived substances in the manufacture of any product intended for use in humans. In fact, it is not only bovine products that are of concern, but all animal-derived substances might harbor pathogens or might elicit unwanted immune responses in exposed people. For a short review of these issues, see Jan van der Valk, "Fetal Bovine Serum—A Cell Culture Dilemma," *Science*, vol. 375, no. 6577, 2022, pp. 143–44. https://doi.org/10.1126/science.abm1317.

test that you will perform simply requires that you transfer samples of your Ham's F-12 medium into culture dishes using aseptic technique. You will then incubate these dishes for several days to see if any contamination appears.

Just before use, you will supplement your Ham's F-12 medium with calf serum. Calf serum is rich in growth factors that help cells survive in culture. Calf serum is **undefined**, *in that it contains a mixture of substances that are not completely identified*. Scientists have devised methods to avoid the use of calf serum, partly because it is undefined and partly because it is expensive. However, the use of calf serum aids cell growth and makes cell culture easier for beginners, so you will use it. Calf serum has already been filter-sterilized by the manufacturer and needs only to be thawed and added to the Ham's F-12 medium under aseptic conditions in a laminar flow hood. Once the calf serum has been added, the Ham's F-12 medium is called a "complete" medium. Some culturists perform another sterility test after calf serum is added.

Planning Your Work: Preparing Ham's F-12 Medium

Draw a sketch of how you will organize your materials in the BSC to perform this exercise.

LABORATORY PROCEDURE

1. At least 25 minutes before you are ready to begin work, turn on the germicidal lamp in the BSC.
   1.1. Make sure the hood's sash (window) is down.
   1.2. Leave the room.
   1.3. At the end of 15 minutes, turn off the germicidal lamp and turn on the fluorescent light and the blower.
   1.4. Raise the sash to the proper level.
   1.5. Swab the interior surface of the hood. Allow 10 more minutes before beginning work in the hood.
2. Wash your hands; put on gloves; swab your gloved hands with 70% ethanol.
3. In a 2 L flask, combine 10.64 g of Ham's F-12 powder and 1.176 g of sodium bicarbonate (cell culture grade). You can mix the powder on an open lab bench because you will sterilize it.
4. Add 800 mL of cell-culture-grade water to the flask.
5. Add stir bar and stir until fully dissolved. The solution should be pink.
6. Transfer dissolved medium to a 1 L graduated cylinder.
7. Bring to volume (BTV) to 900 mL with cell-culture-grade water.
8. Put on your cell culture lab coat and gloves.
9. Swab your gloved hands with 70% ethanol.
10. Swab the BSC with 70% ethanol (EtOH).
11. In the BSC, filter-sterilize the medium (see Figure 8.23).
    11.1. Obtain a filter apparatus that is intended for sterilizing cell culture medium and wipe the bag in which it is stored with 70% EtOH.
    11.2. Obtain a 1 L sterile tissue culture bottle and wipe the outside of it with 70% EtOH.
    11.3. Introduce the filter and culture bottle into the hood.
    11.4. Unscrew the cap of the culture bottle. You can set the cap open-side down in the cell culture hood.
    11.5. Open the bag containing the filter and remove the filter.
    11.6. Screw the filter onto the bottle.
    11.7. Attach a piece of laboratory vacuum tubing to the side of the filter. Attach the tubing to a source of vacuum.
    11.8. Open the filter cap and pour medium into the filter reservoir.
    11.9. Turn on the vacuum. You should see the pink filtrate going into the bottle.

FIGURE 8.23   **Filter-sterilizing cell culture medium.** (a) The filter apparatus comes in a sterile package. (b) Inside a laminar flow cabinet, the filter is removed from its package, screwed onto a sterile culture storage bottle, and connected to a vacuum source. (c) The medium is poured into the filter's reservoir, being careful not to let the filter surface dry. (d) When all the media is sterilized, the filter is removed, and the cap is returned to the sterile culture bottle. The medium can now be removed from the BSC and stored in a refrigerator.

11.10.	Keep pouring the contents of the graduated cylinder into the filter reservoir; take care not to let the filter dry.
11.11.	When all of the medium is transferred, unscrew the filter and rescrew the original cap onto the bottle. This is now sterile Ham's F-12 medium that is incomplete. Label the bottle.
11.12.	Record the appearance of the Ham's F-12 medium immediately after filtration, including color and clarity.
12.	Perform a sterility test.
12.1.	You will need a 10 mL sterile pipette, a marker, a pipette aide, and a cell culture dish. Bring these into the hood by using proper technique.
12.2.	Pipette 10 mL of the incomplete Ham's F-12 medium into a labeled cell culture dish. Place the cell culture dish in a cell culture incubator to check sterility.

12.3. Incubate the cell culture dish for 24 hours. If the medium is sterile, it will neither look turbid nor change color. You may want to check its appearance under a microscope.

12.4. Record the appearance of the medium.

13. Store the prepared medium away from fluorescent light in a 4°C refrigerator.

14. Clear the laminar flow hood. Wipe down the hood with 70% EtOH.

15. Turn off the blower and fluorescent light, lower the hood sash, and switch on the germicidal UV lamp for 15 minutes if you are the last person to use the hood. Turn off the UV lamp after 15 minutes.

16. Wash your hands.

17. When it is time to use the Ham's F-12 medium, make it complete.

17.1. Add 100 mL of calf serum to the 900 mL of Ham's F-12 medium in the hood under aseptic conditions.

17.2. Perform another sterility test on the complete medium, if your instructor directs you to do so.

18. The complete medium may be aliquoted into 100 mL sterile bottles for ease of use and storage. Perform this manipulation in the laminar flow hood.

LAB MEETING/DISCUSSION QUESTIONS

1. List the precautions that you took in this laboratory procedure to minimize the risk of introducing contaminants into your culture vessels.

2. Explain why a change in the color and clarity of your media can indicate contamination.

3. Examine the labels and/or technical information that came with the ingredients for LB broth and Ham's F-12 dehydrated medium. Compare and contrast the ingredients in these two media. They are quite different—why?

Laboratory Exercise 45: Examining, Photographing, and Feeding CHO Cells

OVERVIEW

In this laboratory exercise, you will familiarize yourself with Chinese hamster ovary (CHO) cells. Your instructor will provide you with cultures of healthy cells so you can see what they are supposed to look like. You will learn to assess their appearance, aided by a microscope, and will learn how to document the growth of your cells using a camera mounted on your microscope (if available). You will feed your cells with fresh medium so they can continue to grow and reproduce. You will thus perform these tasks:

- Examine the morphology of healthy CHO cells.
- Practice photographic documentation of cell culture (if equipment is available).
- Learn to recognize signs of contamination or poor health in a culture.
- Feed your cells.

BACKGROUND

A. Basic Features of CHO Cell Culture

i. Anchorage Dependence

CHO cells are **anchorage dependent**, which means that *they require an appropriate solid surface to which to adhere.* This makes sense because mammalian cells normally grow inside the body where they are anchored in a particular location in a tissue. Cultured CHO cells grow in a **monolayer**, meaning *they do not grow on top of each other.* Given suitable conditions, CHO cells will multiply in culture until they reach **confluence**, that is, *completely cover the surface of their vessel.*

Anchorage-dependent cells are grown in the laboratory in disposable plastic plates, flasks, or bottles that are chemically treated to promote cell adherence. **Tissue culture** dishes look slightly taller than bacterial Petri plates. Anchorage-dependent cells, however, will not adhere and grow in untreated, less expensive plastic Petri dishes. In laboratories where both bacterial and mammalian cell culture are practiced, it is important not to confuse the plates used for bacteria with those used for mammalian cells.

ii. Buffering with CO_2/Bicarbonate and Cell Culture Incubators

Mammalian cells are very sensitive to the pH of their environment. The cells generate metabolic waste products as they live, grow, and reproduce. These waste products cause their medium to become acidic and inhospitable to the cells. It is, therefore, important to include a buffering system in cell culture media to help stabilize the pH. The usual buffer system is a bicarbonate/CO_2 buffer, much like the natural system that buffers blood in mammals. For this buffer system to work, the cell cultures must be maintained in a CO_2-rich environment. This is accomplished by connecting the cell culture incubator to a tank of CO_2 and adjusting a valve on the tank to maintain a 5% to 10% CO_2 atmosphere (at 37°C). Cell culture incubators are designed to be used with carbon dioxide and will often provide a gauge that indicates the level of this gas.

Because of the need for CO_2 gas exchange, tissue culture containers cannot be tightly sealed during incubation. They are designed in such a way as to allow gas exchange yet not to allow microbes to enter. The caps of cell culture flasks and bottles should be loosened while they are in the incubator and tightened when they are removed.

It is important to maintain a high humidity within CO_2 incubators to prevent cell culture media from evaporating, so most of these incubators have a bottom tray that is partially filled with water. This water can be fertile ground for the growth of microorganisms, so it is essential that the pan be emptied, cleaned, and disinfected regularly even if chemicals are added to inhibit microbial growth.

iii. Feeding and Passaging Cells

Cells deplete the nutrients in their medium as they grow and reproduce. Also, cell culture medium is buffered, but the buffering capacity can be exceeded if cells remain in the same medium for too long. Therefore, after a time, *used culture medium must be replaced with fresh medium*; this is called **feeding the cells**. Feeding cells is a simple process that involves removing the old medium by aspiration and replacing it with fresh medium. CHO cells should generally be fed every 2 or 3 days, although they may need to be fed sooner if the cells are approaching confluence or a drastic pH change is observed. You will feed your cells in this laboratory exercise.

Cells that are confluent or nearly confluent need a new vessel to which to adhere. They are therefore **passaged**, or **subcultured**, *meaning they are removed from the original vessel and plated into new plates, flasks, or bottles*. When cells are passaged, they are also divided. It would not make sense to put all the cells from one confluent plate onto a new plate because the new plate would then be filled. Instead, the *cells from one vessel are divided into two or more fresh vessels*; this is called **splitting the cells**. You will passage and split your cells in Laboratory Exercise 47.

B. Inverted-Phase Microscopy

Microscopy for cultured mammalian cells is a little different than the microscopy you performed with microbial cultures. For one thing, mammalian cells are larger, so lower-power objectives can be used. Second, the cells will be alive and unstained. Unstained cells are difficult to see because they lack contrast. Therefore, you will be using **phase microscopy**, *a method of manipulating light to increase the contrast of the image*. Another difference is that the cells must not be contaminated, so you will look at them through their culture dish. **Inverted phase microscopes for cell culture** *are equipped with phase contrast lenses and with adequate space on the stage for a cell culture vessel* (Figure 8.24). Your instructor will show you how to operate the model of inverted microscope in your laboratory. Your instructor will also show you how to operate the associated camera if your microscope is equipped with one.

C. Assessing the Condition of Your Cultures

It is important to routinely check that your cell cultures are doing well. You do not want to invest time and expensive reagents on contaminated or otherwise unhealthy cells. There are several characteristics that you can monitor including the following:

- **Cell density and growth rate.** Each healthy cell line has a predictable growth rate and population doubling time. You can periodically count the cells, as you will practice in Laboratory Exercise 46, to assess their growth rate. It is also sometimes possible to follow the culture microscopically, roughly assessing each day what percentage of the culture dish the cells have covered. In this laboratory exercise, you will not actually assess the cells' growth rate because you are only examining the cells at one timepoint. You will, however, try to judge their percent confluence to become accustomed to this practice.
- **Appearance of the growth medium.** Contamination of monolayer cell cultures can be assessed in part by observing the appearance of the medium floating above the monolayer. If the medium looks turbid, there is quite possibly microbial contamination of the cell culture. Cell death also indicates possible contamination.
- **Cell morphology.** Cell lines have a characteristic morphology (Figure 8.25). Once you are familiar with your specific cell line, you will know what cellular morphology to expect. If the cells exhibit an unusual morphology, you may suspect that they are either contaminated with another cell line or are not growing in optimal conditions. Cellular morphologies are generally classified as "epithelial-like," "lymphoblast-like," or "fibroblast-like."

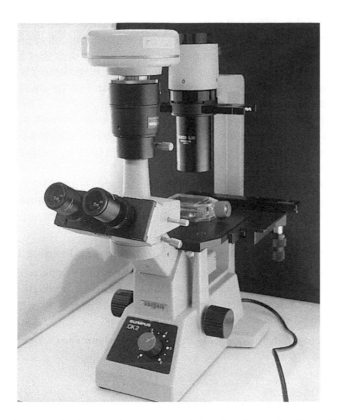

FIGURE 8.24   Inverted-phase microscope with cell culture flask on the stage.

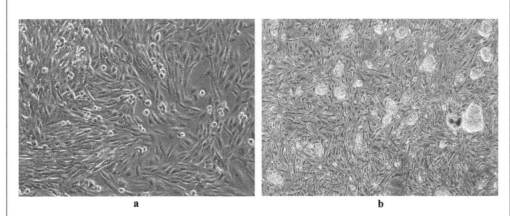

a   b

FIGURE 8.25   Cell morphology. (a) CHO cells. This image was taken using an inverted-phase microscope with the 20× objective. (b) Human embryonic stem cells and fibroblasts photographed at low magnification with a scanning objective. The stem cells grow in round colonies, each consisting of hundreds of cells. The stem cell colonies appear as the lighter "blobs" in this micrograph. Individual stem cells are not discernable at this low magnification. Fibroblasts surround the colonies and exhibit a characteristic elongated, sometimes swirling morphology.

- *Epithelial-like cells* generally grow as an attached monolayer with a flat, polygonal shape, sometimes called "cobblestone" in appearance. CHO cells are epithelial.
- *Fibroblast-like cells* tend to be elongated and attached to the substrate, and they often grow in patterns that resemble swirls.
- *Lymphoblast-like cells* are generally round and grow in suspension rather than attached to a substrate.

- **Uniformity of the culture.** If cells are contaminated by bacteria or yeast or other fungi, you may be able to directly observe the contaminating organisms using high-power microscopy. This will generally require cultures that are highly contaminated. If the cell line harbors a low level of contamination, it is likely that you will miss this contamination using microscopy because of the small size and small number of offending organisms.
- **Attachment to substrate.** Monolayer cultures should be well attached to the substrate. A large percentage of unattached cells could indicate a problem. If cells appear to be "rounding up" or looking like they are getting ready to detach, they may be unhealthy.
- **Other signs of poor health.** Sometimes cell cultures are not harboring any contamination yet are growing suboptimally. Poor growth can be caused by many factors, including problems with medium components, pH, temperature, or aeration.

Planning Your Work: Examining, Photographing, and Feeding CHO Cells

1. Bacterial cells are about 1.5 μm in length; mammalian cells are often on the order of 15 μm (or more) in diameter. Based on your previous experience with bacteria, how do you expect CHO cells to look under low power? Think about the relative size of the two cell types and the amount of detail you might expect.

2. Sketch how you will arrange your work space to feed your culture. Show all supplies that you will need to place in the BSC.

LABORATORY PROCEDURE

Part A: Examining Your Cells

A.1.   Wash your hands and put on your cell culture lab coat and gloves. Swab your hands with 70% alcohol.

A.2.   Learn how to use the inverted-phase microscope.

A.3.   Remove your CHO cells from the incubator and describe the color and appearance of the medium.

A.4.   Place the culture plate on the microscope stage and examine the cells using low power. Check around the plate and estimate the **percent confluence** of your culture; *the percent of the plate that is covered by cells*. Have several classmates and your instructor make similar estimates and try to reach consensus. Figure 8.26 shows a plate of CHO cells that are estimated to be about 90% confluent.

A.5.   Describe the morphology of your cells at low power and photograph the cells if your microscope is equipped with a camera. Draw the cells if you do not have a camera. Label any structures that you see. You should be able to make out the nuclei in some of the cells. Switch to the higher-power objective(s) and describe the morphology of the cells. Photograph them, if possible.

A.6.   Presumably this is a healthy culture provided by your instructor. Pay attention to the morphology and appearance of healthy CHO cells. Take time to study them so you know how they should look in the future.

A.7.   Return the cells to the incubator or feed them as described in Part B.

A.8.   Try to examine and photograph or draw this culture every day or two for the remainder of the week. Check the appearance of the culture medium and the cells. How long does it take for the cells to reach confluence? Get in the habit of examining your cells every time you feed them.

Part B: Feeding Your Cells

You will be feeding a plate of CHO cells with 10 mL of fresh Ham's F-12 medium. Refer to Figure 8.27.

B.1.   Prepare to work aseptically in a BSC, as was described in Laboratory Exercise 43, steps 1–5.

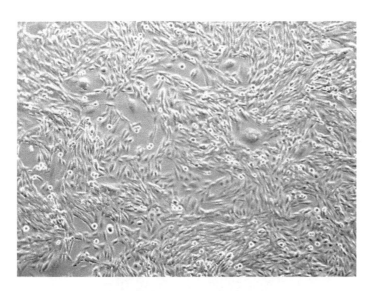

FIGURE 8.26   A plate of CHO cells that is about 90% confluent. This image was taken with the 10× objective in place. The cells were evenly dispersed on this plate so most fields of view looked similar.

B.2.   While the germicidal lamp is on, warm your bottle of Ham's F-12 medium in a water bath at 37°C.

B.3.   Gather your supplies: sterile pipettes (10 mL size), sterile Pasteur pipettes, and pipette aid.

B.4.   When it is warmed, swab the outside of your cell culture medium bottle, including its cap, with 70% ethanol. Be sure to clean the bottle surface next to the cap where microbes from the water bath can accumulate. Place the bottle in the hood.

B.5.   Place your other materials inside the hood, first swabbing each with 70% ethanol.

B.6.   Remove your plate of CHO cells from the incubator and place it in the hood.

B.7.   Loosen, but don't remove, the cap of the bottle of sterile medium.

B.8.   Aseptically insert a sterile Pasteur pipette into the hose leading to the vacuum line of the hood (see Figure 8.27b). Do not allow the tip to touch any surfaces.

B.9.   Remove the lid of the CHO cell culture vessel and aspirate off the medium.

   B.9.1.   You can place the lid face-down on the surface of the hood if you want.

   B.9.2.   You will need to tilt the dish or flask slightly to remove all of the old medium.

   B.9.3.   Do not touch the cells with the tip of the pipette.

B.10. Pipette 10 mL of warmed Ham's F-12 medium onto your cells using aseptic technique.

B.11. Look at your plate under the microscope and take a photo of your fed cells if a camera is available. If possible, photograph your cells over the next couple of days to see how they have grown.

B.12. Return your cells to the incubator. Clean up the hood; put supplies away or properly dispose of them.

B.13. Clean out the vacuum line by aspirating 70% ethanol through it.

B.14. Wipe down the hood interior with 70% ethanol.

B.15. If you are the last person using the hood, turn off the blower and fluorescent light. Turn on the UV lamp for 15 minutes with the hood sash closed.

B.16. Feed your cells every 2–3 days until they are nearly confluent, at which time you will perform Laboratory Exercise 47.

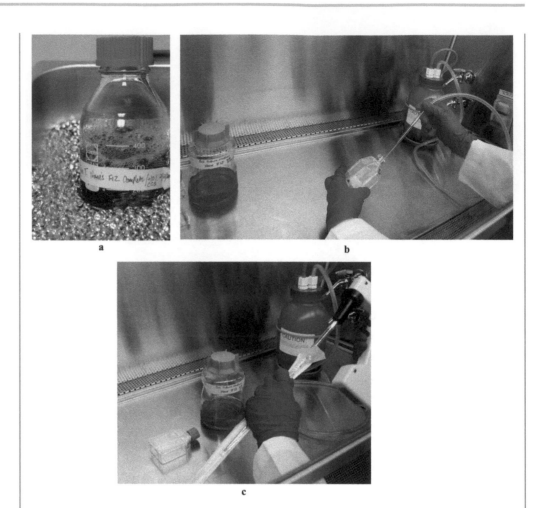

FIGURE 8.27 **Feeding cells.** (a) The complete (with serum) cell culture medium is warmed to 37°C. The warming bath in this photo is a bead bath. It serves the same function as a water bath but is filled with beads instead of water. A bead bath minimizes contamination that can occur with standing water. A water bath, however, is suitable as long as it is routinely cleaned. (b) In this example, the cells are adhered to the surface of a small, plastic cell culture flask. Some culturists find flasks easier to handle than culture dishes, like those shown in Figure 8.22. The culturist has taken a sterilized glass Pasteur pipette and inserted the pipette into plastic tubing. The tubing is connected to a vacuum line that provides vacuum in the BSC. There is a blue bottle between the pipette and the vacuum valve. The bottle contains disinfectant. The culturist has turned on the vacuum and positioned the pipette tip just above the cells in the flask. She is thus aspirating off the old cell culture medium, which flows into the disinfectant bottle to be disposed of later according to the institution's procedure. The cells remain attached to the surface of the flask. The spent medium that is being aspirated is yellowish in color. (c) The culturist now needs to add fresh medium to the flask. She has already loosened the cap of the cell culture medium bottle. She removes 10 mL from the bottle and gently pipettes it into the flask with the cells. She works quickly but carefully to minimize the time the cells are not covered by medium.

LAB MEETING/DISCUSSION QUESTIONS

1. Suppose that you suspect that you may have introduced contamination into your flask during the feeding process. What signs can you look for in your cell culture that can help you address this concern?

2.   Think about the process of feeding your cells. During what parts of the procedure do you think you were most likely to introduce contamination? Can you think of any practices you could modify or introduce to minimize the likelihood of contaminating your culture?

3.   How long does it take your culture to become confluent?

4.   Does performing this laboratory exercise feel different than performing Laboratory Exercise 38? Most people do find that working with mammalian cell cultures feels different than working with microbial cultures.

Laboratory Exercise 46: Counting Cells Using a Hemacytometer

OVERVIEW

You estimated the concentration of bacterial cells in your cultures in Laboratory Exercises 41 and 42. In Laboratory Exercise 41, you counted colonies growing on agar plates. In Laboratory Exercise 42, you used turbidity to estimate cell concentration. As is the case for bacterial cells, an individual mammalian cell is too small to see with the naked eye, and tens of thousands of cells are likely to be present in each milliliter of growth medium. Mammalian cells, however, are larger than bacterial cells, and they can, with reasonable care, be counted using a microscope. It is not necessary to resort to indirect enumeration methods as we did for bacteria. In this exercise, you will count CHO cells using a special glass microscope slide designed for this purpose. To do so, you will perform the following tasks:

- Release cells from their culture substrate.
- Prepare "countable" dilutions of your culture.
- Use a stain to differentiate between live and dead cells in your culture.
- Count cells using a hemacytometer and a microscope.
- Calculate the cell density of your culture.

BACKGROUND

A. Counting Mammalian Cells with a Hemacytometer

In Laboratory Exercise 45, you learned how to assess the growth of your cells by looking at them with a microscope each day and estimating their percent confluence. Depending on the type of cells and the situation, this method of observing growth may be adequate for routine maintenance. There are, however, situations when culturists need a more accurate estimate of cell concentration. If, for example, the cells are to be used in an assay, the assay method will specify how many cells are required. In this case, it is necessary to count the cells before using them. Another time culturists count cells is when expanding a culture to use the cells for production. Production technicians want to seed the right number of cells into the production vessel. This is because mammalian cells are "social"; if too few cells are present, they do not thrive. If, however, too many cells are present, they use up the nutrient medium too quickly.

Mammalian cells are counted in a special counting chamber. A **Neubauer hemacytometer** (Figure 8.28) is *a special modified microscope slide that is used to determine the concentration of cells in a suspension.*

A hemacytometer contains two identical wells, or chambers, into which a small volume of the cell suspension is pipetted (see Figure 8.28a). Each chamber is manufactured

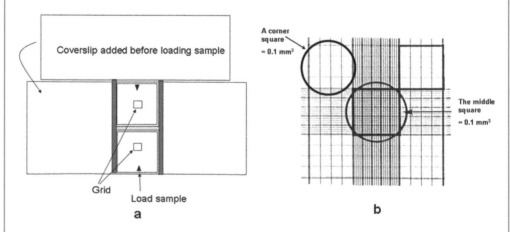

FIGURE 8.28   The hemacytometer is used for counting cells. (a) Top view of hemacytometer showing two identical chambers. (b) The grid as it appears under a microscope.

to hold an exactly specified volume of liquid. When you view the hemacytometer under a microscope, you see that the bottom surface of each well is etched with a grid pattern (see Figure 8.28b). The grid consists of nine larger squares, each of which is further subdivided into smaller squares. Each of the nine larger squares has dimensions of 1 mm by 1 mm by 0.1 mm. Therefore, the volume of fluid over each of these large squares is 0.1 mm^3 because Volume = (Length)(Width)(Height). The units of mm^3 are not very helpful, so they are converted to units of milliliters where 0.1 mm^3 = 10^{-4} mL.

The process of counting cells with a hemacytometer is described in the following steps.

Step 1: Remove a Sample of Cell Suspension; Dilute It with Trypan Blue

A volume of the cell culture suspension is removed and diluted with the dye trypan blue and possibly also with sterile diluent. Living cells exclude trypan blue and appear clear, but the dye readily passes through the membrane of dead cells, which thus appear blue.

Step 2: Apply the Mixture to the Hemacytometer

A coverslip is placed on top of the hemacytometer slide. The cell suspension is mixed well and then is carefully applied under the coverslip into both wells of the hemacytometer in such a way that each chamber is exactly filled.

Step 3: Count the Cells

Using a microscope, all the cells located over the four large corner squares and the center square in one of the chambers are counted. A separate tally is kept of viable and nonviable cells. Sometimes the cells in both chambers are counted, either to make sure that the counts are comparable or to count more cells. (Some analysts try to count enough squares to total at least 100 cells; others aim for at least 200 cells total.)

Step 4: Calculate Percent Viability and the Concentration of Viable Cells

The percent of viable cells is calculated. This value can be revealing. If there is a high concentration of dead cells, then the culture may not be healthy and should not be used for further experiments (though it may be diluted and scaled back up for future use). Once the percent of viable cells is known, the concentration of viable cells in the culture is calculated. These calculations are discussed in Formula 8.1.

B. Removing Adherent Cells from Their Vessel

CHO cells adhere to the surface of their vessel. You will need to dislodge them in order to count them. You will similarly need to remove cells from their plate to passage them in Laboratory Exercise 47. We use an enzyme, trypsin, to disrupt the attachments that bind the cells to the plastic of their vessel. Trypsin is inhibited by serum; therefore, the first step in dislodging the cells is to wash away the serum. You will use the PBS prepared in Laboratory Exercise 40 for this purpose. Trypsin-EDTA, dissolved in medium without serum (and without Ca^{+2} or Mg^{+2} ions), is then added to the cells. EDTA binds divalent cations that may be inhibitory to the enzyme. After you add the trypsin-EDTA solution to the culture vessel, you must look at the cells every minute or so under the microscope. You will see them becoming rounder, and, after a few minutes, most of the cells will begin to float in the medium (see Figure 8.29). Trypsin will eventually damage the cells, so the trypsin must be inhibited as soon as the cells have detached. This is accomplished by adding cell culture medium containing serum. Once the cells are floating, they can be pipetted from their vessel and be counted or used for other purposes.

Formula 8.1

Calculating the Number of Cells/mL from Hemacytometer Counts

$$\frac{\text{\# Viable cells}}{1\,\text{mL}} = \frac{\text{\# Viable cells in 5 large squares}}{5}\left(\text{Reciprocal of dilution}\right)\left(1\times10^4\right)$$

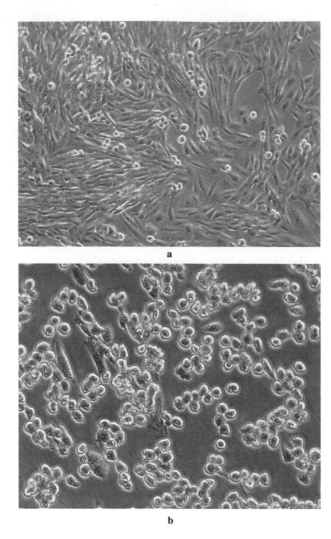

FIGURE 8.29   Trypsin causes cells to detach from the substrate. (a) Shortly after trypsin-EDTA is added to the cells, a few cells can be seen that have become round and begun to detach from the surface of the vessel. (b) When most of the cells are round and floating, as in this photo, trypsinization is stopped.

Consider the parts of this formula:

- The number of viable cells per milliliter is what you want to know. The total number of viable cells in five large squares divided by five gives the average number of cells per large square. Note that if you count more or fewer than five squares, then you need to adjust the formula. For example, if you count only the four corner squares and not the center square, then you divide the total number of viable cells by four to get the average number of cells per large square.
- The dilution must be considered when calculating the number of cells per milliliter. Example 8.1 shows how this is done.
- Finally, each of the nine larger squares has a volume of 10^{-4} mL, but we want to know the number of cells per 1 mL. Therefore, we multiply by 10^4 to calculate how many cells there are in 1 mL.

Example 8.1

- 200 µL of a cell suspension is removed from a plate of cells.
- 300 µL of diluent is added to the suspension.

- 500 µL of trypan blue is added.
- The total volume is thus 1000 µL.
- The total number of viable cells in the four corner squares plus the center square is 245.
- The total number of nonviable cells in these squares is 55.

What is the number of viable cells per milliliter?

Answer

Calculations:

1. The total number of cells counted = 245 + 55 = 300.
2. The percent of viable cells = 245/300 ≈ 82%.
3. The average number of viable cells per square = 245/5 = 49.
4. The dilution is 200 µL/(200 µL + 300 µL + 500 µL) = 200 µL/1000 µL = 1/5.
5. The reciprocal of the dilution is 5/1 or 5.
6. Using Formula 8.1, the concentration of viable cells in the original dish is estimated to be

$$\frac{\text{\# Viable cells}}{1 \text{ mL}} = \frac{245}{5}(5)\left(1 \times 10^4\right) = 2,450,000 \text{ cells / mL}$$

Planning Your Work: Counting Cells Using a Hemacytometer

Sketch in your laboratory notebook how you will organize your materials on your lab bench to perform this exercise. Your sketch should include all needed materials.

LABORATORY PROCEDURE

1. Prepare yourself and the BSC for cell culture work as described in the previous laboratory exercises.
2. Warm Ham's F-12 complete medium and trypsin-EDTA solution in a 37°C water bath.
3. Clean a hemacytometer and its coverslip with 70% alcohol so that they are dust-free and without fingerprints.
4. Remove your culture plate from the incubator and place in the hood. Aspirate old medium from the plate.
5. Rinse with sterile PBS.
    5.1. Add 2 mL of PBS to the plate. Briefly swirl the plate to remove dead cells, which do not firmly attach to the substrate. This step will also help remove medium, serum, and cations that may inhibit the enzymatic action of the trypsin.
    5.2. Aspirate off the PBS.
6. Trypsinize the cells.
    6.1. Add 1 mL of trypsin-EDTA to the plate.
    6.2. Incubate the plate at 37°C.
    6.3. Remove the plate from the incubator after about a minute. Tap the plate to help dislodge the cells from the plastic surface of the dish.
    6.4. Look at the plate using the inverted microscope. If the cells have not been dislodged from the substrate, return the plate to the incubator for another minute.
    6.5. Repeat steps 6.2 to 6.4 until all, or nearly all, of the cells are floating. Leave the trypsin on the cells only long enough to dislodge most of the cells. If you wait too long, the trypsin will begin to damage the cells. The length of time required depends on the activity of the trypsin and the type of cells you have. Generally 1–10 minutes will work. Record how long you keep the trypsin on the cells.

7.   Add 5 mL of Ham's F-12 complete medium to the culture plate; this will stop the effect of trypsin. Pipette the cells up and down gently once or twice to rinse the plate and dislodge cells that might be loosely adhering to the plate. The cells are now ready to be counted.

8.   In a small tube, combine the following:
     0.5 mL of 0.4% trypan blue
     0.3 mL of PBS
     0.2 mL of cell suspension from step 7

9.   Once you have removed cells from the plate for counting, your instructor will tell you whether to keep or discard the rest of the cells on the plate.

10. Pipette the cell/stain suspension gently up and down to break up cell aggregates and suspend the cells evenly. At this point, you no longer need to work aseptically because the stained cells will be discarded.

11. Apply cells to hemacytometer chamber.
     11.1.   Place the coverslip on top of the hemacytometer slide.
     11.2.   Use a sterile Pasteur pipette to transfer the suspension to the slit on the edge of the chamber:
             11.2.1.   Hold the tip of the pipette in the slit on the edge of the chamber and gently expel the liquid until the chamber is just filled. The liquid is drawn under the coverslip into the chamber by capillary action.
             11.2.2.   Avoid an air space in the chamber because this means the chamber is not perfectly filled. Also, do not allow the coverslip to rise because of excess liquid.
     11.3.   Remember that the hemacytometer is nonsterile. Do not return cells from the hemacytometer to your culture.
     11.4.   Fill the second counting chamber in the same way as the first.

12. Count viable and nonviable cells in five squares.
     12.1.   Keep separate tallies of the viable and nonviable cells.
     12.2.   Use a low-power objective such that the total magnification (objective times ocular) is 100×.
     12.3.   A handheld tally counter is helpful for keeping count.
     12.4.   Repeat for the second counting chamber.

13. **Data analysis**
     13.1.   Calculate the percent of viable cells in each chamber.
     13.2.   Calculate the number of viable cells per milliliter in each chamber.
     13.3.   Average the results of the two chambers.

14. Remove all items from the hood.
15. Clean out the vacuum lines by aspirating 70% ethanol through them.
16. Wipe down the hood with 70% ethanol and turn off the fluorescent lamp and blower.
17. When everyone is done in the room, turn on the germicidal lamp for 15 minutes.

LAB MEETING/DISCUSSION QUESTIONS

1.   Discuss the variation you observed in cell numbers per square on your hemacytometer. What factors can you suggest that would cause variability in the number of cells per square in a chamber?

2.   Why take the time to count more than one square? How does counting more squares improve the accuracy of your counts?

3.   You calculated the cell density of your culture in this lab. Suppose you want to set up a culture plate containing 10 mL of your cells diluted with Ham's F-12 media, so that your final culture plate will contain 100,000 cells/plate. How will you prepare this plate? (Indicate the volume of cell suspension you will use, as well as the volume of medium you will need.)

Laboratory Exercise 47: Subculturing Cho Cells

OVERVIEW

When cells use up the space in their culture vessel, they need to be divided and reseeded into two or more fresh vessels. This is called **passaging**, **splitting**, or **subculturing** the cells. In this lab you will split your CHO cells. To do this, you will perform the following tasks:

- Observe why cells need to be subcultured.
- Consider how to select the ratio to subculture the cells.
- Subculture (split) cells.

BACKGROUND

Cultured cells eventually use up the space in their vessel and must be passaged. This involves using trypsin-EDTA to remove the cells from the plastic, as you did in Laboratory Exercise 46. The cells from one plate are then divided into two or more fresh plates.

The **split ratio** *expresses how the cells should be diluted.* The cells can be split so that their final concentration is equal to some predetermined concentration, such as 10^5 cells/mL. To split cells by concentration, the cell density must first be determined, as you did in Laboratory Exercise 46. Another way to determine how the cells should be split is to take cells that are confluent, or nearly so, and to split them so that the cells will be ready on whatever day they may be needed. For instance, a one into eight split means that one plate of cells is divided into eight fresh plates. Cells that are split one into eight will need to pass through three doubling times to return to their original density. This method relies on estimating the percent confluence of a plate and so is less accurate than the method of actually counting the cells with a hemacytometer.

Planning Your Work: Subculturing CHO Cells

Draw a sketch of how you will organize your materials on your lab bench to perform this exercise.

LABORATORY PROCEDURE

Refer to Figure 8.30.

1. Prepare for cell culture as you have in previous exercises.
2. Microscopically observe your cells.
    2.1. Record the general appearance of the cells. Estimate their percent confluence.
    2.2. Decide how to split your cells. For example, you may want a one into eight split, or a one into four split, depending on the condition of your cells and the length of time until your next passage. (In our laboratory, CHO cells have a dividing time on the order of 33 hours.)
3. Label fresh cell culture plates with your initials, date, type of cells, and passage number. The passage number is one more than the previous passage number. Your instructor will tell you the passage number of the plate of cells with which you are currently working.
4. Aspirate old medium from cells.
5. Rinse with sterile PBS.
    5.1. Add 2 mL of PBS to the plate. Briefly swirl the plate to remove dead cells, which do not firmly attach to the substrate. This step will also help remove media, serum, and cations that may inhibit the enzymatic action of the trypsin.
    5.2. Aspirate off the PBS.
6. Trypsinize the cells.
    6.1. Add 1 mL of trypsin-EDTA to the plate.
    6.2. Incubate the plate at 37°C.

FIGURE 8.30   Example of passaging cells. (a) The culturist has prepared for this procedure by warming the trypsin-EDTA solution and fresh cell culture medium in the bead bath, as was shown in Figure 8.27a. He brings all required materials into the hood, including pipettes, a pipette aid, sterile Pasteur pipettes, and fresh cell culture dishes. The warmed cell culture medium and tube of trypsin are visible on the left side of the hood. (b) The culturist removes his plate of cells from the cell culture incubator and examines the cells microscopically to decide on the split ratio. (c) He moves the plate into the BSC and aspirates off the old medium from the plate of cells. Not shown: He briefly washes each plate with PBS, adding the PBS and then aspirating it off. (d) He adds 1 mL of trypsin-EDTA to his plate with cells. (e) He then periodically views the plate under the microscope, returning the cells to the incubator between views. (f) He continues until the cells round up and float. (g) The culturist is now using a slightly different strategy than described in your laboratory procedure to split the cells. He is pipetting the cells from his plate directly into his bottle of fresh cell culture medium. He has previously calculated how many milliliters of fresh medium he needs to achieve the proper split ratio. (Observe that there are different cell culture strategies for splitting cells, all of which provide fresh plates with the right number of cells.) (h) The culturist swirls the bottle of medium containing his cells to create a homogeneous suspension. (i) The culturist pipettes 10 mL of cell suspension into a series of fresh cell culture plates that were previously labeled. He will distribute the cells on the fresh plates by moving them forward and backward, left and right, and diagonally. He will finally return the plates to the incubator.

6.3.   Remove the plate from the incubator after about a minute. Tap the plate to help dislodge the cells from the plastic surface of the dish.

6.4.   Look at the plate using the inverted microscope. If the cells have not been dislodged from the substrate, return the plate to the incubator for another minute.

6.5.   Repeat steps 6.2 to 6.4 until all, or nearly all, of the cells are floating. Leave the trypsin on the cells only long enough to dislodge most of the cells. If you wait too long, the trypsin will begin to damage the cells. The length of time required depends on the activity of the trypsin and the type of cells you have. Generally 1–10 minutes will work. Record how long you keep the trypsin on the cells.

7.   Add 7 mL of Ham's F-12 complete medium to the culture plate. (This will halt trypsinization).

7.1.   Pipette the cells up and down once or twice.

7.2.   Rinse the plate gently to dislodge cells that might be loosely adhering to the plate.

8.   Split cells.

8.1.   For a one into eight split, place 1 mL of your cells into each of eight fresh plates. Add 9 mL of Ham's F-12 complete medium into each of the plates.

8.2.   For a one into four split, place 2 mL of your cells into four fresh plates. Add 8 mL of Ham's F-12 complete medium into each of the plates.

9.   Evenly suspend the cells throughout the plate.

9.1.   Place the plate on the hood bench. Slide the plate away from you and then back toward you several times. The plate is always resting on the bench.

9.2.   Move the plate to the left and right, several times, again with the plate resting on the bench.

9.3.   Move the plate on a diagonal away from yourself and toward yourself several times.

10.   Your instructor may have each student make only one plate, in which case, discard the extra cells. Flood any extra cells with 10% bleach, wait at least 15 minutes, and then dispose of the cells.

11.   Check your plate(s) under the microscope, record its appearance, and place in the incubator.

12.   Return your cell culture medium bottle to the refrigerator. Clean up in the usual manner. Do not forget to run alcohol through the vacuum line.

13.   Maintain, observe, and feed your cells until they must be passaged again.

LAB MEETING/DISCUSSION QUESTIONS

Suppose that you are ready to perform a procedure with your CHO cells today. Unfortunately, as you prepare to work in the hood, you receive a call that requires you to postpone the procedure. You will not be able to attempt this procedure again for 4 days. Quickly, you perform a calculation to determine a split ratio that will allow you to keep your cells growing and ready to work with in 4 days. What split ratio will you use? (Assume that your cells will not experience much, if any, lag phase following the subculturing process. Also, assume they have a 33 hour doubling time.)

Laboratory Exercise 48: Preparing a Growth Curve for CHO Cells

OVERVIEW

In this laboratory exercise, you will prepare a growth curve for your CHO cells. To do so, you will subculture the cells, perform cell counts daily for a week, and use the data to generate a growth curve. From these data, you should be able to estimate the population doubling time of your CHO cell line. You will thus perform the following tasks as you complete this laboratory exercise:

- Practice splitting and counting cells.
- Generate a growth curve for CHO cells.
- Calculate population doubling time.

BACKGROUND

CHO cells follow the same stages of growth as bacterial cells (see Figure 8.31). CHO cells experience a lag phase when they are first seeded into a new vessel with fresh medium. During this time, the cells are metabolically active, but the culture does not experience a net increase in cell number. Once the cells are accustomed to their new conditions and firmly attached to the substrate, they begin to divide and soon enter a period of rapid, exponential growth. The cells will remain in exponential stage for several generations, as long as they are fed fresh medium every 2 or 3 days.

The population doubling time can be determined from the growth curve as shown in Figure 8.31. Pick any two points in the exponential phase where the cell concentration of one point is twice that of the other point, and see how much time has elapsed. It is useful to know the population doubling time so that you can predict when cells will be ready for a given manipulation.

When cells are fed, the old nutrient-depleted medium, along with any accumulated metabolic wastes, is removed and fresh medium is added. Even so, the cells will eventually expand to fill the surface of the vessel. At this point, the cells will reach stationary stage. Eventually, if the cells are not passaged, they will all die.

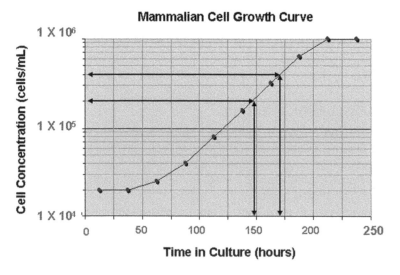

FIGURE 8.31 **Mammalian cell growth curve and determination of doubling time.** Like bacterial cells, these cells also go through lag phase, exponential phase, stationary phase, and then death phase (not shown). The cell concentration in this example doubles in about 24 hours from 2×10^5 cells/mL to 4×10^5 cells/mL.

Planning Your Work: Preparing a Growth Curve for CHO Cells

1. This exercise requires combining the procedures in Laboratory Exercises 45, 46, and 47. Write a protocol in a step-by-step format to guide your work.

2. Prepare a daily schedule. Teams can share the work, but they should be careful to ensure consistency.

3. Prepare a list of all materials you will need and check to make sure they are available.

OVERVIEW

1. If possible, on Monday, follow the directions in Laboratory Exercise 47 to get eight fresh plates of cells.

2. Count the cells to determine their concentration at time = 0.

3. Feed the cells every 2 or 3 days, as usual.

4. Count the cells each day of this week, as well as Monday and Tuesday of next week. Each day, trypsinize and count cells from one of your plates. You can discard the plate afterward.

5. Copy Table 8.4 into your laboratory notebook and record your results.

6. **Data Analysis**

   6.1   Plot cells per milliliter (logarithmic Y-axis) versus time in hours (X-axis). Plot the results by hand on semilog paper or use a graphing program, such as Excel.

   6.2.   Draw a best-fit line through those points that appear to represent the exponential stage of growth.

   6.3.   Determine the population doubling time by finding the time it takes the population concentration to double during the exponential phase.

LAB MEETING/DISCUSSION QUESTIONS

1. Compare class results for doubling time. Discuss any differences you observe.

2. Describe the factors that limit the length of time a cell line can spend in exponential growth.

3. Suppose you are given a new line of cultured cells. Describe, mathematically, how you would determine the population doubling time of this new line.

TABLE 8.4   Cell Counting Results

Day	Hours Post-Seeding	Cell Concentration (cells/mL)	Appearance of Cells
Monday	0		
Tuesday			
Wednesday			
Thursday			
Friday			
Monday			
Tuesday			

UNIT DISCUSSION: GROWING CELLS

MAKING CONNECTIONS

1. Growing cells requires many of the skills that have been introduced throughout this manual. Fill in the following table to show how the exercises in the earlier part of the manual relate to cell culture.

Unit	Laboratory Exercise or Classroom Activity	Relationship to Cell Culture
Unit I Safety	Laboratory Exercise 2	
Unit II Documentation	Classroom Activity 7	
	Classroom Activity 8	
	Laboratory Exercise 3	
Unit III Metrology	Laboratory Exercise 5	
	Laboratory Exercise 6	
	Laboratory Exercise 7	
	Laboratory Exercise 8	
	Laboratory Exercise 9	
Unit IV Spectrophotometry	Laboratory Exercise 12	
Unit V Solutions	Laboratory Exercise 14	
	Laboratory Exercise 16	
	Laboratory Exercise 20	
Unit VI Assays	Laboratory Exercise 25	
Unit VII Separations	Classroom Activity 13	

2. Discuss similarities and differences between mammalian and bacterial cell culture.
3. "Science is facts; just as houses are made of stone, so is science made of facts; but a pile of stones is not a house, and a collection of facts is not necessarily science," stated Jules Henri Poincaré (1854–1912), a French mathematician. As we discussed in the Introduction to this manual, science is more than a pile of facts, and working in the laboratory is more than learning to perform a series of steps in a "cookbook" fashion. Fill in the following table, considering what you have learned in terms of thinking through your laboratory work.

Good laboratory practices
DOCUMENT
PREPARE
REDUCE VARIABILITY AND INCONSISTENCY

ANTICIPATE, PREVENT, CORRECT PROBLEMS
ANALYZE, INTERPRET RESULTS
VERIFY PROCESSES, METHODS, AND RESULTS

Appendix 1

ABBREVIATIONS AND ACRONYMS USED IN THIS LABORATORY MANUAL

ß-gal	ßeta-galactosidase
λ	Greek letter lambda, used to represent the wavelength of light
µS/cm	microsiemens/cm
A_{600}	absorbance at a wavelength of 600 nm (any wavelength can be substituted)
ACS	American Chemical Society
Ag/AgCl₂ electrode	silver/silver chloride electrode
ASTM	American Society for Testing and Materials
ATC	automatic temperature compensation (probe)
ATCC	American Type Culture Collection
BSA	bovine serum albumin
BSC	biological safety cabinet
BSL	Biosafety Level
BTV	bring to volume
CDC	Centers for Disease Control and Prevention
CGMP	current Good Manufacturing Practices
CFR	Code of Federal Regulations
CHO	Chinese hamster ovary
CHP	Chemical Hygiene Plan
CM Sepharose	carboxymethyl Sepharose
DEAE Sepharose	diethylaminoethyl Sepharose
DNA	deoxyribonucleic acid
DTT	dithiothreitol
E. coli	*Escherichia coli*
EDTA	ethylenediaminetetraacetic acid
EPA	Environmental Protection Agency
FDA	Food and Drug Administration
FW	formula weight
GLP	Good Laboratory Practices
GOI	gene of interest
HCl	hydrochloric acid
HEPA	high-efficiency particulate air (filter)
IE	ion exchange
IR	infrared
ISO	International Organization for Standardization
LAI	laboratory-acquired infection
MgAc	magnesium acetate
MgCl₂	magnesium chloride
mS/cm	millisiemens/cm
MW	molecular weight
NaCl	sodium chloride
Na₂CO₃	sodium carbonate
NaOH	sodium hydroxide
OD	optical density
ONP	ortho-nitrophenol
ONPG	ortho-nitrophenyl-β-galactoside
OSHA	Occupational Safety and Health Administration
PCR	polymerase chain reaction
PPE	personal protective equipment
ppm	parts per million

QA	quality assurance
QC	quality control
R&D	research and development
RCF	relative centrifugal force
Rf	retention factor
RFLP	restriction fragment length polymorphism
RNA	ribonucleic acid
SDS	Safety Data Sheet
SI	Système International d'Unités (International System of Units)
SOP	standard operating procedure
TBE	Tris and borate and EDTA
TE	Tris-EDTA
Tris	tris(hydroxymethyl)aminomethane
U	unit (of enzyme activity)
USP	*United States Pharmacopeia*
UV	ultraviolet
V/V	volume per volume
VIS	visible
WFI	water for injection
W/V	weight per volume
W/W	weight per weight
X g	times the force of gravity

Appendix 2

Absolute error: The difference between the true value and the measured value. Absolute error = True value – Measured value.

Absorbance (A): A measure of the amount of light absorbed by a sample defined as $A = \log_{10} (1/t) = -\log_{10} t$, where t is transmittance. Also called "optical density."

Absorbance spectrum: The plot of absorbance of a particular sample when exposed to light of various wavelengths, typically plotted with wavelength on the X-axis versus absorbance on the Y-axis.

Absorption (in spectrophotometry): The loss of light of specific wavelengths as the light passes through a material and is converted to heat energy.

Absorptivity: The inherent tendency of a material to absorb light of a certain wavelength. (Note that absorbance is a measured value that depends on the instrument used to measure it whereas absorptivity is an intrinsic property of a material.)

Absorptivity constant (a or α): A value that indicates how much light is absorbed by a particular substance at a particular wavelength under specific conditions (such as temperature and solvent).

Absorptivity constant (molar): The absorptivity constant when the concentration of analyte is expressed in units of moles per L.

Accuracy: Closeness of agreement between a measurement or test result and the true value or the accepted reference value for that measurement or test.

Acid: A chemical that dissociates and releases H^+ ions when dissolved in water.

Acidic solution: An aqueous solution with a pH of less than 7.

Adapter (in centrifugation): An insert placed in a rotor compartment that permits the use of a smaller-sized tube than could otherwise be accommodated.

Adherent cells (in mammalian cell culture): Cells that grow in a single layer attached to the surface of their culture vessel.

Aerosols: Fine liquid droplets that are suspended in the air.

Agar: A hardening agent derived from seaweed, used to make a solid substrate for bacterial growth.

Air displacement micropipette: Device for measuring microliter volumes, designed so that there is an air cushion between the micropipette and the sample.

Aliquot: The portions that result from subdividing a homogeneous solution or substance into smaller units.

American Type Culture Collection: A global, nonprofit organization that provides and distributes biological resources, particularly cells and tissues.

Amino acids: A class of naturally occurring molecules that are the building blocks of proteins. Every amino acid contains a carbon atom bonded to an amino group ($-NH_2$), a carboxyl group ($-COOH$), a hydrogen atom, and a side chain. Each amino acid has a different side chain that gives it distinctive chemical properties.

Amino group: Part of the core structure of amino acids, NH_3.

Amount: How much of a substance is present.

Analog (also analogue): Smoothly changing; measurement values that are continuous (e.g., as is displayed by a meter with a needle that can point to any value on a scale.)

Analyte: A substance of interest whose presence and/or level is evaluated in a sample by using an instrument, assay, or test.

Analytical balance: An instrument that can accurately determine the weight of a sample to at least the nearest 0.0001 g.

Analytical method: A test used to analyze, identify, or characterize a mixture, compound, chemical, or unknown material.

Analytical wavelength (in spectrophotometry): The wavelength at which absorbance measurements are made in a particular assay.

Aseptic technique: A system of laboratory practices that helps prevent environmental contaminants from entering a culture and also helps prevent cultured cells from escaping to the environment.

Assay: A test of a sample or system. An assay might measure a characteristic of a sample (e.g., identity, purity, activity), a biological response, or an interaction between molecules. The terms "assay," "test," and "method" are often used interchangeably, although the term "assay" is not generally applied to a test of an instrument's performance and generally refers to an analysis of a sample. "Assay" can also be used as a verb, meaning "to determine," for example, "the technician assayed the sample for protein."

ASTM International: An organization that prepares and distributes standards to promote consistent procedures for measurement.

Atomic weight: See **Gram atomic weight**.

Audit: A process in which a person (or team) collects objective evidence to permit an informed judgment about whether an organization is complying with its quality system.

Autoclave: A laboratory pressure cooker that sterilizes materials with pressurized steam.

Automatic temperature compensating probe: A probe placed alongside pH electrodes that automatically reports the sample temperature to the meter.

Background (in spectrophotometry): Light absorbance caused by anything other than the analyte.

Bacterial broth: Aqueous mixtures of nutrients to support the growth and reproduction of bacterial cells; prepared without a hardening agent.

Balance: An instrument used to measure the weight of a sample by comparing the effect of gravity on the sample to the effect of gravity on objects of known mass.

Base (in the context of pH): (1) A chemical that causes H^+ ions to be removed when dissolved in aqueous solutions. (2) A chemical that releases OH^- ions when dissolved in water.

Basic solution (in the context of pH): An aqueous solution with a pH greater than 7.

Batch: "[A] specific quantity of a drug or other material that is intended to have uniform character and quality, within specified limits, and is produced according to a single manufacturing order during the same cycle of manufacture," according to GMP regulations.

Batch record: The document used to direct the manufacture of a product and to record information about manufacturing activities.

Beam: See **Lever**.

Beer's law: A rule that states that the absorbance (A) of a homogeneous sample is directly proportional to both the concentration (C) of the absorbing substance and to the thickness of the sample in the optical path (b): $A = \alpha bC$, where α is the absorptivity constant. Also called "Beer-Lambert law" or "Beer-Bouguer law."

Bioaerosols: Small liquid droplets containing a biological agent that are capable of remaining suspended in the air for long periods of time.

Biological macromolecule: A large and complex molecule that has biological function.

Biological safety cabinets: A specially designed hood that protects people and the outside environment from bioaerosols.

Biological solution: A laboratory solution that supports the structure and/or function of biological molecules, intact cells, or microorganisms in culture.

Biopharmaceutical: (1) A drug product that is manufactured using genetically modified organisms as a production system. (2) Most broadly, any drug manufactured by living cells or organisms (whether or not recombinant DNA techniques are involved), whole cells or tissues used therapeutically, or any drug that is a large biological molecule. These are usually proteins, but they can be RNA or DNA (e.g., DNA for gene therapy).

Bioreactor: A specialized growth chamber used for producing a product in cells (usually refers to mammalian cells) in which conditions of temperature, nutrient level, aeration, pH, and mixing are controlled.

Biosafety levels: A classification scheme managed by the Centers for Disease Control and Prevention that classifies biological agents according to their risk. There are four classes that represent progressively higher risk: BSL1, BSL2, BSL3, and BSL4. BSL1 agents are nonpathogenic and well characterized.

Blank (in spectrophotometry): A reference that contains no analyte but does contain the solvent (for a liquid sample) and any reagents that are intentionally added to the sample. The blank is held in a cuvette that is identical to that used for the sample, or the blank is alternately placed in the same cuvette as the sample. A spectrophotometer compares the interaction of light with the sample and with the blank in order to establish the absorbance caused by the analyte.

Bring to volume: The procedure in which solvent is added to solute (typically in a volumetric flask or graduated cylinder) until the total volume of the solution is exactly the final volume desired.

Buffer: A substance or combination of substances that, when in aqueous solution, resists a change in H^+ concentration even if acids or bases are added.

Buffered salt solution: Solution that is intended to maintain living cells for short periods (minutes to hours) in an isotonic, pH-balanced environment.

Calibrate: To bring the readings of an instrument into accordance with an external standard.

Calibration: (1) Adjustment of a measuring system to bring it into accordance with external values. (2) A process that establishes, under specified conditions, the relationship between values indicated by a measuring instrument or measuring system and values of a trustworthy standard. Calibration permits the estimation of the uncertainty of the measuring instrument or measuring system. The result of a calibration is recorded in a document.

Calibration (of a pH meter): The use of standards of known pH to determine the relationship between the voltage response of the electrodes and the pH of the sample.

Calibration (of a spectrophotometer): (1) The use of standards with known amounts of analyte to determine the relationship between light absorbance and analyte concentration. (2) Bringing the transmittance and wavelength values of a spectrophotometer into accordance with externally accepted values.

Calibration (of a volume-measuring device): (1) Placement of capacity lines, graduations, or other markings on the device so that they correctly indicate volume. (2) Adjustment of a dispensing device so that it dispenses accurate volumes.

Calibration error: Situation where there is an error (usually small) in the upper calibration of a balance.

Calibration standard (for a balance): Objects whose masses are established and documented.

Calomel electrode: A type of reference electrode containing calomel, a paste consisting of mercury metal and mercurous chloride, Hg/Hg_2Cl_2. (Seldom used anymore due to toxicity of mercury.)

Capacity: Maximum load that can be weighed on a particular balance as specified by the manufacturer.

Capacity line: A line marked on an item of glassware to indicate the volume if the item is filled to that mark.

Carcinogen: A type of toxic chemical that is known to cause cancer in animals and is assumed to also cause cancer in humans.

Carryover: Material from one sample that is carried to and contaminates another sample.

Casting tray (in electrophoresis): A tray for molten agarose that shapes the agarose into a rectangular slab with small depressions for the samples.

Cell culture: The process in which living cells derived from the tissue of multicellular animals are maintained in nutrient medium inside dishes, flasks, or other vessels.

Centrifugal force: The force that pulls a particle away from the center of rotation.

Centrifuge: An instrument that generates centrifugal force, commonly used to help separate particles in a liquid medium from one another and from the liquid.

- **Microcentrifuge.** A small centrifuge intended for small volume samples in the microliter to 1 or 2 mL range.
- **Desktop, or clinical centrifuge.** A centrifuge that spins samples at rates slower than 10,000 rpm.
- **Superspeed centrifuge.** A centrifuge that spins samples at speeds from 10,000 to 30,000 rpm.
- **Ultracentrifuge.** A centrifuge that spins samples at speeds of up to and sometimes exceeding 80,000 rpm.

Chelator: An agent that binds to and removes metal ions from solution.

Chemical hygiene plan: An extensive written manual that contains procedures and policies regarding safety issues for a laboratory or institution.

Chemically defined growth medium: An aqueous medium to support the growth and reproduction of cells that contains only known ingredients in known quantities.

Chinese hamster ovary cells: A cell line derived from the ovary of a Chinese hamster that is commonly used for biopharmaceutical production.

Chromatography: A large class of methods used to separate the components of a mixture from one another and identify them based on differences in their relative affinities of the components for two different phases.

Code of Federal Regulations: Published annually, this document contains U.S. federal government regulations.

Colony-forming unit: A unit that measures the number of bacteria in a sample based on the number of visible colonies.

Colorimetric assay: An assay in which a colorless analyte is exposed to another compound(s) and/or to conditions that cause it to develop color. The amount of color that occurs is proportional to the concentration of the analyte.

Comb (in electrophoresis): A toothed device that is placed in molten agarose before it sets that leaves behind depressions, called "wells." Samples are loaded into the wells.

Combination electrode: A measuring electrode and reference electrode that are combined into one housing.

Complete medium (in mammalian cell culture): Medium that has all its components, including supplements and components that are added just before use (e.g., serum).

Compound: A substance composed of atoms of two or more elements that are bonded together.

Concentration: An amount per volume; a ratio. The numerator is the amount of the solute of interest, and the denominator is usually the volume of the entire solution, the solvent and the solute(s) together.

Conductivity: The inherent ability of a material to conduct electrical current.

Conductivity meter: A measuring instrument that measures the conductivity of a solution.

Control: (1) *Evaluation.* An evaluation to check, test, or verify; an item used to evaluate or verify a process, method, or experiment. (2) A known substance. (3) *Authority.* The act of guiding, directing, or managing. (4) *Stability.* A state in which the variability in a process is attributable only to chance (i.e., to the normal variability inherent in the process). (Based on definitions from ASTM International, "Standard Terminology Relating to Quality and Statistics," Standard E 456-06, West Conshohocken, PA: ASTM, 2006.)

Cross-contamination: A situation where cells from one culture accidentally enter another culture.

Cubic centimeter (cc, cm³): A unit of volume that is equal to 1 mL.

Culture media: See **Growth media**.

Current (electrical): The flow of charge. Current may involve the movement of electrons in a conductor or the flow of ions in a solution.

Current Good Manufacturing Practices (CGMPs): The currently accepted minimum standards and requirements for the manufacture, testing, and packaging of pharmaceutical products. "Current" means that a practice may be enforced, even if it is not in the published GMP regulations, if the practice has become accepted by industry.

Cuvette: A sample "test tube" that is designed to fit a spectrophotometer and is made of an optically defined material that is transparent to light of specified wavelengths.

Data: Observations of a variable (singular, datum).

Defined medium: See **Chemically defined growth medium**.

Deionization: A process in which dissolved ions are removed from a solution by passing the solution through a cartridge containing ion exchange resins.

Detector (in a spectrophotometer): Device used to measure the amount of light transmitted through a sample.

Detergent: Substances that have both a hydrophobic "tail" and a hydrophilic "head." Detergents, therefore, have two natures; they can be both water soluble and lipid soluble.

Diffraction grating: Device consisting of a series of evenly spaced grooves on a surface that is used to separate polychromatic light into its component wavelengths.

Digital: Discontinuous measurement values (in contrast to continuous, analog values); for example, digital values are stored and displayed by a computer.

Diluent: A substance used to dilute another. For example, when concentrated orange juice is diluted, the diluent is water.

Dilution: Addition of one substance (often but not always water) to another to reduce the concentration of the original substance.

Dilution series: A group of solutions that have the same components but at different concentrations.

Disinfection: Destruction of most, but not all, microorganisms by means of heat, chemicals, or ultraviolet light.

Dispersion (in spectrophotometry): The separation of light into its component wavelengths.

Distillation: A water purification process in which impurities are removed by heating the water until it vaporizes. The water vapor is then cooled to a liquid and collected, leaving nonvolatile impurities behind.

DNA fingerprinting: A technique for distinguishing individuals based on differences in their DNA sequences.

Documentation: Written records that guide activities and that record what has been done.

Electrode: A metal and an electrolyte solution that participate in an electrochemical reaction.

Electromagnetic radiation: A form of energy that travels through space at high speeds.

Electronic balance: Instrument that determines the weight of a sample by comparing the effect of the sample on a load cell with the effect of standards on that load cell.

Electronic records: Text, graphics, data, audio, or pictorial information that is created, modified, maintained, archived, retrieved, or distributed by a computer system.

Electrophoresis: A class of techniques in which molecules are separated from one another based on differences in their mobility when placed in a gel matrix and subjected to an electrical field.

Equilibrate a column (in ion exchange chromatography): The process of adjusting the pH of the liquid surrounding the beads throughout a column and stabilizing the charge on the beads' surface.

Error: (1) The difference between a measured value and the "true" value (see also **Percent error** and **Absolute error**). (2) The cause of variability in measurements.

Ethidium bromide: A compound that inserts itself into the structure of DNA and fluoresces when it is activated by ultraviolet light.

Field of view (in microscopy): The portion of the sample that is visible.

Filling hole: An opening in a reference electron by which filling solution is introduced into the electrode.

Filling solution: A solution of defined composition within an electrode.

Filter: A device used to separate components of samples on the basis of size; particles smaller than a certain size pass through a porous filter material; particles larger than a certain size are trapped by the filter.

Filtrate: The fluid and any associated particles that have passed through a filter.

Filtration: A separation method in which particles are separated from a liquid or gas by passage through a porous material.

Fixed angle rotor: A rotor that holds tubes at a fixed angle.

Flash point: The temperature at which enough vapor is emitted that the chemical will burn in the presence of an ignition source.

Formula weight (FW): The weight, in grams, of one mole of a given compound. The FW is calculated by adding the atomic weights of the atoms that compose the compound.

Glass electrode: See **pH-measuring electrode**.

Good Laboratory Practices: (1) FDA—The procedures and practices that must be followed when performing laboratory studies in animals or in vitro in order to investigate the safety and toxicological effects of new drugs. (2) EPA—The procedures and practices that must be followed when investigating the effects of pesticides and other agrochemicals. (3) With lowercase letters (i.e., good laboratory practices [glp])—A general term used to refer to all quality practices in any laboratory.

Good Manufacturing Practices: See **Current Good Manufacturing Practices**.

Graduated cylinder: Cylindrical vessel calibrated to deliver various volumes.

Graduations: Lines marked on glassware, plasticware, and pipettes that indicate volume.

Gram atomic weight: The weight, in grams, of 6.022×10^{23} (Avogadro's number) atoms of a given element.

Gravimetric method: A method that involves the use of a balance; gravimetric methods are used to calibrate the volume-measuring devices. The volume of the liquid is calculated from its weight and its density at a particular temperature. This method takes advantage of the high accuracy and precision attainable with modern balances.

Gravity (in centrifugation): The unit of measure for the rate of acceleration of gravity. The Earth's normal gravity is defined as $1 \times g$.

Growth media: Solutions that support the survival and growth of cells in culture. These solutions include nutrients, vitamins, salts, and other required substances.

Hazard: The equipment, chemicals, and conditions that have a potential to cause harm.

High-efficiency particulate air filter: A filter used to remove particulates, including microorganisms, from air.

Horizontal rotor: A rotor in which tubes or sample bottles swing into a horizontal position when centrifugal force is applied. Also called "swinging-bucket rotor."

Hydrates: Compounds that contain chemically bound water. The weight of the bound water is included in the formula weight of hydrates.

Hydrogen ion: H^+.

Hydroxide ion: OH^-.

Hygroscopic compound: A compound that absorbs moisture from the air.

Incident light: Light that strikes or shines on a substance.

International Organization for Standardization (ISO): A network of the national standards institutes of many countries, on the basis of one member per country, with a central secretariat in Geneva, Switzerland, that coordinates the system.

Ion: An atom or group of atoms with a net charge as a result of having lost or gained electrons.

Ion exchange: The process in which ions in solution are exchanged with similarly charged ions associated with ion exchange resins.

Ionic strength: A measure of the charges from ions in an aqueous solution.

ISO 9000: A set of internationally accepted standards that outline a system for quality management aimed at ensuring the quality of a product or service.

Isotonic solution: A solution that has the same solute concentration as the interior of the cell.

Junction: A part of a reference electrode that allows filling solution to flow into the solution whose pH is being measured.

Knife edge: In mechanical balances, a support on which the beam rests that allows it to swing freely.

Laboratory notebook: A bound ledger assigned to an individual in which that individual keeps a chronological log of everything done and observed in the laboratory.

Leveling (a balance): Adjusting the balance to a level, horizontal position.

Leveling bubble: A bubble that is used to determine whether the balance is level or not. When the balance is level, the bubble is centered in a window.

Lever: A rigid bar used in mechanical balances on which the sample to be weighed is balanced against objects of known mass. Also called "beam."

Light: Electromagnetic radiation with wavelengths from 180 to 2500 nm; includes the ultraviolet, visible, and infrared regions of the electromagnetic spectrum.

Light-emitting diode (LED): A diode that emits light when current flows through it.

Light scattering: An interaction of light with small particles in which the light is bent away from its initial path.

Light source (in spectrophotometry): The bulb or lamp that emits the light that shines on the sample.

Limit of detection: The lowest level of the material of interest that a method or instrument can detect.

Limit of quantitation: The lowest level of the material of interest that a method or instrument can quantify with acceptable accuracy and precision.

Line of best fit: A line connecting a series of data points on a graph in such a way that the points are collectively as close as possible to the line.

Linear relationship: A relationship between two properties such that when they are plotted on a graph, the points form a straight line.

Linearity: (1) A characteristic of a direct reading device. If a device is linear, calibration at 2 points (e.g., 0 and full scale) calibrates the device. (2) A measure of how well an instrument follows an ideal, linear relationship.

Linearity (for a balance): The ability of a balance to give readings that are directly proportional to the weight of the sample over its entire weighing range.

Linearity check (in pH measurement): A check of whether the meter's response is linear through its entire range.

Lot number: An identifying number that is used to distinguish the product of one manufacturing run from another.

Mass: The amount of matter in an object, expressed in units of grams. Mass is not affected by the location of the object or the effect of gravity.

Measurement: Quantitative observation; numerical description.

Measuring pipette: A type of pipette calibrated with a series of graduation lines to allow the measurement of more than one volume.

Mechanical balance: A balance that uses a lever and that does not generate an electrical signal in response to a sample. The object to be weighed is placed on a pan attached to the lever and is balanced against standards of known mass.

Method: The means of performing an analysis. A method describes the steps necessary to perform an analysis and related details, such as how the sample should be obtained and prepared, the reagents that are required, the setup and use of instruments, comparisons with reference materials, calculations, and so on.

Metrology: The study of measurements.

Microbe: See **Microorganism**.

Microliter pipette: or micropipette: A term used to refer to various styles of device that measure volumes in the 1 to 1000 μL range.

Microorganism: An organism that can be seen only through a microscope; includes bacteria, protozoans, algae, and fungi. Some people would also consider viruses and prions to be microorganisms, although these substances are not living and, therefore, are not technically organisms.

Microscope lens: A specially shaped piece of glass that bends light in specific ways to create a magnified image of a specimen.

- **Ocular.** Lens through which the viewer looks at the specimen. Oculars usually magnify the image 4X or 10X.
- **Objective.** Lenses used to magnify the image.
  - **Scanning-objective.** Objective with a magnification of 4× that is useful for quickly surveying the whole slide.
  - **Low-power objective.** Objective with a magnification of 10×.
  - **High-power objective.** Objective with a magnification of 40×.
  - **Oil-immersion objective.** Objective with a magnification of 100× that requires the presence of oil between the lens and the specimen.
- **Condenser lens.** Lens that helps to focus the light from the lamp onto the slide.

Mohr pipette: A type of serological pipette that is calibrated so that the liquid in the tip is not part of the measurement.

Molarity: An expression of concentration of a solute in a solution that is the number of moles of solute dissolved in 1 L of total solution.

Mole: A mole of any element contains 6.022×10^{23} (Avogadro's number) atoms. Because some atoms are heavier than others, a mole of one element weighs a different amount than a mole of another element. A mole of a compound contains 6.022×10^{23} molecules of that compound.

Molecular weight: See **Formula weight**.

Monochromatic: Light that is of one wavelength.

Monochromator: Device used to separate polychromatic light into its component wavelengths and to select light of a certain wavelength (or narrow range of wavelengths).

Mutagen: A type of toxic chemical that alters the genetic makeup of a cell. All mutagens are suspected to also be carcinogens.

National Institute of Standards and Technology: A federal agency that works with industry and government to advance measurement science and develop standards.

Negative control: A sample known not to contain the analyte of interest used to ensure that an assay is working correctly.

Nominal volume: The desired volume for which a pipetting device is set.

Optical density: See **Absorbance**.

Osmotic equilibrium: A condition where the rate of water flow into and out of a cell are equal.

Packing a column (in chromatography): The process of pouring stationary-phase beads into a column and causing them to settle into an even bed.

Parts per million: An expression of concentration of solute in a solution that is the number of parts of solute per 1 million parts of total solution. Any units may be used but must be the same for the solute and the total solution.

Pasteur pipette: A type of pipette used to transfer liquids from one place to another but not to measure volume.

Path length: The distance light passes through the sample; typically measured as the length of the cuvette in centimeters or millimeters.

Pathogen: An agent that is both infectious (able to infect a host) and pathogenic (can cause disease in the host).

Pellet (in centrifugation): Components of a sample that have settled to the bottom of a container after centrifugation.

Percent: A fraction whose denominator is 100.

Percent error: An indication of the accuracy of a measurement as calculated by this equation:

$$\frac{\text{True value} - \text{Measured value}(100\%)}{\text{True value}}$$

Percent solution: An expression of the concentration of solute in a solution where the numerator is the amount of solute (in grams or millimeters), and the denominator is 100 units (usually mL) of total solution.

- **Weight-per-volume percent.** The grams of solute per 100 mL of solution.
- **Volume percent.** The milliliters of solute per 100 mL of solution.
- **Weight percent.** The grams of solute per 100 g of solution.

Performance verification: A process of checking that an instrument is performing properly.

Personal protective equipment: Specialized clothing or equipment worn to protect an individual.

pH: A measure representing the relative acidity or alkalinity of a solution; expressed as the negative log of the H^+ concentration when concentration is expressed in moles per liter. pH = $-\log$ [H^+].

pH-measuring electrode: An electrode whose voltage depends on the concentration of H^+ ions in the solution in which it is immersed.

pH meter: A term that is commonly used to refer to a specialized voltage meter and its accompanying electrodes.

Pipette: A hollow tube that allows liquids to be drawn in and dispensed from one end; generally used to measure volumes in the 0.1–25 mL range. Also spelled "pipet."

Pipette aid: A device used to draw liquid into and expel it from pipettes.

pKa: The pH at which a buffer stabilizes and is least sensitive to additions of acids or bases.

Plunger (in volume measurement): Part of a manual micropipette that is compressed by the operator as liquids are taken up and expelled.

Polychromatic light: A combination of light of many wavelengths.

Polymerase chain reaction: An enzymatic method of amplifying the amount of a specific sequence of DNA many thousands of times.

Positive control: A sample known to contain the analyte of interest. Used to ensure that an assay is working correctly.

Positive displacement micropipette: A volume-measuring device, such as a syringe, where the sample comes in contact with the plunger and the walls of the pipetting instrument.

Positively charged: A material in which electrons are depleted.

Precision: (1) The consistency of a series of measurements or tests obtained under stipulated conditions. (2) The fineness of increments of a measuring device; the smaller the increments, the better the precision.

Preparative method: Method that produces a material or product for further use (perhaps for commercial use or perhaps for further experimentation). Compare to **Analytical method**.

Procedure: A specified way to perform a process.

Proportion: Two ratios that have the same value but different numbers. For example: 1/2 = 5/10.

Protein assay: A test of the amount or concentration of protein in a sample.

Proteins: Diverse biological molecules composed of amino acid subunits that control the structure, function, and regulation of cells.

Protocol: A step-by-step outline that tells an operator how to perform an experiment that is intended to answer a question.

Purified water: A general term for laboratory water from which impurities have been removed.

Qualitative analysis: The analysis of the identity of substance(s) present in a sample.

Quality: All the features of a product or service that are required by the customer or user, according to ISO.

Quality assurance: An organizational unit in a company that provides confidence that product quality requirements are fulfilled, in part through the effective distribution and management of documentation.

Quality control: A department responsible for monitoring processes and performing laboratory testing to ensure that products are of suitable quality.

Quantitation limit: The lowest amount of analyte in a sample that can be quantitatively determined with suitable precision and accuracy.

Quantitative analysis: The measurement of the concentration or amount of a substance of interest in a sample.

Radius of rotation: The distance from the center of rotation to the material being centrifuged.

Range: (1) A range of values, from the lowest to the highest, that a method or instrument can measure with acceptable results. (2) A statistical measure that is the difference between the lowest and the highest values in a set of data.

Ratio: The relationship between two quantities (e.g., 25 miles per gallon).

Raw data: The first record of an original observation.

Readability (for a balance): Value of the smallest unit of weight that can be read.

Reference electrode: An electrode that maintains a stable voltage for comparison with the pH-measuring electrode or an ion-sensitive electrode.

Relative centrifugal field: The ratio of a centrifugal field to the earth's force of gravity.

Research and development: The organizational unit in a company that finds ideas for products, performs research and testing to see if the ideas are feasible, and develops promising ideas into actual products.

Resistivity: The reciprocal of conductivity. The units are expressed in ohm-centimeters.

Restriction endonucleases: Enzymes that cleave DNA at sites with specific base pair sequences.

Restriction fragment length polymorphism analysis (RFLP): A method of analyzing DNA that can be used to identify individuals based on slight differences in the sequence of their DNA in specific regions of the genome.

Retention factor (in chromatography): A measure of the affinity of a solute for the stationary phase versus the mobile phase, defined as (Distance moved by solute)/(Distance moved by solvent).

Revolutions per minute: A measure of the speed of rotation in a centrifuge.

"Right-to-know" law: Law that regulates hazardous materials use in the workplace and emphasizes the responsibility of employers to provide safety information to their employees.

Rotor: The device that rotates in the centrifuge and holds the sample tubes or other containers.

Safety Data Sheet: A document that provides information about the properties, safety, handling, and storage of a chemical. These documents are required by law to be provided with every chemical that is sold in the United States.

Sample: A subset of the whole that represents the whole (e.g., a blood sample represents all the blood in an individual).

Sample chamber: The location in which the sample is placed.

Sash: Window of a hood constructed of impact-resistant materials, which can be raised and lowered by the user.

Scanning (in spectrophotometry): Process of determining the absorbance of a sample at a series of wavelengths.

Semilog paper: A type of graph paper that has a log scale on one axis and a linear scale on the other axis.

Sensitivity (for a laboratory balance): The smallest value of weight that will cause a change in the response of the balance that can be observed by the operator.

Sensitivity (of a detector or instrument): Response per amount of sample.

Separations: Methods used to separate specific substances from one another.
- **Analytical separations.** Separations whose goal is to produce information.
- **Bioseparations.** Separations that involve biological macromolecules.
- **Preparative separations.** Separations whose goal is to produce a tangible item for future use.

Serial dilutions: Dilutions made in series (each one derived from the one before) and that all have the same dilution factor. For example, a series of 1/10 dilutions of an original sample would be 1, 1/10, 1/100, 1/1,000, etc.

Serological pipette: Term that usually refers to a calibrated glass or plastic pipette that measures in the 0.1 mL to 25 mL range. These pipettes are calibrated so that the last drop is in the tip. This drop needs to be "blown out" to deliver the full volume of the pipette.

Serum: The liquid component of blood from which blood cells and most clotting factors have been removed.

SI system: A standardized system of units of measurement, adopted in 1960 by a number of international organizations, that is derived from the metric system.

Siemen (S): A unit of conductance.

Significant figure: A digit in a number that is a reliable indicator of value.

Silver/silver chloride electrode: A type of reference electrode that contains a strip of silver coated with silver chloride and immersed in an electrolyte solution of KCl and silver chloride.

Slope (of a line): Given a straight line plotted on a graph, a numerical indication of how steeply the line rises or falls. Slope is the ratio of vertical change to horizontal change between any two points on the line. Slope can be calculated by choosing any two points on the line and dividing the change in their y values by the change in their x values.

Solute: A substance that is dissolved in some other material.

Solution: A homogeneous mixture in which one or more substances is (are) dissolved in another.

Solvent: A substance that dissolves another.

Specific activity: A value that communicates the amount of a specific protein per milligram of total protein.

Specification: The defined limits within which physical, chemical, biological, and microbiological test results for a product should lie to ensure its quality.

Specificity (of an analytical method): A method's ability to measure accurately and specifically the analyte in the presence of components that may be expected to be present in the sample.

Spectrophotometer: An instrument that measures the effect of a sample on the incident light beam.

Spectrum: Electromagnetic radiation separated or distinguished according to wavelength.

Standard (relating to measurement): (1) Broadly, any concept, method or object that has been established by authority, custom, or agreement to serve as a model in the measurement of any property. (2) A physical object, the properties of which are known, with sufficient accuracy to be used to evaluate another item; a physical embodiment of a unit. (3) In chemical or biological measurements, a substance or a solution that is used to establish the response of an instrument or an assay method to the analyte. (4) A document established by consensus and approved by a recognized body that establishes rules or guidelines to make a procedure consistent among various people. (5) In spectrophotometry, a mixture including the analyte of interest dissolved in solvent and used to determine the relationship between absorbance and concentration for that analyte.

Standard curve (in spectrophotometry): A graph that shows the relationship between absorbance (on the Y-axis) and the amount or concentration of a substance (on the X-axis).

Standard deviation: A measure of the dispersion of observed values or results expressed as the positive square root of the variance. A statistical measure of variability.

Standard method: A technique to perform a measurement or an assay that is specified by an external organization to ensure consistency.

Standard microbiological practices: Rules that should always be followed to minimize the risk of working with microbes.

Standard operating procedure: A set of instructions for performing a routine method, manufacturing operation, administrative process, or maintenance operation.

Standard weight: Any weight whose mass is known with a given uncertainty.

Sterile: A situation in which no living substance is present.

Sterile technique: See **Aseptic technique**.

Sterilization: The destruction of virtually all viable organisms.

Sterilizing filter: A filter that does not release fibers and produces a filtrate containing no demonstrable microorganisms.

Stock solution: A concentrated solution that is diluted to a working concentration.

Supernatant: The liquid medium above a pellet after centrifugation.

Swinging-bucket rotor: See **Horizontal rotor**.

Tare: A feature that allows a balance to automatically subtract the weight of the weighing vessel from the total weight of the sample plus container.

Teratogen: A type of toxic chemical that is known to cause defects in a developing fetus.

Transfection: Introduction of foreign DNA into host cells; usually refers to eukaryotic host cells.

Transformation: Introduction of foreign DNA into host cells; usually refers to prokaryotic host cells.

Transilluminator (in electrophoresis): A box with a UV light source used to visualize DNA stained with ethidium bromide.

Transmittance: The ratio of the amount of light transmitted through the sample to the amount of light transmitted through the blank.

Transmitted light: Light that passes through an object.

Tris: One of the most common buffers in biotechnology laboratories. Tris buffers over the normal biological range (pH 7 to pH 9), is nontoxic to cells, and is relatively inexpensive.

Tris base: Unconjugated Tris; has a basic pH when dissolved in water.

Tris-HCl: Tris buffer that is conjugated to HCl.

True value (of a measurement): The actual value for a measurement that would be obtained in the absence of any error.

Turbid solution: One that contains numerous small particles that both absorb and scatter light.

Two-point calibration: If the response of a device is linear, calibration at two points (e.g., 0 and full scale) calibrates the device.

Ultraviolet radiation: The region of the electromagnetic spectrum from about 180 to 380 nm.

United States Pharmacopeia: (1) An organization that promotes public health by establishing and disseminating officially recognized standards for the use of medicines and other health care technologies. (2) The compendium containing drug descriptions, specifications, and standard test methods for such parameters as drug identity, strength, quality, and purity. This compendium is recognized as a legal authority by the FDA.

Validation (of an analytical procedure): A process used by the scientific community to evaluate a test method or assay. This involves determining the ability of the method to reliably obtain a desired result, the conditions under which such results can be obtained, and the limitations of the method.

Verification: Confirmation that an instrument, method, or process is functioning properly.

Vertical rotor: A rotor in which tubes are held upright in the sample compartments.

Viable organism: A living organism that can reproduce.

Visible radiation: The region of the electromagnetic spectrum from about 380 to 780 nm. Visible light of different wavelengths in this range is perceived as different colors.

Volume: The amount of space a substance occupies, defined in the SI system as Length × Length × Length = Meters^3. More commonly, the liter (dm^3) is used as the basic unit of volume.

Volumetric glassware: A term used generally to refer to accurately calibrated glassware intended for applications where high-accuracy volume measurements are required.

Voltage: The potential energy of charges that are separated from one another and attract or repulse one another.

Volts: Units of potential difference between two points in an electrical circuit.

Wavelength: The distance from the crest of one wave to the crest of the next wave.

Weigh boat: A plastic or metal container designed to hold a sample for weighing.

Weighing paper: Glassine-coated paper used to hold small samples for weighing.

Weight: The force of gravity on an object.

Wells (in electrophoresis): Depressions in a gel, formed by a comb, and into which samples are loaded.

Working distance (in microscopy): The distance between the objective lens and the specimen.

X-axis: The main horizontal line on a graph.

"X solution": A stock solution where X means how many times more concentrated the stock is than normal. A 10X solution is 10 times more concentrated than the solution is normally prepared.

Y-axis: The main vertical line on a graph.

Y-intercept: The point at which a line passes through the Y-axis, where X = 0, on a two-dimensional graph.

Appendix 3

SELECTED BIBLIOGRAPHY

RESOURCES SPECIFICALLY ASSOCIATED WITH THIS MANUAL

Seidman, Lisa A. *Basic Laboratory Calculations for Biotechnology.* 2nd ed. Boca Raton, FL: CRC Press, 2022. (Includes background on the math calculations in this manual along with problem sets with answers.)

Seidman, Lisa A., Moore, Cynthia J, and Mowery, Jeanette. *Basic Laboratory Methods for Biotechnology: Textbook and Laboratory Reference.* 3rd ed. Boca Raton, FL: CRC Press, 2022. (Includes more in depth discussions of the techniques covered in this laboratory manual.)

GENERAL RESOURCES

Ballinger, Jack T., and Shugar, Gershon. *Chemical Technicians' Ready Reference Handbook.* 5th ed. New York, NY: McGraw-Hill Book Co., 2011.

Lundblad, Roger L., and Macdonald, Fiona M. ed. *The Practical Handbook of Biochemistry and Molecular Biology.* 5th ed. Boca Raton, FL: CRC Press, 2018. (Contains extensive information about biological materials including buffers for biological systems.)

The Merck Index On-line: An Encyclopedia of Chemicals, Drugs, and Biologicals. https://www.rsc.org/merck-index. (An encyclopedia with information about thousands of biologically relevant chemicals including their structures, densities, molecular weights, and solubilities.)

U.S. Pharmacopeial Convention. *The United States Pharmacopeia--National Formulary (USP–NF).* http://www.usp.org/USPNF/. (A book of public standards for medicines, dosage forms, drug substances, excipients, medical devices, and dietary supplements. The methods in the USP are accepted by the FDA.)

UNIT I: SAFETY IN THE LABORATORY

Centers for Disease Control and National Institutes of Health, U.S. Department of Health and Human Services, Public Health Service. *Biosafety in Microbiological and Biomedical Laboratories.* 5th ed. Washington, DC: CDC, 2009. https://www.cdc.gov/labs/pdf/CDC-BiosafetymicrobiologicalBiomedicalLaboratories-2009-P.pdf

Cold Spring Harbor Laboratory. *Safety Sense: A Laboratory Guide.* 2nd ed. Cold Spring Harbor, MA: CSH Laboratory Press, 2007. (This is a short basic guide for students and individuals new to the laboratory.)

Flinn Scientific Company. "The Flinn Scientific Online Catalog." Batvia Il: Flinn Scientific, Inc. http://www.flinnsci.com/. (This catalog includes extensive, practical information about chemical safety topics such as storage and disposal. The web site is also an excellent source of information.)

Furr, Keith. *CRC Handbook of Laboratory Safety.* 5th ed. Cleveland, OH: CRC, 2000. (This is a comprehensive reference for laboratory safety topics.)

Wooley, Dawn, and Byers, Karen, eds. *Biological Safety: Principles and Practices.* 5th ed. Washington, DC: ASM Press, 2017. (An excellent general reference on biological safety.)

UNIT III: METROLOGY IN THE LABORATORY AND UNIT IV: SPECTROPHOTOMETRY AND THE MEASUREMENT OF LIGHT

ASTM International. *Standard Specification for Piston or Plunger Operated Volumetric Apparatus.* ASTM E-1154. West Conshohocken, PA http://www.ASTM.org. (Recommended for anyone responsible for calibrating micropipettes. Contains the definitive procedure for pipette calibration and verification.)

ASTM International. *Standards Provide Specific, Technical Information on a Number of Metrology-related Topics. Many of the Methods in This Manual Are Ultimately Based on ASTM Standards.* Standards are available for purchase at www.ASTM.org and at some libraries.

Freedman, David, Pisani, Robert, and Purves, Roger. *Statistics*. 4th ed. New York, NY: W.W. Norton, 2007. (A general statistics book with information about standard deviation, accuracy, precision, and other important concepts in metrology.)

Kenkel, John. *Analytical Chemistry for Technicians*. 4th ed. Boca Raton, FL: CRC Press, 2019. (Provides readable explanations of instrumental analysis, pH, and other measurement-related topics.)

Mettler-Toledo. *pH Theory Guide: Theory and Practice of Laboratory pH Applications*. https://www.mt.com/us/en/home/library/guides/process-analytics/ph-measurement-guide.html

National Institute of Science and Technology (NIST) publications provide information on specific measurement topics and are available through NIST at www.NIST.gov.

UNIT V: BIOLOGICAL SOLUTIONS

For general information about molarity and percent solutions, consult any basic chemistry textbook. Manufacturers' catalogs are a valuable source of information about chemicals. The sigma-aldrich company catalog, for example, contains formula weights, compound names, and formulas for many chemicals. www.sigmaaldrich.com

Ausubel, Fred M., Brent, Roger, Kingston, Robert E., Moore, David D., Seidman, J.G., Smith, John A., and Struhl, Kevin eds. *Current Protocols in Molecular Biology*. New York, NY: John Wiley and Sons, 2010. (Contains information that helps to understand the purposes of the components of biological solutions.)

Burgess, Richard K., and Deutscher, Murray P. ed. *Guide to Protein Purification*. 2nd ed. San Diego, CA: Academic Press, 2009. (Contains information that helps to understand the purposes of the components of biological solutions.)

Perkins, John J. *Principles and Methods of Sterilization in the Health Sciences*. 2nd ed. Springfield, Il: Charles C. Thomas Publisher, 1983. (A classic in-depth reference on sterilization methods.)

Scopes, Robert K. *Protein Purification: Principles and Practice*. Robert K. Scopes, ed. 3rd ed. New York, NY: Springer-Verlag, 1994. Reprinted 2010. (Contains information that helps readers to understand the purposes of the components of biological solutions.)

Sambrook, Joseph, and Russell, David W. *Molecular Cloning: A Laboratory Manual*. 3rd ed. Cold Spring Harbor, NY: Cold Spring Harbor Laboratory Press, 2001. (Contains information that helps readers to understand the purposes of the components of biological solutions.)

Thermo Scientific. *Thermo Scientific Waterbook*. 2016. https://www.thermofisher.com/us/en/home/products-and-services/promotions/life-science/thermo-scientific-waterbook.html. (Contains information that helps readers to understand different types of water purification equipment.)

UNIT VI: ASSAYS

SPECTROPHOTOMETRIC ASSAYS

Glasel, Jay A. "Validity of Nucleic Acid Purities Monitored by 260 nm/280 nm Absorbance Ratios." *BioTechniques* 18, no. 1 (1995): 62–3.

Lundblad, Roger L., and Price, Nicholas C. "Protein Concentration Determination: The Achilles' Heel of cGMP?" *BioProcess International* (January 2004): 38–47.

Manchester, Keith L. "Use of UV Methods for Measurement of Protein and Nucleic Acid Concentrations," *BioTechniques* 20, no. 6 (1996): 968–70.

Manchester, Keith L. "Value of A_{260}/A_{280} Ratios for Measurement of Purity of Nucleic Acids." *BioTechniques* 19, no. 2 (1995): 208–10.

Mettler-Toledo. *UV Spectrophotometry Applications and Fundamentals*. https://www.mt.com/us/en/home/library/guides/laboratory-division/1/uvvis-spectrophotometry-guide-applications-fundamentals.html

Willfinger, William W., Mackey, Karol, and Chomczynski, Piotr, "Effect of pH and Ionic Strength on the Spectrophotometric Assessment of Nucleic Acid Purity." *BioTechniques* 22, no. 3 (1997): 474–81.

METHOD VALIDATION

Various organizations provide information relating to the validation of assays, analytical procedures, methods, and tests. The International Conference on Harmonisation of Technical Requirements for Registration of Pharmaceuticals for Human Use (ICH) has been helpful in harmonizing definitions and in determining the basic requirements for validation in the pharmaceutical area (although some differences still exist among organizations). A few key documents and interesting articles are listed here.

International Conference on the Harmonisation of Technical Requirements for the Registration of Pharmaceuticals for Human Use. *Validation of Analytical Methods* (Definitions and Terminology). ICH Guideline Q2A, 1994.

International Conference on the Harmonisation of Technical Requirements for the Registration of Pharmaceuticals for Human Use. *Analytical Validation–Methodology*. ICH Guideline Q2B, 1996.

Kanarek, Alex D. "Method Validation Guidelines," *BioPharm International*, 2005. http://www.biopharminternational.com/biopharm/ (A good summary of the status of method validation in the pharmaceutical industry and related terminology.)

United States Pharmacopeia. "Validation of Compendial Methods," Chapter 1225 in the *United States Pharmacopeia*. Rockville, MD: The United States Pharmacopeial Convention, Inc., 2006.

UNIT VII: BIOLOGICAL SEPARATION METHODS

Beckman Coulter, Inc. Fullerton, CA. https://www.beckman.com/resources/technologies/centrifugation/principles. (A manufacturer of centrifuges, this company has excellent publications regarding centrifuge principles, applications, and rotor safety.)
Cutler, Paul. *Protein Purification Protocols (Methods in Molecular Biology)*. 2nd ed. Totowa, NJ: Humana Press, 2003.
Miller, James M. *Chromatography: Concepts and Contrasts*. 2nd ed. Hoboken, NJ: Wiley Interscience, 2022.
Rickwood, David, and Graham, John. *Biological Centrifugation*. Berlin, Germany: Springer Verlag, 2001.
Westermeier, Reiner. *Electrophoresis in Practice: A Guide to Methods and Applications in DNA and Protein Separations*. 4th ed. Hoboken, NJ: Wiley VCH, 2005.

UNIT VIII: GROWING CELLS

The American Type Culture Collection (ATCC) Supplies Cells and Provides a Number of Resources Relating to the Culture of all Types of Cells, Including Published References, Media Formulations, and Web-based Technical Information http://www.atcc.org
Atlas, Ronald M. *Handbook of Microbiological Media*. 3rd ed. Boca Raton, FL: CRC Press, 2004. (Contains the formulations, methods of preparation, sources, and uses for several thousand different media.)
BD Diagnostic Systems. *Difco and BBL Manual of Microbiological Culture Media*. Sparks, MD: BD Diagnostic Systems. http://catalog.bd.com/bdCat/viewProduct.doCustomer?productNumber=220225 (A classic reference with information and microbial media formulations for a variety of applications.)
BioProcess International. Culture Media: A Growing Concern in Biotechnology. *Supplement to BioProcess International Magazine*, June 2005. http://www.bioprocessintl.com/journal/supplements/2005/June/ (Contains ample information about culture media from a production standpoint. Also contains basic cell biology background.)
Darling, D.C., and Morgan, S.J. *Animal Cells Culture and Media: Essential Data*. Chichester, UK: Wiley, 1994. (Small handbook with extensive technical information.)
Fresheny, R. Ian. *Culture of Animal Cells: A Manual of Basic Technique*. 7th ed. Hoboken, NJ: Wiley-Liss, 2016. (A commonly referenced manual of mammalian cell culture techniques with extensive information for both beginners and more experienced culturists.)
Martin, Bernice M. *Tissue Culture Techniques*. Boston, MA: Birkhäuser Press, 1994. (A handy student-oriented introduction to the basic principles of cell culture.)
Sigma-Aldrich. *Fundamental Techniques in Cell Culture: Laboratory Handbook*. 3rd ed. https://www.sigmaaldrich.com/deepweb/assets/sigmaaldrich/marketing/global/documents/425/663/fundamental-techniques-in-cell-culture.pdf
Thermo Fisher Scientific. *Introduction to Cell Culture*. https://www.thermofisher.com/us/en/home/references/gibco-cell-culture-basics/introduction-to-cell-culture.html
The following websites have photographs and text that will help you gain an appreciation of the different cell types you may encounter in cell culture laboratories.
Molecular Expressions, http://micro.magnet.fsu.edu/primer/techniques/fluorescence/gallery/cells/index.html
Nikon Microscopy U, http://www.microscopyu.com/galleries/fluorescence/cells.html

Appendix 4

BRIEF METRIC REVIEW

To complete the laboratory exercises and classroom activities in this manual, it is necessary to make measurements and perform calculations using the metric system. The metric system uses units such as meters, grams, and liters (Table A4.1). As you know, the United States uses a different system of measurement that includes units such as miles, pounds, gallons, inches, and feet. If you are like most people in the U.S., using the metric system will at first seem like trying to use a foreign language. The information in this appendix will hopefully help you to use this system in your laboratory work.

The units in the U.S. system are not related to one another in a systematic fashion. For example, for length, 12 in = 1 ft; 3 ft = 1 yd. There is no pattern to the relationships. In contrast, in the metric system, there is only one basic unit to measure each property. The basic unit is then modified systematically by the addition of prefixes (Table A4.2). For example, the prefix *milli-* means 1/1000 so a millimeter is a meter/1000. Similarly, the unit of volume is a liter, therefore, a milliliter (mL) is a liter/1000. The same prefixes are used to modify the meaning of each of the basic units, whether it be a liter, a gram, or an ampere.

It is important to be familiar with conversions between common units. For example, biologists often need to convert between liters (L) and milliliters (mL) or between milliliters and microliters (µL). For this, you need to know that 1 L = 1000 mL and that 1 L = 1,000,000 µL. To use micropipettes in molecular biology, you must know that 1 mL = 1000 µL. The conversions in Tables A4.3 and A4.4 are very commonly used, and you should practice these conversions until you have them memorized and can think of them quickly. Other unit conversions will be used less frequently and may be looked up as needed. To practice conversions using example problems and calculations, refer to the book *Basic Laboratory Calculations for Biotechnology*, as referenced in the Bibliography, Appendix 3.

- A microliter is 1/1,000,000 liters and is normally abbreviated µL, but sometimes biologist refer to 1 µL as 1 "lambda" (λ).
- The terms "cc" and "cm³" (in reference to volume) are the same as a mL. Both stand for "cubic centimeter," the space occupied by a mL.
- Micrograms may be abbreviated as µg or mcg where "mc" stands for "micro." The abbreviation mcg is often used when drug dosages are reported.

TABLE A4.1   A Selected List of Basic Units in the Metric System

Property Measured	Unit of Measurement	Abbreviation
Length	meter	m
Mass	gram	g
Amount of substance	mole	mol
Electrical current	ampere	A
Volume	liter	L

TABLE A4.2   Prefixes Used in the Metric System

Decimal	Prefix	Symbol	Power of 10
1,000,000,000,000,000	peta-	P	10^{15}
1,000,000,000,000	tera-	T	10^{12}
1,000,000,000	giga-	G	10^{9}
1,000,000	mega-	M	10^{6}
1,000	kilo-	k	10^{3}
1	Basic unit, no prefix		
0.1	deci-	d	10^{-1}
0.01	centi-	c	10^{-2}
0.001	milli-	m	10^{-3}
0.000001	micro-	μ	10^{-6}
0.000000001	nano-	n	10^{-9}
0.000000000001	pico-	p	10^{-12}
0.000000000000001	femto-	f	10^{-15}

TABLE A4.3   Commonly Used Conversions to Memorize

Kilo- (k) 10^3	Property	Basic Unit	Milli- (m) 10^{-3}	Micro- (μ) 10^{-6}	Nano- (n) 10^{-9}
	Volume	liter (L)	1 L = 1000 mL	1 L = 1,000,000 μL	1 L = 1,000,000,000 nL
1000 g = 1 kg	Weight	gram (g)	1 g = 1000 mg	1 g = 1,000,000 μg	1 g = 1,000,000,000 ng
1000 m = 1 km	Length	meter (m)	1 m = 1000 mm	1 m = 1,000,000 μm	1 m = 1,000,000,000 nm
NA	Amount of material	mole (mol)	1 mol = 1000 mmol	1 mol = 1,000,000 μmol	1 mol = 1,000,000,000 nmol

TABLE A4.4   Commonly Used Conversions to Memorize

Kilo- (k) 10^3	Property	Basic Unit	Milli- (m) 10^{-3}	Micro- (μ) 10^{-6}	Nano- (n) 10^{-9}
	Volume	liter (L)	1 mL = 10^{-3} L	1 μL = 10^{-6} L	1 nL = 10^{-9} L
1 g = 10^{-3} kg	Weight	gram (g)	1 mg = 10^{-3} g	1 μg = 10^{-6} g	1 ng = 10^{-9} g
1 m = 10^{-3} km	Length	meter (m)	1 mm = 10^{-3} m	1 μm = 10^{-6} m	1 nm = 10^{-9} m
NA	Amount of material	mole (mol)	1 mmol = 10^{-3} mol	1 μmol = 10^{-6} mol	1 nmol = 10^{-9} mol

Notes: Although not used in this manual, biologists sometimes use terms that are not part of the "regular" metric system or that are not as common in other disciplines:

Appendix 5

CALCULATING STANDARD DEVIATION

Standard deviation is a *calculated value that describes how much the values in a data set vary from one another.* Consider this simple example. Assume the data below are lengths (in millimeters) of eight insects:

4   5   6   7   7   7   9   11

The mean for these data is 7 mm. An intuitive way to calculate how much the data are dispersed is to take each data point, one at a time, and see how far it is from the mean (see Table A5.1). *The distance of a data point from the mean is its* **deviation**. There is a deviation for each data point; sometimes the deviation is positive, sometimes negative, and sometimes the deviation is zero.

It is useful to summarize all the deviations with a single value. An obvious approach would be to calculate the average of the deviations. However, for any data set, the sum of the deviations from the mean is always zero, as shown in Table A5.1. Therefore, mathematicians devised the approach of squaring each individual deviation value to result in all nonnegative numbers. *The squared deviations can be added together to get the* **total squared deviation**, also called the **sum of squares**. In this example, the sum of squares is 34 mm² (see Table A5.2). The sum of squares is always a nonnegative number that indicates how much the values in a data set deviate from the mean.

The **population variance** *is the total squared deviation divided by the number of values* (see Formula A5.1). In this example, the sum of squares is 34 mm², n is 8, and the population variance is as follows:

$$\text{variance} = \frac{34 \text{ mm}^2}{8} = 4.25 \text{ mm}^2$$

The standard deviation is calculated by taking the square root of the variance:

$$\text{standard deviation} = \sqrt{\text{variance}}$$

In this example:

$$\text{standard deviation} = \sqrt{4.25 \text{mm}^2} \approx 2.06 \text{ mm}$$

Note that the standard deviation has the same units as the data.

The larger the variance and the standard deviation, the more dispersed are the data.

FORMULA A5.1

CALCULATING THE VARIANCE

$$\text{Variance (of a population)} = \frac{\text{Sum of squared deviations from the mean}}{n}$$

$$= \frac{\Sigma (X - \text{Mean})^2}{n}$$

where n is the number of values in the data set.

TABLE A5.1   Calculation of Deviation from the Mean

(Value – Mean)	Deviation
1st value – Mean	= (4 – 7) = –3 mm
2nd value – Mean	= (5 – 7) = –2 mm
3rd value – Mean	= (6 – 7) = –1 mm
4th value – Mean	= (7 – 7) = 0 mm
5th value – Mean	= (7 – 7) = 0 mm
6th value – Mean	= (7 – 7) = 0 mm
7th value – Mean	= (9 – 7) = +2 mm
8th value – Mean	= (11 – 7)= +4 mm
Sum	= 0 mm

TABLE A5.2   Calculation of the Sum of Squares

(Deviation)	(Deviation^2)
–3	9 mm^2
–2	4 mm^2
–1	1 mm^2
0	0 mm^2
0	0 mm^2
0	0 mm^2
+2	4 mm^2
+4	16 mm^2
Total squared deviation (sum of squares)	= 34 mm^2

The preceding discussion skipped over a significant detail regarding the calculation of the variance and standard deviation. Statisticians distinguish between the variance and standard deviation of a population and the variance and standard deviation of a sample. The variance of a population is called σ^2 (sigma squared). The variance of a sample is called S^2. The standard deviation of a population is called σ (sigma). The standard deviation of a sample is sometimes abbreviated S and sometimes SD.

The formulas used to calculate the variance and standard deviation of a sample are slightly different than those for a population. In the previous discussion, we showed how to calculate the variance and standard deviation of a population. For a sample, the denominator is n 1 rather than n. The equations for variance and standard deviation of a sample are thus shown in Formulas A5.2 and A5.3. In this manual, we are generally looking at data from a sample and so routinely use n – 1 in the denominator of the equation.

Many scientific calculators are set up to easily calculate the standard deviation. Often the operator can enter the data and then press one key to get the standard deviation for a population and a different key to get the standard deviation for a sample. (Refer to the calculator instructions.)

FORMULA A5.2

CALCULATING THE VARIANCE OF A SAMPLE

$$\text{Variance}\left(\text{of a sample}\right) = \frac{\text{Sum of squared deviations from the mean}}{n-1}$$

Alternatively, this equation can be expressed as:

$$S^2 = \frac{\sum (X - \overline{X})^2}{n - 1}$$

where n is the number of values in the data set.

FORMULA A5.3

CALCULATING THE STANDARD DEVIATION OF A SAMPLE

$$SD = \sqrt{\frac{\sum (X - \overline{X})^2}{n - 1}}$$

where n is the number of values in the data set.

Example

The heights (in cm) for a sample of seven college women are as follows:

| 162.5 | 166.7 | 155.6 | 159.7 | 163.4 | 164.2 | 160.1 |

For this sample, calculate the mean (X), the deviation (d) of each value from the mean, the squared deviation (d²), the variance (S^2), and the standard deviation (SD).

Answer

The mean is 161.74 cm.

Measured Value	d (cm)	d²(cm²)
162.5	0.757	0.573
166.7	4.96	24.6
155.6	−6.14	37.7
159.7	−2.04	4.16
163.4	1.66	2.76
164.2	2.46	6.05
160.1	−1.64	2.69
	Sum =	78.53 cm²

$S^2 = 78.53/6 \text{ cm}^2 \approx 13.09 \text{ cm}^2$ $SD = \sqrt{13.09 \text{cm}^2} = 3.618 \text{cm} \approx 3.618 \text{ cm}$

Appendix 6

EQUIPMENT, SUPPLIES, AND REAGENTS REQUIRED FOR EACH UNIT

I. EQUIPMENT

A (✓) indicates that the following equipment is needed to prepare or perform the activities and exercises in that unit. Key items will also be listed in Appendix 7.

Equipment Needed	Unit Number							
	I	II	III	IV	V	VI	VII	VIII
Access to Excel or other graphing program					✓			✓
Access to internet	✓			✓	✓		✓	
Access to word processing computer program			✓		✓			
Autoclave								✓
Balance standard weight set for student use			✓					
Balances, include at least one analytical balance/student group	✓	✓	✓	✓	✓	✓	✓	✓
Beakers (50 mL, 100 mL, 200 mL, 500 mL)	✓	✓	✓	✓	✓	✓	✓	✓
Biohazard waste container								✓
Blender						✓		
Block-building set			✓					
Bunsen burner and striker								✓
Carbon dioxide (CO_2) incubator								✓
Carboy with spigot, 10 L							✓	
Cell spreaders								✓
Ceramic spot plate							✓	
Chromatography columns (0.8 × 4 cm) (e.g., Bio-Rad Poly-Prep Chrom Col. #731-1550)							✓	
Clinical centrifuge							✓	✓
Column clamp							✓	
Complete set of Safety Data Sheets	✓	✓	✓	✓	✓	✓	✓	✓
Conductivity meter with probe (similar to Fisher # 15-077-977 traceable expanded range digital conductivity meter, https://www.fishersci.com/shop/products/fisher-scientific-traceable-expanded-range-conductivity-meter/15077977)					✓			✓
Cuvette racks				✓		✓		✓
Cuvette washer (Fisher #14-385-946A)				✓		✓		
Digital camera	✓						✓	
Dropper bottles						✓		
Electrophoresis gel box, casting tray, comb, and power supply							✓	
Erlenmeyer flasks: 250 mL, 2 L, with cotton or sponge plugs								✓
Fume hood					✓		✓	
Funnel							✓	
Gel spatula							✓	
Graduated cylinders (10 mL, 50 mL, 100 mL, 500 mL, 1L)	✓	✓	✓	✓	✓	✓	✓	✓
Handheld counters								✓
Heat-resistant gloves							✓	

(Continued)

Equipment Needed	Unit Number							
Hemacytometers and coverslips								✓
Ice bucket						✓	✓	
Incubator, 37°C								✓
Incubator/shaker, 37°C								✓
Inoculating loop								✓
Laminar flow hood or biological safety cabinet w/ germicidal lamp								✓
Magnetic stir bars	✓	✓	✓	✓	✓	✓	✓	✓
Microcentrifuge			✓				✓	
Micropipettes, 1 set/student or group	✓		✓			✓	✓	✓
Microscope, inverted phase								✓
Microscope, light								✓
Microscope camera (optional)								✓
Microwave oven							✓	
Mortar and pestle							✓	
Pasteur pipette sterilizer canister or container								✓
pH meter and pH probe, 1/student or group			✓		✓		✓	✓
Pipette aids (e.g., bulbs)	✓	✓	✓	✓	✓	✓	✓	✓
Plastic spray bottles								✓
Prepared microscope slides								✓
Protractors		✓						
Quartz cuvettes					✓	✓		✓
Refrigerator			✓					
Rulers	✓	✓	✓				✓	
Ring stand					✓		✓	
Scissors							✓	
Spectrophotometer, UV light					✓	✓		
Spectrophotometer, visible light					✓	✓		✓
Staining trays							✓	✓
Stirring hot plates (or separate stirrers and hot plates)	✓	✓	✓	✓	✓	✓	✓	✓
Stopcock or tubing clamp							✓	
Storage bottles with caps	✓	✓	✓	✓	✓	✓	✓	✓
Test-tube racks							✓	✓
Thermometers			✓				✓	
Timer						✓		
UV lamps (handheld)	✓							
UV transilluminator							✓	
Volumetric flasks, 100 mL and 500 mL						✓		
Vortex mixer					✓		✓	✓
Water bath						✓		✓
Weighing spatulas	✓	✓	✓	✓	✓	✓	✓	✓
Wide-mouthed glass jars								✓
Vacuum aspirator bottle (e.g., BelArt #199170001)								✓
Vacuum pump and vacuum tubing								✓

II. CONSUMABLE SUPPLIES

The consumable items in the following table are assumed to be available (✓) and are not always listed in Appendix 7 in the setup notes for individual activities and exercises.

Consumable Supplies Needed	I	II	III	IV	V	VI	VII	VIII
Absorbent paper (e.g., Benchkote)	✓						✓	
Aluminum foil								✓
Autoclave tape								✓
Bibulous paper								✓
Calibration buffers, pH 4, 7, and 10				✓	✓			
Cell-culture-treated culture dishes (e.g., Corning #430167)								✓
Centrifuge tubes, 15 mL and 50 mL								✓
Chinese hamster ovary cells, BSL1 (ATCC #CCL-61)								✓
Conductivity standards (e.g., Fisher, #09-328-11, 1417 microsiemen/cm)					✓			
Cuvettes, plastic or glass for visible spectrophotometry				✓		✓		✓
Cuvettes, preferably quartz, for UV spectrophotometry. Alternatively, UV compatible plastic						✓		
Disposable cups		✓						
E. coli bacteria, K-12 strain, (e.g., Carolina Biological Supply #155065)								✓
Ethanol, 70% for cleaning lab benches, in spray bottles								✓
Filter paper, (e.g., Whatman 3M)	✓						✓	
Glass cleaner (e.g., Sparkle)								✓
Glass wool							✓	
Gloves, disposable in assorted sizes, nitrile	✓	✓	✓	✓	✓	✓	✓	✓
Graph paper (semilog graph paper, Unit 8)	✓			✓		✓		✓
Ice						✓	✓	
Immersion oil, (e.g., Sigma #56822)								✓
Laboratory markers, permanent (e.g., Sharpies)	✓	✓	✓	✓	✓	✓	✓	✓
Laboratory tissue (e.g., Kimwipes)	✓	✓	✓	✓	✓	✓	✓	✓
Labeling tape	✓	✓	✓	✓	✓	✓	✓	✓
Lens paper								✓
M&M's candy				✓				
Marshmallows		✓						
Microcentrifuge tubes, 1.5 mL	✓					✓	✓	
Micropipette tips	✓		✓			✓	✓	✓
Microscope glass slides								✓
Microscope prepared slides, ideally mammalian cell slides and stained bacterial cell preparations								✓
Paper towels	✓	✓	✓	✓	✓	✓	✓	✓
Parafilm (to cover glassware)	✓	✓	✓	✓	✓	✓	✓	✓
Pasteur pipettes, without cotton plugs	✓		✓	✓	✓	✓	✓	✓
Petri dishes								✓
pH paper							✓	
Pipettes (e.g., serological pipettes), 1 mL, 5 mL, 10 mL, and 25 mL	✓	✓	✓	✓	✓	✓	✓	✓
Plastic lids from 96 well plates or plastic Petri dishes	✓							

(Continued)

Consumable Supplies Needed	Unit Number							
	I	II	III	IV	V	VI	VII	VIII
Purified water (e.g., distilled or run through a commercial water purifying system. When the manual calls for "purified water," it means the quality of distilled water or better. If necessary, house deionized water can probably be substituted in most cases, but it may be problematic and may introduce inconsistency into solutions.) Cell-culture-grade water (e.g., 18 MΩ/cm @ 25°C filtered through 0.22 μm filter) is needed for Unit VIII if working with mammalian cells.	✓	✓	✓	✓	✓	✓	✓	✓
Stericup filter unit 1000 mL, 0.22 μm, PVDF membrane (e.g., Millipore #SCGVU10RE)								✓
Steritop filter unit 1000 mL, 0.22μm, PES membrane, 45 mm neck size (e.g., Millipore #SCGPT10RE)								✓
Test tubes (various sizes)							✓	
Toothpicks, flat		✓						✓
Weigh boats and weighing paper	✓	✓	✓	✓	✓	✓	✓	✓
Wet-erase markers (e.g., Vis-à-Vis)							✓	
White chalk					✓			
Wooden tongue depressors, for stirring and as spatulas		✓			✓			

III. REAGENTS AND REAGENT COMPONENTS AND CULTURE MEDIA

A (✓) indicates that the following equipment is needed to prepare or perform the activities and exercises in that unit.

Reagent Components Needed	Unit Number							
	I	II	III	IV	V	VI	VII	VIII
Acetic acid, glacial, [CAS #64-19-7]* (e.g., Sigma #338826)				✓				
Agar, granulated (e.g., BD Difco #214530)								✓
Agarose, gel electrophoresis grade (e.g., Sigma #A9539)							✓	
Apple						✓		
Artist water-based paint (e.g., acrylic or watercolor paint)							✓	
Artist oil-based paint							✓	
Beta-galactosidase Grade VI from E. coli, [CAS # 9031-11-2] (e.g., Sigma G6008)						✓		
Beta-galactosidase Grade VIII from E. coli, [CAS # 9031-11-2] (e.g., Sigma #G5635)						✓	✓	
Bio-Rad Protein Assay, Dye Reagent Concentrate Bio-Rad 500-0006 https://www.bio-rad.com/webroot/web/pdf/lsr/literature/LIT33.pdf						✓		
Biuret test solution [CAS #108-19-0] (e.g., Flinn #B0050 or Sigma #15280)					✓			
Bleach, sodium hypochlorite [CAS #7681-52-9]								✓
Boric acid, FW 61.83 [CAS #10043-35-3] (e.g., Sigma #B6768)							✓	
Bovine serum albumin (BSA), [CAS #9048-46-8] (e.g., Sigma #A2153, Sigma #A7906, or Flinn #A0300)					✓	✓		
Bromophenol blue, FW 669.96 [CAS #115-39-9] (e.g., Sigma #114391)							✓	
Calcium chloride (CaCl), FW 110.98 [CAS #10043-52-4] (e.g., Sigma #C1016)					✓			
Calf serum, (e.g., Hyclone cosmic calf serum SH30087)								✓
Carrot						✓		
CM Sepharose, Sigma CCF-100							✓	
DayGlo (e.g., #D-282 UV-Blue Color Corp, Cleveland, Ohio)	✓							
Deoxyribonucleic acid sodium salt from herring testes, genomic [CAS #9007-49-2] (e.g., Sigma #D6898)						✓		

Reagent Components Needed	Unit Number							
	I	II	III	IV	V	VI	VII	VIII
DEAE Sepharose (e.g., Sigma #DCL-6B-100)							✓	
Dibasic sodium phosphate anhydrous (Na$_2$HPO$_4$), FW 141.96 [CAS #7558-79-4] (e.g., Sigma #S7907)			✓	✓	✓	✓	✓	✓
Dibasic sodium phosphate heptahydrate (Na$_2$HPO$_4$ 7H$_2$O), FW 268.07 [CAS #7782-85-6] (e.g., Sigma #S9390)					✓			
Dibasic sodium phosphate dihydrate (Na$_2$HPO$_4$ 2H$_2$O), FW 177.990 [CAS #10028-24-7] (e.g., Sigma #71643)						✓	✓	
Dithiothreitol (DTT), FW 154.25 [CAS # 3483-12-3] (e.g., Sigma #43817)						✓		
Egg, chicken						✓		
Ethanol, 70%, [CAS # 64-17-5] (e.g., Fisher #25-467-01)								✓
Ethidium bromide, FW 394.31 [CAS # 1239-45-8] (e.g., Sigma 46047)							✓	
Ethyl acetate, FW 88.11 [CAS # 141-78-6] (e.g., Sigma 319902) Note that ethyl-acetate is EPA U-Listed (U012). This means it must be disposed of properly. Purchase small amounts for effective inventory control. Consult local sanitary sewer regulations before disposal.							✓	
Ethylenediaminetetraacetic acid (EDTA) disodium salt dihydrate (EDTA), FW 371.24 [CAS # 6381-92-6] (e.g., Sigma #E4884)					✓		✓	
Fluorescein, FW 332.31 [CAS # 2321-07-5] (e.g., Sigma 46955)	✓							
Food coloring, liquid dye, red, blue, green, & yellow, obtain from grocery store, McCormick brand, if available		✓	✓	✓				
Glucose (dextrose), FW 180.16 [CAS # 50-99-7] (e.g., Fisher D16-500)						✓		
Glycerol, FW 92.09 [CAS # 56-81-5] (e.g., Sigma #G6279)					✓			
Glycine, FW 75.07 [CAS # 56-40-6] (e.g., Fisher #G46-500)						✓		
Gram stain kit, (e.g., Carolina Biological Supply #821050)								✓
Ham's F-12 medium, w/l-glutamine, w/o sodium bicarbonate, (e.g., Cellgro, 50-040-PB)								✓
Honey						✓		
Hydrochloric acid solution 6 N (HCl) [CAS # 7647-01-0] (e.g., Fisher SA56-500)					✓			
Iodine-potassium iodide solution (Lugol's solution), (e.g., Flinn 10027 or Sigma L6146)					✓			
Isopropanol, (2-propanol), FW 60.10 [CAS # 67-63-0]							✓	
Lambda DNA/*Hind*III marker (e.g., Promega #G1711)							✓	
Lambda DNA/*Eco*RI/*Hind*III marker (e.g., Promega #G1731)							✓	
Magnesium acetate, FW 214.45 [CAS # 16674-78-5] (e.g., Sigma #M2545)					✓		✓	
Magnesium chloride anhydrous (MgCl$_2$) FW 95.21 [CAS # 7786-30-3] (e.g., Sigma #M8266)					✓			
Magnesium chloride hexahydrate (MgCl$_2$ • 6H$_2$O), FW 203.3 [CAS # 7791-18-6] (e.g., Sigma M2670)					✓			
Magnesium sulfate anhydrous (MgSO$_4$) FW 120.37 [CAS # 7487-88-9] (e.g., Sigma #208094)				✓	✓	✓		
Magnesium sulfate heptahydrate (MgSO$_4$ 7H$_2$O), FW 246.47 [CAS # 10034-99-8] (e.g., Sigma #230391)				✓		✓		
Methylene blue, FW 373.90 [CAS # 7220-79-3] (e.g., Sigma #M9140)							✓	
Monobasic potassium phosphate anhydrous (K H$_2$PO$_4$) FW 136.09 [CAS # 7778-77-0]								✓
Monobasic sodium phosphate anhydrous (NaH$_2$PO$_4$), FW 119.98 [CAS # 7558-80-7] (e.g., Sigma #S8282)				✓		✓	✓	

(Continued)

Reagent Components Needed	Unit Number							
	I	II	III	IV	V	VI	VII	VIII
Monobasic sodium phosphate monohydrate (NaH$_2$PO$_4$ H$_2$O), FW 137.99 [CAS # 10049-21-5] (e.g., Sigma S9638)			✓		✓	✓	✓	
Niacin reference standard (nicotinic acid), meets USP testing standards, FW 123.11, [CAS # 59-67-6] (e.g., Sigma #N5410)						✓		
Niacin test sample, FW 123.11 [CAS # 59-67-6] (e.g., Sigma #N4126)						✓		
Niacin test sample, drug store niacin supplement tablets						✓		
o-Nitrophenol or 2-Nitrophenol (ONP), FW 139.11 [CAS # 88-75-5] (e.g., Sigma #N19702)				✓				
2-Nitrophenyl β-D-galactopyranside or o-Nitrophenyl β-D-galactopyranside (ONPG), FW 301.25 [CAS # 369-07-03] (e.g., Sigma #N1127)						✓	✓	
Onion						✓		
Orange G, FW 452.37 [CAS # 1936-15-8] (e.g., Sigma #01625)							✓	
pBR322 plasmid DNA, 10 µg (e.g., Promega #D1511)							✓	
Polyvinyl alcohol (PVA), [CAS # 9002-89-5] (e.g., Sigma P1763)		✓						
Potassium chloride (KCl), FW 74.55 [CAS # 7447-40-7] (e.g., Sigma #P3911)				✓	✓			✓
Potato						✓		
Restriction buffer, 10X, supplied with restriction enzyme							✓	
Restriction enzyme, either *Eco*RI, *Hind*III or *Bam*HI, (e.g., Promega)							✓	
Rhodamine B, FW 479.01 [CAS # 81-88-9] (e.g., Sigma #R6626)	✓							
Risk Reactor e.g., PXT-07-Invisible Blue (www.riskreactor.com)	✓							
Roccal-D Plus disinfectant (e.g., Pfizer #1487)								✓
Safranin O, FW 350.84 [CAS # 477-73-6] (e.g., Sigma #S2255)							✓	
Salad oil						✓		
Sodium bicarbonate (NaHCO$_3$), FW 84.01 [CAS # 144-55-8] (e.g., Sigma #S6014)								✓
Sodium carbonate anhydrous (Na$_2$CO$_3$) FW 105.99 [CAS # 497-19-8] (e.g., Sigma #S7795)				✓		✓		
Sodium chloride (NaCl) FW 58.44 [CAS # 7647-14-5] (e.g., Sigma #S9888)					✓		✓	✓
Sodium dodecyl sulfate (SDS), also known as lauryl sulfate or sodium lauryl sulfate, FW 288.38 [CAS # 151-21-3] (e.g., Sigma #436143)						✓		
Sodium hydroxide (NaOH), FW 40.00 [CAS # 1310-73-2] (e.g., Sigma #221465)					✓		✓	✓
Sodium tetraborate decahydrate (Na$_2$B$_4$O$_7$ 10H$_2$O) FW 201.22 (Borax) [CAS # 1303-96-4] (e.g., Sigma #S9640)		✓						
Starch from potato, soluble [CAS # 9005-25-8] (e.g., Flinn #S0122 or Sigma #S2004)						✓		
Sucrose, FW342.30 [CAS # 57-50-1] (e.g., Fisher #S5-500)						✓	✓	
Tide, powder detergent	✓							
Tris base, FW 121.14 [CAS # 77-86-1] (e.g., Sigma #T4661)			✓		✓		✓	
Tris-HCl, FW 157.60 [CAS # 1185-53-1] (e.g., Sigma #T3253)			✓				✓	
Trypan blue solution 0.4%, sterile filtered and cell culture tested, FW 960.81 [CAS # 72-57-1] (e.g., Sigma #T8154)								✓
Trypsin-EDTA 1X (0.5 g porcine trypsin + 0.2 g EDTA•4Na per liter of Hank's balanced salt solution w/phenol red, sterile filtered, cell culture tested) (e.g., Sigma #T3924 or Invitrogen #25300-054)								✓
Tryptone								✓
Xylene cyanol, FW 538.61 [CAS # 2650-17-1] (e.g., Sigma #335940)							✓	
Yeast extract, (e.g., BD BBL #211929)								✓

Chemical Abstract Services (CAS), a division of the American Chemical Society, assigns these CAS registry numbers to every chemical element, compound, polymer, biological sequence, mixture, and alloy identified in the literature. This will help you find a specific chemical if the manufacturer and/or catalog number noted in this table is no longer available.

Appendix 7

RECIPES AND PREPARATION NOTES

UNIT I SAFETY

CLASSROOM ACTIVITY 3

Equipment	Per Student or Group	Consumables	Per Student or Group
Rulers	1	Graph paper	2 sheets
		Pens or markers in various colors	5 or 6 to share

CLASSROOM ACTIVITY 5

Consumables	Per Student or Group
Boxes of gloves in various sizes; S, M, and L	1 box of each type available

LABORATORY EXERCISE 1

Equipment	Per Student or Group	Consumables	Per Student or Group
UV lamps	1 for each group, ideally	Fluorescein (Save concentrated solutions for Lab Ex. 2.)	0.5 g
Glassware	16 containers/group Reagent bottles are ideal because they can be capped. Beakers, flasks, or large test tubes with caps will also work.	Rhodamine B (Save concentrated solutions for Lab Ex. 2.)	0.5 g
Lab coat and UV-safe safety glasses	Each person	DayGlo D-282 UV-Blue Dye or Risk Reactor PXT-07-Invisible Blue Powder	0.5 g
SDS for all chemicals	Copy for each person or can be shared	Tide powder detergent	0.5 g
Balance	1	Lab bench paper, preferably white	1 sheet
		3M filter paper 4″ × 4″	Several

LABORATORY EXERCISE 2

Equipment	Per Student or Group	Consumables	Per Student or Group
Handheld UV lamps	Ideally 1 per group	Serological pipettes, Pasteur pipettes, and pipette aids	Bag to share
Digital camera; do not use flash	Optional	Absorbent light-colored paper	
Micropipettes; optional depending on students' skills		Any fluorescent solution from Laboratory Exercise 1. More-concentrated stocks are preferable to the dilutions.	
		Microcentrifuge tubes	
		Micropipette tips	

UNIT II DOCUMENTATION

LABORATORY EXERCISE 3

Consumables	Per Student or Group
Lab notebook	1
Disposable cups and wooden tongue depressors for stirring	Several/student
Food coloring, optional	A few drops
4% polyvinyl alcohol (PVA)	50–10000 mL/student
5% sodium tetraborate decahydrate (Borax)	20–50 mL/student

RECIPES

4% PVA	5% Borax
40 g of PVA	5 g Borax
900 mL purified water between 80°C–90°C	90 mL purified water
Slowly dissolve PVA in hot, purified water with constant stirring; do not allow to boil.	Dissolve Borax in purified water.
	BTV 100 mL.
BTV 1 L.	

CLASSROOM ACTIVITY 8

Equipment		Consumables	
Protractors (to determine angles between molecules, if desired)	2 or 3 to be shared	Model building materials such as marshmallows, both small and large	At least 40 per student
Rulers	2 or 3 to be shared	Colored markers	Several students can share 5 or 6 markers
		Flat toothpicks	Several students can share a box

UNIT III METROLOGY

CLASSROOM ACTIVITY 9

Equipment	Per Group	Consumables	Per Group
Block-building set, ideally including some that are rounded and some that are triangular; include rulers to act as beams	Assorted blocks/ group	M&M's candy	Several M&M's per group
Quarter coin	1	Weighing dishes	2 per group

LABORATORY EXERCISE 4

1. Place thermometers in a beaker of water. Label each thermometer with a number, using tape. Immerse the thermometers to their immersion line, the black line toward the lower part of the thermometer. Allow them to equilibrate to room temperature.
2. Place items on the balances in the classroom. Turn on and adjust the balances so they are ready to be read.
3. Fill graduated cylinders of various sizes with varying amounts of colored water.
4. Place various items next to rulers to be measured.

LABORATORY EXERCISE 5

Equipment	Per Student or Group	Consumables	Per Student or Group
Balances	As many different types as available; ideally, some analytical and some not; both mechanical and electronic	Unknowns with answer key	3 in weight ranges to match several different balances in the lab
100 g standard weight and/ or standard weight set	1 per group, ideally		

LABORATORY EXERCISE 6

Equipment	Per Student or Group
Balances	1
Standard weight sets	1 set
Four items that are about one-fourth the capacity of the balance to be used; rubber stoppers and corks have worked; also see Figure 3.14	1 set

LABORATORY EXERCISE 7

Equipment	Per Student or Group	Consumables	Per Student or Group
Micropipettes with tips	1 set ideally	Part A: Food coloring (100 mL each) with the following labels: "Simulated 1 M Tris" "Simulated 1 M KCl" "Simulated DNA Sample" "Simulated PCR Mix" "Simulated MgCl$_2$"	Can be shared
Instructions that come with micropipettes	1 set	Unknowns of colored water One less than 20 μL One 20–100 μL One >100 μL Answer key	For each student
A variety of pipette aids and measuring pipettes	To be shared		

LABORATORY EXERCISE 8

Equipment	Per Student or Group	Consumables
Micropipettes	1 set ideally	Micropipette tips
Analytical balance	1	

LABORATORY EXERCISE 9

Equipment	Per Student or Group	Consumables	Per Student or Group
pH meters; ATC probes are helpful but not necessary. If ATC probes are not used, provide thermometers.	As many as possible of different styles should be available on a rotation basis.	Part C: Step C.1 Tris buffer at high and low pH	≈ 300 mL
Calibration buffers: pH 4, 7, and 10	To be shared	Part C: Step C.3 Refrigerated Tris buffer 0.1 M, pH between 7 and 8 when at room temperature at room temperature (RT)	≈ 300 mL
At least 4 standards, ideally others in addition to those with pH 4, 7, and 10	To be shared	Part C: Step C.4 1 M Tris, pH between 7 and 9 (RT)	≈ 300 mL
		Part C: Step C.4 1 M phosphate buffer, pH between 7 and 9 (RT)	≈ 300 mL
		Part D: Two labeled bottles of any buffer, any pH	≈ 300 mL All groups share the same bottles.

RECIPES

Part C: Step C.1	Part C: Step C.1
<u>0.1 M Tris Buffer ≈ pH 10 at RT</u>	<u>0.1 M Tris Buffer ≈ pH 4 at RT</u>
12.1 g Tris base	15.76 g Tris-HCl
Dissolve in 900 mL purified water.	Dissolve in 900 mL purified water.
Check pH; solution should be ≈ pH 10.9.	Check pH; solution should be ≈ pH 4.0.
BTV 1 L.	BTV 1 L.
Part C: Step C.3	Part C: Step C.4
<u>0.1 M Tris Buffer, pH 7–8 at RT</u>	<u>1 M Tris Buffer, pH 7–9 at RT</u>
12.11 g Tris base	121.14 g Tris base
Dissolve in 800 mL purified water.	Dissolve in 800 mL purified water.
Adjust pH to 7–8 with HCl	Adjust pH to 7–9 with HCl.
BTV 1 L.	BTV 1 L.
Store at 4°C.	

Part C: Step C.4
<u>1 M Phosphate Buffer ≈ pH 7 at RT</u>
54 g monobasic sodium phosphate monohydrate ($NaH_2PO_4\ H_2O$)
87 g dibasic sodium phosphate anhydrous (Na_2HPO_4)
Dissolve in 1 L purified water.
pH should be ≈ 7.

UNIT IV SPECTROPHOTOMETRY

LABORATORY EXERCISES 10, 11, 12

The materials and preparations for Laboratory Exercises 10, 11, and 12 are in this single table because it is convenient to prepare all the food coloring materials at the same time.

Equipment	Per Student or Group	Consumables	Per Student or Group
Visible light spectrophotometer, (e.g., Spec-20 instrument and a scanning spectrophotometer)	1	Piece of white chalk 15–20 mm long	1
Cuvettes for each spectrophotometer	At least 1	Laboratory Exercise 10: McCormick brand food coloring: Green diluted 1/1500 Red diluted 1/1500 Yellow diluted 1/1500	50 mL of each color
Vortex mixer, optional	1 or 2 for class	Laboratory Exercise 11: McCormick brand food coloring (We read once that the initial concentration is 25,000 ppm. We have not been able to confirm this, but use it as a starting point.) Red food coloring standards: 1 ppm, 5 ppm, 7 ppm, 10 ppm, 15 ppm	250 mL of each standard
Access to Excel or other graphing program, optional	1 computer	Laboratory Exercise 12: McCormick brand food coloring Stock of red food coloring, not diluted	2.5 mL
Cuvette washer with cuvette washing solution, optional	1 for class	Laboratory Exercise 12: Unknowns of red food coloring at concentrations between 1 and 100 ppm in 5 mL aliquots Answer key for unknowns	2 samples of unknowns
		Graph paper or access to graphing program	

RECIPES

Laboratory Exercise 10 Food Coloring Solutions, 1/1500 Dilution Green, red, and yellow each: 0.33 mL liquid food coloring 500 mL purified water	Laboratory Exercise 11 Red Food Coloring Standards,1–15 ppm Assume that original bottle of liquid red food coloring is 25,000 ppm. Make 250 mL of red food coloring standards at the following concentrations: 1 ppm, 5 ppm, 7 ppm, 10 ppm, and 15 ppm When diluting food coloring, use the $C_1V_1 = C_2V_2$ equation. Example: $C_1V_1 = C_2V_2$ $(25{,}000 \text{ ppm})(?) = (1 \text{ ppm})(250 \text{ mL})$ $? = 0.01 \text{ mL} = 10\ \mu L$
Laboratory Exercise 12 Red Food Coloring Unknowns Prepare red food coloring stock solution at 1000 ppm from the original 25,000 ppm bottle. Prepare unknowns in the 1–50 ppm range with purified water. Aliquot 2 unknowns of 5 mL per student. It is also possible to use unknowns with a concentration greater than 50 ppm, thus requiring students to prepare dilutions.	
Cuvette Washing Solution Use cuvette washing solution for situations where detergent and water is not sufficient. To make 1 L: 200 mL purified water 750 mL 70% ethanol 50 mL acetic acid, glacial Handle with caution; the cuvette washing solution is acidic. Place liquid in squirt bottles for easier handling. Do not allow acid to remain in cuvettes more than 1 hour. Rinse thoroughly with purified water.	

LABORATORY EXERCISE 13

Equipment	Per Student or Group	Consumables	Per Student or Group
Spectrophotometer, any type	1	10 mM o-nitrophenol (ONP)	3 mL
Cuvettes for each spectrophotometer	At least 1	1 M sodium carbonate	10 mL
		Z Buffer	40 mL

RECIPES

10 mM ONP	Z Buffer	1 M Sodium Carbonate
0.2782 g ONP BTV 200 mL with Z buffer.	4.26 g Na_2HPO_4 2.40 g NaH_2PO_4 0.373 g KCl 0.06 g $MgSO_4$ or 0.0882 g $MgSO_4 \bullet 7H_2O$ BTV 500 mL with purified water. Check the pH and adjust to pH 7.0 if needed.	53.00 g Na_2CO_3 BTV 500 mL.

UNIT V BIOLOGICAL SOLUTIONS

CLASSROOM ACTIVITIES 11, 12

Equipment	Per Student or Group	Consumables
Computer with access to the internet and/or chemical catalogs to review	1	Reagent containers for the following: bovine serum albumin (BSA), NaCl, $CaCl_2$, and $MgCl_2$

LABORATORY EXERCISE 14

Equipment	Per Student or Group	Consumables	Per Student or Group
Balance	1	BSA	10 mg
Conductivity meter		NaCl	4.55 g
1 conductivity meter with standards recommended by the manufacturer of your meter.		$CaCl_2$	5.6 g
We like to use a single meter for the whole class to ensure that any differences between groups is due to differences in the solutions and not the meters. However, using a single meter can slow the process.		$MgCl_2$	1.5 g

LABORATORY EXERCISE 15

Equipment	Per Student or Group	Consumables	Per Student or Group
Balance	1	NaCl Ideally, provide several containers to avoid delays or put a container at each balance.	51 g
Conductivity meter	1 conductivity meter for the class (see note in Exercise 14)	Conductivity standards	1 bottle
Glassware This exercise requires a lot of glassware, but a mix of beakers, flasks, and other vessels can be used.	17		

LABORATORY EXERCISE 16

Equipment	Per Student or Group	Consumables	Per Student or Group
Fume hood	1 fume hood	6 M HCl	50 mL
Usual equipment for preparing solutions (e.g., glassware, graduated cylinders, pH meters with standards, conductivity meters with standards, balances with weighing supplies)		Gloves	One box of each size per group
		NaOH	8 g
		Tris base	15 g
		Monobasic sodium phosphate monohydrate	27.6 g
		Dibasic sodium phosphate anhydrous or dibasic sodium phosphate heptahydrate	28.4 g or 53.6 g

LABORATORY EXERCISE 17

Equipment	Consumables	Per Student or Group
Usual equipment for preparing solutions (e.g., glassware, graduated cylinders, pH meters with standards, conductivity meters with standards, balances with weighing supplies)	NaCl	1.2 g
	pH standards	pH 4, 7, 10
	Magnesium acetate (MgAc)	0.214 g
	Glycerol	5 mL
	Tris base	2.5 g
	HCl and NaOH for adjusting pH	As prepared in Laboratory Exercise 16

LABORATORY EXERCISE 18

Equipment	Consumables	
Usual equipment for preparing solutions (e.g., glassware, graduated cylinders, pH meters with standards, conductivity meters with standards, balances with weighing supplies)	EDTA disodium salt dihydrate	19 g
	Tris base	13 g
	NaOH pellets	10–15
	HCl and NaOH for adjusting pH	As prepared in Laboratory Exercise 16

LABORATORY EXERCISE 19

Equipment	Consumables	Per Class
Usual equipment for preparing solutions (e.g., glassware, graduated cylinders, pH meters with standards, conductivity meters with standards, balances with weighing supplies)	Acid and base for adjusting pH	As prepared in Laboratory Exercise 16
	Tris base	1 container
	NaCl	1 container
	$MgCl_2$	1 container
	KCl	1 container
	$CaCl_2$	1 container

LABORATORY EXERCISE 20

Equipment	Consumables	Per Student Or Group
Usual equipment for preparing solutions (e.g., glassware, graduated cylinders, pH meters with standards, conductivity meters with standards, balances with weighing supplies)	Monobasic sodium phosphate monohydrate	27.6 g
	Dibasic sodium phosphate anhydrous or dibasic sodium phosphate heptahydrate	28.4 g or 53.6 g

UNIT VI ASSAYS

LABORATORY EXERCISE 21

Equipment	Per Student or Group	Consumables	Per Student or Group
Dropper bottles	Several	Iodine-potassium iodide solution	10–20 mL
Blender, optional	1	Biuret test solution	10–20 mL
Mortar and pestle	1	Starch extract, low concentration	10–20 mL
Micropipettes, optional		Starch extract, high concentration	10–20 mL
		Protein extract, low concentration	10–20 mL
		Protein extract, high concentration	10–20 mL
		Onion juice	5 mL
		Potato juice	5 mL
		Sucrose, dissolved	5 mL
		Glucose, dissolved	5 mL
		Carrot juice	5 mL
		Apple juice	5 mL
		Egg albumin, dissolved	5 mL
		Honey	5 mL
		Amino acids, dissolved	5 mL
		BSA, dissolved	5 mL
		Salad oil	5 mL

RECIPES

Starch Extract, low concentration (0.1% starch) 0.1 g starch, soluble 100 mL of 80°C purified water Add starch to warm water. Stir until dissolved.	Starch Extract, high concentration (1.5% starch) 1.5 g starch, soluble 100 mL of 80°C purified water Add starch to warm water. Stir until dissolved.	Protein Extract, low concentration (0.1% protein) 0.1 g BSA 100 mL purified water Add BSA to purified water. Stir until dissolved.
Protein Extract, high-concentration (1.5% protein) 1.5 g BSA (Fraction V) 100 mL purified water Add BSA to water. Stir until dissolved.	Onion Juice 80 g onion, chopped 40 mL purified water Grind onion with water using mortar and pestle.	Potato Juice 80 g potato chunks without skin 40 mL purified water Grind potato chunks in mortar and pestle with water.
Sucrose, dissolved 0.5 g sucrose 50 mL purified water. Dissolve sucrose in water.	Glucose, dissolved 0.5 g glucose 50 mL purified water Dissolve glucose in water.	Carrot Juice 40 g shredded carrot 40 mL purified water Grind shredded carrot with water using a mortar and pestle.
Apple Juice 20 g peeled apple chunks 40 mL purified water Grind apple chunks with water using a mortar and pestle.	Egg Albumin, dissolved 1 egg white 100 mL purified water Dissolve egg white in water.	Honey in Water 4 g honey 40 mL purified water Dissolve honey in water.
Amino acid, dissolved 2 g amino acid (e.g., glycine) 40 mL purified water Dissolve amino acid in water.	BSA, dissolved 0.75 g BSA 50 mL purified water Add BSA to water. Stir until dissolved.	Salad Oil Solution 4 g salad oil 40 mL purified water Add oil to water. Shake to mix.

LABORATORY EXERCISE 22

Equipment	Per Student or Group	Consumables	Per Student or Group
UV spectrophotometer	1	0.1 mg/mL DNA solution	5 mL per group
Quartz cuvettes or UV- compatible plastic cuvettes with rack If you use plastic, check to be sure the cuvettes do not absorb light in the 260–280 nm range.	2	DNA samples of unknowns 5–75 µg/mL It is also possible to use unknowns with a concentration >75 µg/mL, thus requiring students to perform a dilution.	1 or more

RECIPES

0.1 mg/mL DNA Solution	DNA Samples of Unknowns
10 mg DNA, genomic 100 mL purified water Dissolve DNA in water. Prepare aliquots for each student or group.	For each student or group, prepare one or more samples of unknowns in the range of 5–75 µg/mL and possibly one or more with a concentration >75 µg/mL. Use the 0.1 mg/mL DNA solution to prepare unknowns.

LABORATORY EXERCISE 23

Equipment	Per Student or Group	Consumables	Per Student or Group
UV spectrophotometer	1	100 µg/mL DNA solution	10 mL
Quartz cuvettes or UV-compatible plastic cuvettes; cuvette rack	2	1 mg/mL protein solution	10 mL

RECIPES

100 µg/mL DNA Solution	1 mg/mL Protein Solution
10 mg DNA, genomic 100 mL purified water Dissolve DNA.	100 mg BSA 100 mL purified water Dissolve protein.

LABORATORY EXERCISE 24

Equipment	Per Student or Group	Consumables	Per Student or Group
Visible spectrophotometer	1	Bio-Rad protein assay, dye reagent concentrate	14 mL
Cuvettes and rack		BSA stock solution, 1 mg/mL	1 mL
		BSA samples of unknowns	2 unknowns, 10 mL of each

RECIPES

1 mg/mL BSA Stock Solution	BSA Samples of Unknowns
30 mg BSA 30 mL purified water Dissolve BSA.	Prepare BSA samples of unknowns in the range of 50–100 µg/mL from the 1 mg/mL BSA stock solution.

We use the Bio-Rad protein assay dye reagent concentrate with good results. An instruction booklet is available from the company that contains important information regarding interferences, standards, and frequently asked questions. There are other protein assay kits available, in addition to the Bio-Rad kit. While we have not tried these other products, we assume that they would work equally well and that the instructions would be similar.

LABORATORY EXERCISE 25

Equipment	Consumables	Per Student or Group
Spectrophotometer with cuvettes and rack; UV capability not required	Bio-Rad protein assay, dye reagent concentrate	10 mL This is an open-ended lab, and amount depends on the number of standards and samples assayed.
	1 mg/mL BSA protein standard	1-2 mL
	Sodium dodecyl sulfate (SDS)	4 g
	0.1 M sodium phosphate buffer	varies
	β-galactosidase	1 k unit vial If doing Laboratory Exercises 26 and 27, order a 3 k or 5 k unit vial.
	β-galactosidase 1 mg/mL stock solution, keep on ice	0.2 mL

RECIPES

1 mg/mL BSA Protein Standard 30 mg BSA 30 mL purified water Dissolve BSA. Prepare aliquots.
1% SDS (SDS is a fine powder and may be difficult for students to weigh; therefore, we suggest giving students a 1% stock solution from which to make their assay dilutions.) 4 g SDS 400 mL purified water Place the SDS in a beaker and add the purified water with gentle stirring to dissolve the SDS. Rapid stirring or agitation may cause the SDS to foam.
0.1 M Sodium Phosphate Buffer, pH 7.5 Refer to Laboratory Exercise 16, Box 5.4, for recipe.

LABORATORY EXERCISE 26

Equipment	Per Student or Group	Consumables	Per Student or Group
37°C water bath	1	β-galactosidase 1 mg/mL solution in 0.1 M sodium phosphate buffer Keep on ice	0.1 mL
Timer	1	β-galactosidase unknown	0.1 mL
Visible spectrophotometer with cuvettes and rack	1	Z buffer	15-20 mL
Ice bucket and ice	1	0.1 M sodium phosphate buffer	
		1 M sodium carbonate (stop solution)	3-6 mL
		2-nitrophenyl β-D-galactopyranside or o-nitrophenyl β-D-galactopyranside, (ONPG substrate solution) 4 mg/mL in 0.1 M sodium phosphate buffer	2-4 mL
		0.5 M dithiothreitol (DTT)	5 mL

RECIPES

0.1 M Sodium Phosphate Buffer, pH 7.5
Refer to Laboratory Exercise 16, Box 5.4 for recipe.

1mg/mL β-Galactosidase Stock Solution in 0.1 M Sodium Phosphate Buffer, pH 7.5

β-galactosidase from Sigma (catalog number G5635) comes as a powder that is sold in "units," not by weight. The powdered enzyme preparation includes solid impurities. Depending on the purity of a particular preparation, the recipe for making a 1 mg/mL solution will vary.

Example of How to Dilute the Enzyme to 1 mg/mL

A β-galactosidase vial has the following statement:

"6.3 mg solid, 830 units/mg solid, 1050 units/mg protein"

For the purposes of preparing a stock solution, assume that all the protein in the vial is β-galactosidase, although this may not be exactly the case.

830 units/mg solid/1050 units/mg protein ≈ 0.79

so the powder preparation is about 79% protein.

6.3 mg X 79% ≈ 4.977 mg of protein

Dissolve the entire contents of the vial in 4.977 mL of 0.1 M sodium phosphate buffer, pH 7.5.

Don't try to weigh out portions of a vial; reconstitute the entire contents of the vial at one time.

Prepare aliquots of the 1 mg/mL solution in 0.1 mL volumes and store at −20°C.

Keep β-galactosidase on ice when it is used in the classroom.

β-Galactosidase Unknown

Prepare a β-galactosidase unknown from the 1 mg/mL stock solution.

Z Buffer, pH 7.0	1 M Stop Solution
4.26 g Na_2HPO_4 anhydrous	53.00 g Na_2CO_3
2.40 g NaH_2PO_4 anhydrous	BTV 500 mL.
0.373 g KCl	
0.06 g $MgSO_4$ or 0.0882 g for $MgSO_4 \bullet 7H_2O$	
BTV 495 mL.	
Adjust pH to 7.0.	
Store at 4°C for up to 2 months.	
Add 5 mL of 0.5 M DTT right before use; wear gloves.	

ONPG Substrate Solution
200 mg 2-nitrophenyl β-D-galactopyranoside (ONPG)
50 mL 0.1 M sodium phosphate buffer, pH 7.5
Dissolve ONPG in 0.1 M sodium phosphate buffer.
Prepare aliquots and store at –20°C for up to one year.
If the solution turns a slight yellow color after thawing, it should be discarded.
0.5 M DTT
0.385 g DTT
BTV 5 mL in purified water.
Prepare aliquots and store at –20°C.
DTT is less pungent and less toxic than 2-mercaptoethanol, but we still recommend that you wear gloves and use a fume hood when working with concentrated DTT. To avoid the necessity of using a balance in a fume hood to weigh the powder, purchase a 1 g vial and reconstitute the contents in 13.0 mL of purified water.

LABORATORY EXERCISE 27

Equipment	Per Student or Group	Consumables	Per Student or Group
Visible spectrophotometer with cuvettes and rack	1	β-galactosidase, grade VIII Sigma #G5635, 600–1,200 units/mg protein	1k unit vial
37°C water bath	1	β-Galactosidase grade VI Sigma #G6008, 250-600 units/mg protein	1k unit vial
Timer	1	Enzyme preparation I	350 µL
Ice bucket with ice	1	Enzyme preparation II	350 µL
Micropipettes	1 set	0.1 M sodium phosphate buffer	
		BSA stock solution, 1 mg/ml	1 mL
		Bio-Rad protein assay, dye reagent concentrate	15–20 mL
		1 M sodium carbonate (stop solution)	3–6 mL
		ONPG substrate solution 4 mg/mL in 0.1 M sodium phosphate buffer	2–4 mL
		0.5 M DTT	5 mL
		Z buffer	15–20 mL
		Gloves	
		Ice	

RECIPES

<u>Enzyme Preparation I: β-Galactosidase solution 100 μg/mL in water (Sigma #G5635)</u> **Step 1.** Reconstitute Sigma #G5635 β-galactosidase solution to 1 mg protein/mL in 0.1 M sodium phosphate buffer, pH 7.5. See Laboratory Exercise 26 recipes for an example of the calculations required. **Step 2.** Dilute β-galactosidase solution from step 1 to 100 μg/mL in purified water (e.g., add 400 μL of the solution from step 1 to 3600 μL purified water to yield 4 mL of 100 μg/mL β-galactosidase [Sigma #G5635]). <u>Enzyme Preparation II: β-galactosidase solution 100μg/mL in water (Sigma #G6008</u> **Step 1.** Reconstitute Sigma #G6008 β-galactosidase solution to 1 mg protein/mL in 0.1 M sodium phosphate buffer, pH 7.5. See Laboratory Exercise 26 recipes for an example of the calculations required. **Step 2.** Dilute β-galactosidase solution from step 1 to 100 μg/mL in purified water.	

0.1 M sodium phosphate buffer, pH 7.5	See Laboratory Exercise 16, Box 5.4.
ONPG substrate solution	See Laboratory Exercise 26 recipes.
0.5 M dithiothreitol (DTT)	See Laboratory Exercise 26 recipes.
Z buffer, pH 7.0	See Laboratory Exercise 26 recipes.
Stop solution	See Laboratory Exercise 26 recipes.
1 mg/mL BSA	See Laboratory Exercise 25 recipes.

LABORATORY EXERCISE 28

Equipment	Per Student or Group	Consumables	Per Student or Group
UV spectrophotometer	1	Niacin reference standard	200 mg
Quartz cuvettes or UV-compatible plastic cuvettes	2	Niacin test samples (e.g., niacin supplement tablets from drug store)	200 mg each
Micropipettes	1 set	Micropipette tips	
500 mL and 100 mL volumetric flasks	1 each		

RECIPES

Niacin reference standard (recipe in Laboratory Exercise 28). Niacin test samples (recipe in Laboratory Exercise 28). Sigma #N5410 is a niacin reference standard which meets USP testing standards. Sources for niacin test samples include: Sigma #N4126, niacin supplements, and vitamin B-complex supplements from the drug store or grocery store. Niacin supplement tablets are not pure nicotinic acid and may contain other fillers making them difficult to dissolve. Try grinding the tablets with a mortar and pestle; then stir or heat the mixture to help dissolve the tablets.

UNIT VII BIOLOGICAL SEPARATION METHODS

CLASSROOM ACTIVITY 13

Equipment	Per Student or Group
Access to Internet or centrifuge catalogs	1

LABORATORY EXERCISE 29

Equipment	Per Student or Group	Consumables	Per Student or Group
Fume hood	1	Absorbent paper	Enough to cover work area in hood
Ethyl-acetate- compatible test tubes	5	Artist water-based paints	2 or more colors, 2 mL of each color
Spatulas	1 or more; can be washed between samples	Artist oil-based paints	2 or more colors, 2 mL of each color
Vortex mixer	1	Paint unknown	1
		Paint standards	3 or 4

RECIPES

Paint Unknown	Paint Standards
Mix one color of water-based paint with another color of oil-based paint.	Supply students with a sample of each paint to test.

LABORATORY EXERCISE 30

Equipment	Per Student or Group	Consumables	Per Student or Group
Assorted colors (e.g., red, green, blue, and orange) Sharpie permanent markers	1 each	Whatman 3MM chromatography paper 70–75 mm × 70–75 mm (A paper cutter works well to cut multiple sheets of chromatography paper at one time.)	1 paper for each ink tested plus unknowns
Ruler and scissors	1	Isopropanol (Isopropanol may be saved and reused for the same lab.)	10 mL
Pencil	1	Unknown	1
Assorted colors Vis-à-Vis wet-erase markers	1 each		
96 well plate lids, plastic	4		

RECIPE

Unknown
Make one unknown for each student; dot one color of the wet-erase marker on top of one color dot of permanent marker. Do not use black and blue permanent marker in the unknown because they are difficult to distinguish.

LABORATORY EXERCISE 31

Equipment	Per Student or Group	Consumables	Per Student or Group
Electrophoresis gel box, casting tray, comb, and power supply	1	Agarose	0.3 g
Micropipettes	1 set	10X Tris/borate/EDTA (TBE) electrophoresis buffer	30 mL
Thermometer	1	1X TBE electrophoresis buffer	200–400 mL depending on gel box
Usual equipment for preparing solutions (e.g., glassware, graduated cylinders, pH meters with standards, balances with weighing supplies)	1	Xylene cyanol, bromophenol blue, orange G, and safranin O loading dyes	10 µL of each dye
10 L carboy with spigot	1	Dye mixture	10 µL
Microwave oven	1	Dye unknown	10 µL
Thermal gloves	1 pair	1 M Tris, pH 8.0	100 mL

RECIPES

<table>
<tr><td colspan="2">

<u>10X TBE Electrophoresis Buffer</u>

Makes 1 L

Add the following dry ingredients to 700 mL of purified water in a 2 L flask:

1 g NaOH (MW = 40.0)

108 g Tris base (MW = 121.10)

55 g boric acid (MW = 61.83)

7.4 g EDTA (disodium salt MW = 372.24)

Stir to dissolve, preferably using a magnetic stir bar and stir plate.

Add purified water to bring the volume to 1 L.

If stored TBE comes out of solution, place the flask in a water bath at 37°C–42°C with occasional stirring until all solid matter goes back into solution.

Store at room temperature indefinitely.

</td></tr>
<tr><td>

<u>1X TBE Buffer</u>

Makes 10 L

Into a spigotted carboy, add 9 L of purified water to 1 L of 10X TBE.

Stir to mix.

Store at room temperature indefinitely.

</td><td>

<u>1 M Tris, pH 8.0</u>

8.88 g Tris-HCl

5.3 g Tris base

Dissolve in 80 mL of purified water.

Check pH and adjust if needed.

BTV 100 mL.

</td></tr>
<tr><td>

<u>Loading-Dye Solvent</u>

50 g sucrose

1 mL 1 M Tris, pH 8.0

BTV 100 mL.

</td><td>

<u>Xylene Cyanol Dye, 0.25%</u>

0.025 g xylene cyanol

10 mL loading-dye solvent

</td></tr>
<tr><td>

<u>Bromophenol Blue Dye, 0.25%</u>

0.025 g bromophenol blue

10 mL loading-dye solvent

</td><td>

<u>Orange G Dye, 0.25%</u>

0.025 g orange G

10 mL loading-dye solvent

</td></tr>
<tr><td>

<u>Safranin O Dye, 0.25%</u>

0.025 g safranin O

10 mL loading-dye solvent

</td><td>

<u>Mixture of Four Dyes, 0.25%</u>

0.025 g xylene cyanol

0.025 g bromophenol blue

0.025 g orange G

0.025 g safranin O

10 mL loading-dye solvent

</td></tr>
<tr><td>

<u>Unknown Mixture, 0.25%</u>

Mix any combination of dyes, 0.025 g each, in 10 mL loading-dye solvent

</td><td>

Save extra loading-dye solvent to make xylene cyanol and bromophenol blue loading-dye for Laboratory Exercise 32.

</td></tr>
</table>

LABORATORY EXERCISE 32

Equipment	Per Student or Group	Consumables	Per Student or Group
Electrophoresis gel box, casting tray, comb, and power supply	1	Lambda DNA/*Hind*III marker Lambda DNA/*Eco* RI and *Hind*III marker	1 vial each for class
Gel spatula	1	1 X TBE buffer	200–400 mL
Micropipettes	1 set	Loading dye	10 mL batch for class
Ice bucket with ice	1	Samples	5
Microwave oven	1	Ethidium bromide stain	40–50 mL
Staining trays	1	Gloves	
UV transilluminator	1	Micropipette tips	
Camera	1	Agarose	0.5 g
Funnel	1		

RECIPES

See Laboratory Exercise 31 recipes shown previously for 1 M Tris pH 8.0, loading-dye solvent, and TBE Buffer recipes.

For this simulation, use two molecular weight markers. Promega's lambda DNA/*Hind*III marker (catalog number G1711) and lambda DNA/*Eco*RI + *Hind*III Marker (catalog number G1731) are shown in the illustration for this exercise (Figure 7.24), but any two molecular weight markers can be used. Some companies will supply 6X loading dye with the markers.

The volumes and concentrations of DNA used in this exercise have been optimized for staining with ethidium bromide. If methylene blue stain is preferred, increase the DNA concentration four to five times, keeping the volumes the same.

Review the Safety Briefing concerning ethidium bromide in Laboratory Exercise 32.

Samples For lanes 2, 4, and 5 each prepare: 500 ng of lambda DNA/*Hind*III marker (Promega #G1711 marker 500 ng = 1 μL) 1 μL loading dye BTV 6 μL with purified water. Mix and centrifuge briefly. For lanes 3 and 6 each, prepare the following: 500 ng of lambda DNA/*Eco*RI + *Hind*III marker (Promega #G1731 marker 500 ng = 1 μL) 1 μL loading dye BTV 6 μL with purified water. Mix and centrifuge briefly.	Loading Dye 0.025 g xylene cyanol 0.025 g bromophenol blue 10 mL loading-dye solvent Ethidium Bromide Stock Staining Solution, 5 mg/mL 25 mg ethidium bromide stain BTV 5 mL in purified water. Store the ethidium bromide stock staining solution in the dark; wrap the tube containing the stain in aluminum foil and label it. Review the Safety Briefing concerning ethidium bromide in Laboratory Exercise 32.
Ethidium Bromide Working Stain, 1 μg/mL 100 μL ethidium bromide stock stain solution BTV 500 mL with purified water. Store in an opaque unbreakable bottle.	Methylene Blue Staining Solution, 0.025% 0.125 g methylene blue BTV 500 mL with purified water. (Use as less hazardous alternative to ethidium bromide.)

LABORATORY EXERCISE 33

Equipment	Per Student or Group	Consumables	Per Student or Group
Electrophoresis gel box, casting tray, comb, and power supply	1	Lambda DNA/*Hind*III marker or Lambda DNA/*Eco*RI and *Hind*III marker or other marker	1 vial for class
Gel spatula	1	1X TBE buffer	200–400 mL
Micropipettes	1 set	Agarose	0.5 g
Funnel	1	Loading dye	10 mL batch for class
Microwave oven	1	Ethidium bromide stain	40–50 mL
Staining trays	1	Gloves	
UV transilluminator	1	Micropipette tips	
Camera	1	10X TBE buffer	500 mL

RECIPES

See Laboratory Exercises 31 and 32 Recipes for 1 M Tris pH 8.0, 10 X TBE buffer, loading-dye, and ethidium bromide stain recipes.

Review the Safety Briefing concerning ethidium bromide in Laboratory Exercise 32.

LABORATORY EXERCISE 34

Equipment	Per Student or Group	Consumables	Per Student or Group
Electrophoresis gel box, casting tray, comb and power supply	1	Lambda DNA/*Hind*III marker	1 μL
Gel spatula	1	Sample DNA	5 μL
Micropipettes	1 set	1X TBE buffer	200–400 mL
Funnel	1	Agarose	0.5 g
Microwave oven	1	Loading dye	10 mL batch for class
Staining trays	1	Ethidium bromide stain	40–50 mL
UV transilluminator	1	Gloves	
Camera	1	Micropipette tips	
		pBR322 plasmid DNA	1 vial for class
		*Eco*RI, *Hind*III, or *Bam*HI restriction enzyme, optional	1 vial for class

RECIPES

See Laboratory Exercises 31 and 32 recipes for 1X TBE buffer, loading-dye, and ethidium bromide stain recipes.

Review the Safety Briefing concerning ethidium bromide in Laboratory Exercise 32.

Sample DNA

The sample DNA may come from any plasmid source. Your laboratory may have plasmid DNA already on hand that could be used. If no DNA is available, purchase 10 μg of pBR322 plasmid DNA (4361 base pairs [bp]) (e.g., Promega catalog number V1511, 1 μg/μL).

Plasmid DNA exists in one of three major conformations: supercoiled, relaxed or "nicked circle," or linear. To avoid the confusion of seeing several bands in a lane known to contain only uncut plasmid DNA, cut the DNA sample with a restriction enzyme so one uniform linear band will be seen on the gel. The plasmid pBR322 contains one restriction site for each of the following restriction enzymes, *Eco*RI, *Hind*III, and *Bam*HI. To perform the restriction digest do the following:

Pipette the following into one microcentrifuge tube:

10 μL of 1 μg/μL pBR322 plasmid

2 μL of restriction enzyme (10 units/μL) of *Eco*RI, *Hind*III, or *Bam*HI (use only one enzyme)

4 μL of 10X restriction buffer, supplied with the enzyme

24 μL of purified water

Incubate in a 37°C water bath for 60 min.

After the restriction digest is complete, the concentration of the pBR322 plasmid will be 0.25 μg/μL or 250 ng/μL.

LABORATORY EXERCISE 35

Equipment	Per Student or Group	Consumables	Per Student or Group
Syringe, 5–10 mL capacity or Bio-Rad column (0.8 cm × 4 cm) Poly-Prep Chrom column	2	Glass wool	1 package for class
Ring stand	1	DEAE Sepharose	2.5 mL
Column clamp	2	CM Sepharose	2.5 mL
Column tubing		pH paper	1 box for class
Test-tube rack		Protein sample (venomX)	0.5 mL
Stopcock or tubing clamp	1	Parafilm	1 roll for class
Column collection tubes	20	Microscope slide	10
Ceramic spot plate	1	Elution buffer	1 L for class
		Equilibration buffer	1 L for class
		Low-salt buffer	1 L for class
		Detection solution	50 mL for class

RECIPES

<u>5 M NaCl Stock Solution</u> 292.2 g NaCl BTV 1 L with purified water.	<u>Elution Buffer (0.5 M NTM)</u> 0.5 M NaCl 100 mL of 5 M NaCl stock 0.01 M Tris 10 mL of 1 M Tris stock, pH 7.6 0.01 M MgAc 10 mL of 1 M MgAc stock BTV 1 L with purified water.
<u>1 M Tris Stock Solution, pH 7.6 at 4°C</u> 140.4 g Tris-HCl 13.4 g Tris base BTV 1 L with purified water.	<u>Low-Salt Buffer (0.2 M NTM)</u> 0.2 M NaCl 40 mL of 5 M NaCl stock 0.01 M Tris 10 mL of 1 M Tris stock, pH 7.6 0.01 M MgAc 10 mL of 1 M MgAc stock BTV 1 L with purified water.
<u>1 M MgAc Stock Solution</u> 85.8 g Mg acetate BTV 400 mL with purified water.	<u>Equilibration Buffer (0.2 M NTM 10X Tris, pH 7.6)</u> 0.2 M NaCl 40 mL of 5 M NaCl stock 0.1 M Tris 100 mL of 1 M Tris stock, pH 7.6 0.01 M MgAc 10 mL of 1 M MgAc stock BTV 1 L with purified water.
<u>Protein Sample venomX</u> For this simulation, use β-galactosidase 1 mg/mL stock solution, in 0.1 M sodium phosphate buffer, pH 7.5, as described in the recipes for Laboratory Exercise 26. Dilute the stock to 0.04 mg/mL using 0.2 M NTM buffer.	<u>Detection Solution</u> For this simulation, the detection solution is ONPG substrate solution, as described in the recipes for Laboratory Exercise 26.

UNIT VIII GROWING CELLS

LABORATORY EXERCISE 36

Equipment	Per Student or Group	Consumables	Per Student or Group
Compound light microscope	1	Glass cleaner (Sparkle brand was recommended by a microscope technician and works well in our laboratory.)	1 bottle for class
Prepared slides	Several, ideally including one of stained bacterial cells	Lens paper	1 pack
		Immersion oil	1 bottle, can be shared

LABORATORY EXERCISE 37

Equipment	Per Student or Group	Consumables	Per Student or Group
Bunsen burner	1	Sterile Petri dish	2
Sterile glass bottle	2	Sterile conical tube	2
Autoclave		Sterile liquid nutrient medium	60 mL
Striker or matches	1	Disinfectant (e.g., 10% bleach or 70% ethanol in spray bottle)	50 mL
Pipette bulb	1	Agar control plate	1
10 mL sterile pipettes	6	Glass microscope slides	1 pack
Compound light microscope	1	Immersion oil	1 bottle
Biohazardous waste container	1		

RECIPES

Sterile Liquid Nutrient Medium-LB Broth	Agar Control Plate-LB agar
10 g tryptone	10 g tryptone
5 g yeast extract	5 g yeast extract
10 g sodium chloride	10 g sodium chloride
0.5 mL 4 N sodium hydroxide	0.5 mL 4 N sodium hydroxide
BTV 1 L with purified water.	15 g agar
Mix to dissolve.	BTV 1 L with purified water.
Divide broth into bottles for each student or group.	Warm and mix to dissolve.
Autoclave for 15 min at 121°C	Autoclave for 15 min at 121°C.
	Cool molten agar to ~60°C and pour plates.
	See Laboratory Exercise 38 for the procedure to make LB agar plates.
4 N Sodium Hydroxide, NaOH	
16 g NaOH pellets	
80 mL purified water	
Add pellets gently and stir.	
The solution will get very hot.	
When the pellets are dissolved, BTV 100 mL with purified water.	

LABORATORY EXERCISE 38

Equipment	Per Student or Group	Consumables	Per Student or Group
Autoclave		Tryptone	10 g
2 L Erlenmeyer flask	1	Yeast extract	5 g
Heat-resistant gloves	1 pair	Sodium chloride	10 g
1 L graduated cylinder	1	Agar	15 g
Hot plate/stirrer	1	Autoclave tape	1 roll for class
Cotton or sponge plug for flask	1	Aluminum foil	1 roll for class
Balance and weighing supplies	1	Steam chemical indicator strips, optional	1 pack for class
Inoculating loop		4 N sodium hydroxide	0.5 mL
Bunsen burner and striker		Disinfectant (e.g., 10% bleach or 70% ethanol in spray bottle)	1
37°C incubator		Sterile Petri dishes	35–40
Biohazardous waste container		*E. coli* culture on an agar plate	1

RECIPES

See Laboratory Exercise 37 for 4 N sodium hydroxide solution recipe.
Prepare *E. coli* cultures on agar plates one day in advance.

LABORATORY EXERCISE 39

Equipment	Per Student or Group	Consumables	Per Student or Group
Bunsen burner and striker	1	Sterile toothpicks	Several
Sterile Pasteur pipette and bulb	1	*E. coli* culture plate from Laboratory Exercise 38	1
Inoculating loop	1	Glass microscope slide	1 pack
Compound light microscope	1	Sterile water	5 mL
Staining trays	1	Gram-stain kit	1
Biohazardous waste container	1	Bibulous paper, optional	1 pack
		Immersion oil	1 bottle

LABORATORY EXERCISE 40

Equipment	Per Student or Group	Consumables	Per Student or Group
Autoclave	1	Autoclave tape	1 roll for class
250 mL storage bottles	1	Potassium chloride	1 bottle for class
10 mL pipettes	1	Dibasic sodium phosphate, anhydrous	1 bottle for class
Screw-cap test-tubes, approx. size 18 × 150 mm	10	Monobasic potassium phosphate, anhydrous	1 bottle for class
Autoclavable test tube rack	1	Sodium chloride	1 bottle for class
Usual equipment for preparing solutions (e.g., glassware, graduated cylinders, pH meters with standards, conductivity meter[s] with standards, balances with weighing supplies)		Steam chemical indicator strips, optional	1 pack for class

LABORATORY EXERCISE 41

Equipment	Per Student or Group	Consumables	Per Student or Group
Bunsen burner and striker	1	*E. coli* culture plate from Laboratory Exercise 38	1
Autoclave		50 mL sterile tube	1
5 mL sterile pipette and bulb	1	PBS dilution tubes; see Laboratory Exercise 40	
Inoculating loop	1	LB agar plates from Laboratory Exercise 38	5
37°C incubator/shaker	1	Sterile LB broth; see Laboratory	
Vortex mixer	1	Exercise 38	
Test-tube rack	1	5–10 mL	
1 mL sterile pipettes or micropipettes and sterile tips	3 or 4		
Cell spreader	1		
Widemouthed glass jar with 70% ethanol	1		
Biohazardous waste container	1		

LABORATORY EXERCISE 42

Equipment	Per student or group	Consumables	Per student or group
Spectrophotometer, visible light	1	Overnight *E. coli* culture; see Laboratory Exercise 41	1 tube
Cuvettes	2	60 mL LB broth in 250 mL-size Erlenmeyer flask	1
1 ml sterile pipette and bulb	Several	LB broth for spectrophotometer blank	5 mL
37°C incubator/shaker		Disinfectant, 10% bleach in spray bottle	1
Access to computer with Excel	1	Semilog graph paper	1 or 2 sheets
Biohazardous waste container	1	9.9 mL and 9 mL PBS dilution tubes, optional	See Laboratory Exercise 40
		LB agar plates, optional	5

LABORATORY EXERCISES 43, 44, 45, 46, 47, 48

The materials and preparations for Laboratory Exercises 43–49 are in this single table because it is convenient to prepare all the reagents and materials at the same time and performing fewer manipulations reduces the chance of contamination.

Equipment	Per Student or Group	Consumables	Per Student or Group
Biological safety cabinet with germicidal lamp	1	Ham's F-12 medium, incomplete	1 bottle containing at least 70 mL
Cell culture CO_2 incubator at 37°C with 5% CO_2	1	Sterile 15 mL centrifuge tubes	1 pkg
Autoclave	1	Sterile calf serum for cell culture (Laboratory Exercise 44)	100 mL
Pipette aid	1	70% ethanol in spray bottle	1
Sterile 125 mL cell culture medium bottles (for Laboratory Exercises 43, 44, and 45)	2	Sterile serological pipettes	One bag each of 1 mL, 5 mL, and 10 mL
Biological hazardous waste containers	1	Cell-culture-treated culture dishes, sterile	1 pkg
Vacuum pump and vacuum tubing	1	Ham's F-12 complete medium made in Lab Exercise 44	1 bottle
Sterile 1 L glass cell culture bottle (for Laboratory Exercises 44 and 45)	1	Cell-culture-grade water	
Test-tube rack	1	Sterile PBS Laboratory Exercises 46–48	2 mL/dish
Vacuum aspirator bottle containing disinfectant solution	1	0.2 μm cellulose acetate (CA) 1000 mL bottle top filter or 0.2 μm PES 1000 mL receiver filter unit (Laboratory Exercise 44)	1
Micropipettes	1 set	Sterile Pasteur pipettes	Varies
Inverted-phase microscope	1	Chinese hamster ovary (CHO) cells	1–8 dishes
Light microscope	1	Trypsin-EDTA (Laboratory Exercises 46–48)	2 mL/dish
Microscope camera, optional	1	0.4% trypan blue (Laboratory Exercises 46 and 48)	1 mL
Hemacytometer with coverslip (Laboratory Exercises 46 and 48)	1		
Handheld counter (Laboratory Exercises 46 and 48)	1		
37°C water bath	1		

RECIPES

Ham's F-12 Cell Culture Medium, incomplete	Ham's F-12 Medium, complete
Combine the following: 10.62 g Ham's F-12 medium powder 1.176 g sodium bicarbonate 800 mL cell culture grade water Mix until dissolved. Bring pH to 7.5. BTV 900 mL with cell-culture-grade water. Filter sterilize with a 0.22 μm CA 1000 mL bottle top filter, or 0.22 μm PES 1000 mL receiver filter unit.	See Laboratory Procedure in Laboratory Exercise 44 for instructions on making Ham's F-12 medium, complete.
Sterile PBS See Laboratory Procedure in Laboratory Exercise 40 for instructions on making sterile PBS.	Cell-culture-grade water can be purchased or produced on site. Typical specifications for cell culture grade water are 18 MΩ/cm resistivity at 25°C, filter-sterilized using a 0.22 μm filter.

Index

9 781032 419916